Nazis, Women AND Molecular Biology

Berlin, 1946

Nazis, Women AND Molecular Biology

Memoirs of a Lucky Self-Hater

Gunther S. Stent

Briones Books
Kensington, California

OTHER BOOKS BY GUNTHER S. STENT

Molecular Biology of Bacterial Viruses (1963)
The Coming of the Golden Age (1969)
Molecular Genetics (1970;1978)
Function and Formation of Neural Systems (1977)
Paradoxes of Progress (1978)
Morality as a Biological Phenomenon (1978; 1981)
A Critical Edition of J.D. Watson's *The Double Helix (1980)*
Max Delbrück's *Mind from Matter (1986)*

Copyright © 1998 by Gunther S. Stent
All rights reserved. No part of this book may be reproduced or transmitted in any form or by any electronic or mechanical means, without the prior written permission of the publisher.
ISBN 0-9664563-0-0

Briones Books
145 Purdue Avenue
Kensington, CA 94708

I thank my sister, Claire Hines, and my brother, Ronald Stent, for helping me reconstruct memories of my childhood and young adulthood. I gratefully acknowledge fellowship appointments at the Wissenschaftskolleg zu Berlin (1985–1990) and at the Fogarty Resident Scholars Center of the U.S. National Institutes of Health (1990–1991), both of which supplied me with financial support, staff assistance, and facilities for writing these memoirs. I am indebted to Alan Rinzler for his overall editing of the text, to my childhood classmate, Lili Cassel-Wronker, for her advice on book design, and to Judith Martin, for her inspiration, literary guidance, criticism, and interpretation throughout this project.

Gunther S. Stent
Berkeley, California
March 1998

To Stefan Loftur

Prologue

I AM A WIDOWER IN HIS SEVENTIES who has written these memoirs to reflect on his previously unexamined emotional development. I went through life like a sleepwalker. For example, it was only while writing these memoirs that, for the first time, I really *read* the letters I received many years ago from my lovers. When those gallantly crafted messages first arrived, I skimmed over their emotional outpourings for such technical information as the time and place of the next date.

I finally realized that as a child I learned to shut down my powers of observation, to isolate myself emotionally from a world in which I faced two awful threats. One was my mother's suicide when I was twelve, after four years of an acute depressive psychosis, during which I was ever aware of the danger of her killing herself and she was unavailable for maternal guidance. The other was Hitler's *Sturm-Abteilungen* (or SA storm troopers), whom I watched marching through the Berlin streets bawling:

> *Wenn's Judenblut vom Messer spritzt,*
> *Dann geht's noch mal so gut.*
> When Jew blood spurts from our knives,
> We'll all have twice-better lives.

For many years after I fled from Nazi Germany in my teens, I dreamed, night after night, that I was back in Berlin, this time for real and not merely dreaming. I dream I'm walking through the city, tremendously excited to see all the familiar places again—the Zoo Station, the KaDeWe department store, the Brandenburg Gate—and running into this one or that one of my old schoolmates. But then I realize that the Nazis will kill me if they catch me. Why was I so dumb as to come back? So I'm desperately devising schemes to re-escape by sneaking across the frontier as I had done in 1938, except that in my dream I'm usually trying to cross a border between Germany and the United States. I never succeed in my getaway and always wake up in a fright, just as the Gestapo is closing in and about to throw me into a concentration camp. I'm infinitely relieved when I realize that I'm safely in America. Very often this canonical dream concluded with an episode of frustrated erotic desire. At last, I'm about to taste of the fruit forbidden, and hence made so alluring, by the 1935 Nuremberg Law for the Protection of German Blood and German Honor. Namely I'm about to make love to a blonde Berlin shiksa. But the act is never consummated because of the untimely arrival of the Gestapo.

My life was not spectacular. Yet I was lucky beyond my wildest dreams. I not only survived the Nazis, whom I had every reason to believe would kill me sooner or later, but I prospered and had a respectable academic career. I had many friends and confidants. Yet all the while, my inner life remained hidden, not just from them but even from me.

To protect the privacy of some persons, I have changed their names and a few inconsequential details of their backgrounds. Furthermore, the account of my two years as Max Delbrück's postdoctoral student is not meant to provide a balanced history of the origins and early development of the discipline that later came to be known as molecular biology. In my story, I focus mainly on the contributions of my friends, while failing to mention many people whose work also played a crucial role in the eventual elucidation of the molecular basis of heredity of which I have previously provided a more complete historical account. (G. S. Stent and R. C. Calendar; *Molecular Genetics,* published by W. H. Freeman, San Francisco, second edition, 1978.)

PART ONE

Following My Nightmare

Waking Up in Berlin

I WOKE UP, ROUSED BY SHOOTING. Soviet soldiers surrounded me. Where was I? What was happening to me? Then I remembered. The war was over. Hitler was dead. I was at the Admiralspalast Theater in the Russian sector. A gun had just been fired on stage. I dozed off during the world premiere of *Postmeister Wyrin*, put on by the State Opera. I hadn't been in the Admiralspalast since my mother took me here about fifteen years ago to see a variety show. Now the State Opera was using this threadbare theater, because nothing but picturesque rubble remained of its own, once magnificent house on the Unter den Linden boulevard, a few blocks away. Except for the Soviet soldiers, the audience was shabbily dressed.

Even without my U.S. Army uniform, it would have been easy to pick me out as an American (or Ami) by my haircut. It was longer than the Soviet Ivans', though shorter than the Krauts', who were wearing their hair much longer than was acceptable in the States. The crew-cuts, fashionable not so long ago in Hitler's days, had disappeared. The Kraut women didn't look very attractive. I didn't spot any who measured up to the kind of sexy Fräulein with whom, according to Stateside press reports about a *Fräulein Wunder*, the city was said to be crawling. The stench of the fetid bodies of unwashed, soap-and-hot-water-less Krauts pervaded the hall.

Yesterday, November 11, 1946, I returned to my native city, after an eternity of eight years. It was the twenty-eighth anniversary of World War I Armistice Day, the day on which, as I had learned in my history class at the Bismarck Gymnasium, the Jew-Bolsheviks had finally succeeded in back-stabbing the valiantly victorious German army and navy. I had come on the U.S. Army's "Berlin Express," made up of red cars—sleepers and a diner—still bearing the MITROPA insignia. Before turning into my berth a couple of hours out of Frankfurt, I went for a late snack in the diner: white table-cloths, Mitropa-emblemed flatware, white-coated, bow-tied Kraut waiters.

I took a seat next to the window and saw the reflection of my jaunty face thrown back at me from the darkened landscape. I was pleased with myself riding in style. As the train stopped at a station, ragged, down-and-out Krauts crowding the dark platform stared at me eating in luxury in the brightly lit diner. I felt good: It served those Nazi bastards and their brats right to be hungry and miserable, shivering out there in the cold!

As we pulled into Berlin's Wannsee station next morning, I heard the railroad hands talking in the vulgar Berlin patois, which I hadn't heard for such a long time. The whole lot had probably been Nazis too, like the Krauts out on the platform last night, but by their voices I heard that I was home. I *saw* that I was home when I spotted the Army-requisitioned, canary-yellow Berlin city bus waiting trainside for the Allied passengers. On its flank, the double-decker was still festooned with an old cigarette advertisement: **BERLIN RAUCHT JUNO,** painted in white letters on a rectangular red ground. No other German, let alone foreign, town ever put the boxy Berlin type of double-decker on its streets, except Moscow, where some liberated specimens circulated for a few years after the war.

There were no competitors among the Berlin Express passengers for a place on the upper-deck front-row bench of the bus that had always been fought over by Berlin juveniles. Yet I instinctively raced upstairs, to make sure that I wouldn't be beaten to a prize front-row seat for the ride that would take us from Wannsee Station through the Grunewald forest to U.S. military government headquarters in the Dahlem district.

The receptionist who checked me in at the Hotel Gossler, a patrician villa converted into a transient officers' billet in Dahlem, had thrown in the State Opera ticket as a bonus when I sold him a carton of Camels for a thousand Reichsmarks in the old swastika-emblemed banknotes still in circulation. I was glad to get the ticket because although I was an opera fan, I had hardly been to an opera for eight years. Even the second balcony at the Chicago Opera was out of the reach of a hard-up student like me.

Too bad they weren't putting on *Don Giovanni* or *Fidelio*, but only *Postmeister Wyrin*, which I found boring. Since the opera's libretto was based on Pushkin's tale of the backwoods postmaster whose young daughter ran away with a passing cavalry officer, I thought it must be a Russian import. Maybe the composer had been unable to persuade the Bolshoi to put it on at home and sent it to the fringe of the Soviet Empire to try out on the vanquished

Krauts? Not exactly; it transpired that the composer, Florizel von Reuter, was an expatriate American from Davenport, Iowa, who had *come to* Nazi Berlin eight years earlier, just when I left it, and he spent the whole war there teaching violin to the Krauts. A lousy, traitorous Nazi collaborator! It served him right that later musical history bore out my callow assessment of his opera's lack of merit. I never heard of *Postmeister Wyrin* being put on again anywhere, and it didn't make it into *The New Kobbé's Complete Opera Book*.

As the crowd surged through the exit into the cold November night after the final curtain, a blonde girl in an olive-drab overcoat pushed up against me. Maybe she is Allied personnel too? No, she is too good-looking. Anyhow, the cut of her coat is too feminine to be government issue; in fact, it seemed to be home-sewn from a couple of GI woolen blankets. There's something to that *Fräulein Wunder* after all! She smiled at me, and I asked,

"How did you like the opera?"

As I meant her to be, she was startled that an Ami spoke to her in Berlin-flavored German.

"It was terrific. How come you talk German like a Berliner?"

"Because I *am* a Berliner. I left Berlin for the States when I was a kid. I grew up in Chicago."

"Did you leave because of Hitler?"

"*Richtig!* I left because of Hitler."

She didn't ask whether, and I didn't feel like confessing that I had had to leave because I was a Jew.

Side by side, we walked down the Friedrichstrasse, lined by lifeless ruins. No one would have guessed that this godforsaken canyon through mountains of rubble had once been a world-class thoroughfare of pleasure, famous for its popular cafés, smart boutiques, and fancy streetwalkers.

"Heading for the U-Bahn subway station, Mister?"

"*Jawohl, Fräulein.*"

"Me too."

"It sure seems like it's going to be a cold winter, doesn't it?"

"It sure does."

Allied personnel rode free on German public transport. Being in the company of an Ami, the girl haughtily ignored the ticket taker at the subway entrance. The unique odor of the U-Bahn, as evocative of home to my nose as the Berlin patois is to my ears, engulfed us. A northbound train pulled in, and neither of us budged to board it. Aha! Like me, she must be southbound. My deduction was confirmed when the southbound train came: Both of us got on the same car, the crowd pressing us against each other. The stench of bodies was even stronger in the jam-packed subway than in the theater.

Yesterday morning, to enact my I'm-back-in-Berlin nightmare, I had gone to look at what was left of the Zoo Station, the KaDeWe department store, the Brandenburg Gate. And tonight, here she was, the blonde shiksa of my dream, standing right next to me in the U-Bahn! As I took in these surreal scenes, I was wondering how I could be so sure that I wasn't merely dreaming, as usual, that I'm back for real *this* time. Maybe I was going to wake up pretty soon and find myself, as usual, in bed at the Tau Epsilon Phi fraternity house in Champaign, Illinois!

No, *she* had to be really real. I had just been lucky running into her, so far, at least. But my luck was going to run out momentarily, because at the second stop, Stadmitte, or City Center, I would have to get off, to change to the westbound Krumme Lanke line. What was I going to do? Not get off at Stadtmitte and stay on to wherever she is going? If I did, what would I do wherever she finally got off? Get off with her and say, "Why, isn't it the strangest coincidence that I happen to be heading for Hermannplatz too!" What else would I say on the Hermannplatz platform, way on the other end of town from Dahlem, where my billet was? I couldn't possibly ask to take her home, where her hulking blond husband or boyfriend was waiting. No, staying on this train with her would be one big waste of time. I better change trains at Stadtmitte, and head straight for the Hotel Gossler. It was a damned pity, all the same. She seemed to be just the kind of Fräulein I had been hoping to run into.

"*Auf Wiedersehen, Fräulein.* I change at the next stop. Nice talking to you."

"Wait! I change there, too!"

Again we got into the same car. For God's sake, I couldn't let this fabulous stroke of luck simply slip through my fingers! So I introduced myself.

"I'm really a civilian. An American scientist. I work for the military government."

"Pleasure to meet you, Mister Stent. Funny coincidence! I'm a civilian too! A German actress. I work at the Schiffbauerdamm Theater."

Her name was Hildegard U. For all I knew, she might get off at the next stop.

"Would you, by any chance, like to have supper with me in Dahlem, Fräulein U.? But maybe Dahlem is too far out of your way?"

"No, it's not too far out of my way. Actually I wouldn't mind a bite to eat."

"*Fantastisch!*"

I was aglow with pride over my skills as one real smooth pickup artist.

"What do you do for the military government?"

"Analyze scientific documents. I'm in Berlin for only a couple of weeks, on special assignment. My headquarters are in Frankfurt."

"Are you really a scientist? You look pretty young."

"I'm twenty-two. A Wunderkind, as they say. And you?"

"Nineteen."

There was one little problem: Where was I going to take her for supper? Because of the shortage of food and tight rationing, there were no legitimate German restaurants where one could just walk in and order a meal. All I could think of was the Harnack Haus, formerly the faculty club of the Dahlem research institutes of the recently defunct Kaiser-Wilhelm-Gesellschaft for the Promotion of the Sciences. According to the receptionist at the Hotel Gossler, it had been turned into an American officers' club with an excellent mess. I wanted to visit the famous Harnack Haus anyway.

As we got off at the Thielplatz stop in Dahlem, I took Hildegard's arm, to walk her to the Harnack Haus. Surfacing from the U-Bahn, we came up into a world very different from that we had left when we had gone underground in the Friedrichstrasse. Instead of the ghostly landscape of the wrecked downtown, suburban Dahlem offered a pleasing panorama of rows of largely intact villas, each standing on its own generous plot of ground. Most of the villas seemed to be in need of a paint job and other maintenance work, but there were no signs of the devastation caused by the war elsewhere in the city.

The Harnack Haus is a conservatively designed three-story mansion with a gabled, red-tiled roof. Although completed in 1929, it bears no trace of the modernist Bauhaus style that characterized so many buildings that went up in Berlin, and especially in Dahlem, in the late 1920s. To signify the

Harnack Haus's social rather than administrative or research function, its main entrance sports a *porte cochère* whose canopy is supported by two stone columns decorated with botanical and zoological motifs in relief. It was taken over by the U.S. Army after the war, in a nearly perfect state of preservation.

That night, Harnack Haus was brilliantly lit, in dramatic visual contrast to the dimly illuminated villas of the neighborhood. Next to its main entrance was a large sign:

KEIN ZUTRITT FÜR DEUTSCHE
NO ADMITTANCE FOR GERMANS

On first sight, this sign brought back scary memories of similar signs posted in prewar Germany, which denied Jews access to restaurants and other public facilities. Such an anti-Semitic advertisement had frightened me as a six- or seven-year-old, before the Nazis even came to power. It was a banner stretched across the highway at the entrance to Zinnowitz, one of the little towns on the island of Usedom, off the Baltic coast, where the popular sea-side beach resorts closest to Berlin were located. The banner enjoined Jews to pass through town without stopping, so as not to pollute the place with their miasma. What if our car happened to have a breakdown in this burg? Why, these Nazi monsters would pull us out of the car and hit us over the head with their clubs, break our legs, or maybe even kill us!

Such signs were rare in Berlin until the near-total exclusion of Jews from public life was decreed in the fall of 1938, in the wake of the Kristallnacht pogrom. What signs had appeared in Berlin were temporarily removed in the summer of 1936, to sanitize the city visually for the Olympic Games. The games, with which Hitler bamboozled the international community into believing that his intentions were wholly benign, had been secured for Berlin by the Weimar Republic's Jewish representative on the International Olympic Committee, Theodor Lewald. To help put over his charade as Mr. Nice Guy, Hitler allowed Lewald to take part in the opening ceremonies.

In the provinces, however, anti-Jewish exclusionary signs had become ubiquitous. I ran into them in the summer of 1938—my last in Germany—when I spent a two-week holiday in Garmisch in the Bavarian Alps with my father. We had arranged to rent a room in a bed-and-breakfast place, whose landlady didn't ask whether we were Jews. But we were dismayed to find that there wasn't a single restaurant in Garmisch that didn't prominently display a sign with the notice:

> JUDEN UNERWÜNSCHT
> (Jews not wanted here.)

Because we intended neither starving nor cutting our Bavarian holiday short, we ignored those signs. At every meal, I thought that if we were found out, we'd probably be beaten up by the SA goons, or arrested by the police, or maybe even sent down the Munich road to the Dachau concentration camp.

All the same, I actually enjoyed our holiday, maybe because of my Berliner's pride in doing the forbidden, or "*daffke*" (one of the many Yiddish words assimilated into general Berlin argot). To provide a socially stabilizing counterforce, the prevalent Berlinesque yen for *daffke* is yoked to a vigilant passion for preventing all the other *Daffkes* in town from getting away with murder. Moreover, thanks to what I eventually realized had become my Nazi-imbued slave mentality, each time we were not found out I gloated over our dumb, overbearing Nazi masters for not being as all-powerful as they thought.

So on second sight, the sign denying Germans access to the Harnack Haus made me feel good, just as did the scene of the down-and-out Krauts on the station platform I saw from the Frankfurt–Berlin train the other night. They are certainly getting a dose of their own medicine!

Yet, on third sight, I realized that, having just picked up one of those Krauts and promised to take her here for supper, the sign applied to me as well. A promise being a promise, what choice was there for me but to ignore the sign and take my guest inside? Maybe her olive-drab coat would fool the sentry. Hildegard, who could read the sign just as well as I, didn't say a word, and I didn't ask her whether she was willing to defy military government decrees. Undoubtedly she had known all along that Germans weren't allowed into Allied messes. So we two *Daffkes* just strode up to the *porte cochère*, me more frightened than she. The black-uniformed, civilian security guard at the door saluted me as we walked past him.

I helped Hildegard take off her coat, fleetingly feeling her shoulders and letting my hands touch her neck, as if by accident. She was wearing a plain dark-blue dress, maybe cut from Naval Stores cloth. With her blue eyes and upswept flaxen hair, she looked even more attractive than in olive-drab. As we were shown to a table, she followed the headwaiter, head lifted high, as if it were the most natural thing for her to be in this place. I could see that it was not for nothing that she was an actress. My fellow Amis looked at her with interest. Did they suspect that she was not Allied personnel? We sat down, and I ordered dinner: soup, steak with vegetables and French fries, ice cream for dessert—plain but hearty American fare.

The dining room with its crystal chandeliers was even more elegant than the restaurant at the Hotel Urbana-Lincoln, the classiest eating place in Champaign-Urbana. The Urbana-Lincoln came to my mind because of the bittersweet memory of having taken a beautiful music graduate student there for dinner on the night I asked her to marry me. She turned me down, kindly but firmly.

Now here I was, a mere doctoral candidate in physical chemistry, about to dine in the very hall where some of the all-time greats of physical science known to me from my textbooks—Max Planck, Max von Laue, Otto Hahn, Werner Heisenberg, and maybe even Albert Einstein himself—used to take their meals. Of course, it wasn't because of my scientific achievements—I had none—that I was in the Harnack Haus tonight. I was here because of Hitler and the war he lost, in which I hadn't even fought. Come to think of it, if there hadn't been any Hitler, there would have been little chance that I, the twenty-two-year-old nonentity, would be taking a girl to dinner in this refectory of the scientific superstars. Not just *any* girl, but a sexy blonde starlet, who wouldn't have given me the time of day in the old days. How the world has changed! And all I had to do was to survive Hitler.

A band of DP musicians, one a Hungarian-style gypsy fiddler, was playing tunes from Austrian and German operettas in the Harnack Haus's dining room, and people were dancing. The men were in uniform, but most of the women wore dowdy civilian clothes. No doubt they were bona fide Allied personnel. No unwashed bodies here! Hildegard and I sat in silence, like a long-married couple: Obviously we couldn't talk in German here. Does Hildegard speak English? Even if she did, her English couldn't be good enough to fool the Americans at the next table. So we just gazed, smiled, and winked. Although certainly good-looking, Hildegard wasn't a major beauty by Stateside glamour-girl norms: I doubted she could be elected Homecoming Queen at Illinois. Maybe her face was a little too round; maybe her nose was turned up a little too much. But her eyes sparkled. I would take her in my arms on the dance floor in between courses, press her body against mine, and squeeze her hand by way of exploring further possibilities.

The waiter brought our soup. Then a burly mess sergeant came up to Hildegard.

"Your military identification card, please, Miss."

"She doesn't have one."

He could see from the shoulder patch sewn on my officer's tunic, which said, "U.S. Scientific Consultant," that I wasn't real Army; merely another goddam, no-count civilian from the goddam military government, with "simulated" commissioned officer's rank

"Get this Kraut woman out of here, Mister!"

"Can't we finish our soup?"

"Nothing doing! If she's not gone on the double, I'll call the MP."

As we got up from the table, Hildegard's eyes lingered on the soup. Out in the street, under the *porte cochère,* I felt that I was back where I had been eight years earlier: the impotent Jewish *schlemiel* kicked around by the *goyim.* Deeply humiliated, I stammered,

"I'm terribly sorry, Fräulein U."

Hildegard laughed. "Never mind, Mr. Stent. It's really funny. Imagine! Being kicked out with an Ami! I've been kicked out of lots of places before, but never with an Ami!"

What next? I couldn't think of any other place in Dahlem where a fraternizing couple could sup at this late hour and was too defeated to want to continue this romantic adventure.

"The last U-Bahn train is gone. I'll walk you home."

She laughed again: "You'd be in for a long hike, my dear Sir. I'm living in the Wedding district, in the French sector. Still remember? It's about ten miles across town."

"OK. So I'll walk you to wherever you were going on the U-Bahn. Didn't you say it wasn't too far out of the way from Dahlem?"

"By now, it's too late for me to go there. I'd better traipse home."

I was puzzled by her story. Hardly an hour had gone by since she accepted my invitation on the U-Bahn. What couldn't she do now at her original destination that she could have done an hour ago? And how would she have gotten home if she hadn't come with me? I didn't try to solve this enigma by quizzing her.

"Let's walk to my billet at the Hotel Gossler, a few blocks down the road. Maybe the receptionist might know how we can get transport to Wedding. He gave me the opera ticket for tonight. So I owe him the good fortune of meeting you in the first place."

Taking my half-joking gallantry at face value, Hildegard blushed. We were in luck. Back at the Gossler, there was an Army jeep taxi service. The receptionist said that if I cared to invest a few cigarettes, he could arrange for a driver to take us to Wedding. We climbed into the back seat of the open jeep; it was very cold. I put my arm around Hildegard, snuggling to keep warm; she snuggled back. My spirits were slowly recovering from the degradation they had suffered at the Harnack Haus. The German driver didn't know his way around Berlin. So we two natives had to show him the way, through the street-lamp-less dark city, whose ruins came into view and vanished again ghostlike, fleetingly illuminated by the jeep's headlights

It turned out that she could speak fairly fluent English, as well as some French (which she had picked up from the military occupiers of Wedding). We kidded around in three languages, switching from one to another in mid-sentence to show off our linguistic savvy.

"Got a wife in the States?"

"*Nein.*"

"*Une fiancée?*"

"*Pas non plus.* Actually, I'm not all that interested in women. My main interest in life is science."

In fact, the opposite was the case. My main interest in life was precisely women and their love, rather than science. Just like Balzac, I sought professional fame and glory mainly to raise my chances of success with the woman of my dreams. I had been falling in love with girls ever since I could remember, from kindergarten through primary and secondary school to college. But in none of these infatuations had my love been requited, let alone consummated, not even by a kiss. Except for a traumatic love affair I had in college with a girl who jilted me after we had not only done plenty of kissing and then some, but were even engaged.

My grotesque lie was intended as another lame attempt at gallantry. I thought I was flattering Hildegard by hinting that had I not been captivated by her striking charms when we bumped into each other at the doors of the Admiralspalast, I would now be cooped up in my room at the Hotel Gossler, poring over some thrilling scientific text. Hildegard did not react. Maybe she didn't take in what I said.

"What about you? Got a husband or fiancé?"

"*Nein.* I'm living with my mother and little brother. My father's gone. He was a Party *Bonze*. The Ivans arrested him right after they took Berlin. We haven't heard from him since. We've got no idea where he is."

As the jeep crossed the Spree from Charlottenburg into the Moabit district, I said to Hildegard,

"We must be close to the Essener Strasse. Before the war, my family owned an apartment house at Essener Strasse 20.

"How come you don't own it any more?"

"The Nazis made us sell it and confiscated the money. It was part of the 1-billion-mark fine Fatso Hermann Göring collected from the German Jews in November 1938."

"What was the fine for?"

"To pay for the damage done to our property by the Nazi mob during the Kristallnacht pogrom."

Hildegard looked puzzled. Obviously she was not well informed on this subject.

"Kristallnacht? And what's a 'pogrom'?"

"That's a Yiddish term for a massacre of helpless Jews."

"Is Yiddish the same as Hebrew?"

I realized I had let slip that I was Jewish. I believed in her ignorance and counted on her sympathy.

"And why did you Jews have to pay for the damage done to *you* by the Nazi mob?"

"Because Fatso said we had provoked this spontaneous outburst of patriotic indignation in the first place. By having one of our Jew boys, Herschel Grünspan, kill a staff member of the German embassy in Paris. Want to hear one of the Jewish gallows humor jokes about the fine? *Question:* Who's the world's greatest alchemist? *Answer:* Hermann Göring; he can turn Grünspan [verdigris, the toxic blue-green pigment formed on outdoor copper and bronze surfaces] into gold."

Hildegard didn't laugh. Maybe she didn't know what an alchemist was. All around us, nothing but ruins.

"Let's have a look at the rubble on our property. It'd be comforting to know that even if the Nazis hadn't stolen our building eight years ago, nothing would be left of it by now anyhow."

I thought that the huge rubble pile ought to prove to Hildegard that my folks were people of substance. I managed to steer the driver to the Essener Strasse. But there was no rubble pile at No. 20. Our former building was completely whole, with all of its forty-or-so flats occupied, except that the kosher butcher shop in the basement had been turned into a garage. The building was almost the only intact structure on this desolate landscape. Hildegard did seem impressed:

"It's really very big! Too bad it isn't yours any more!"

When I dropped her at her apartment house in working-class Wedding, Hildegard said,

"Thanks for an interesting evening, Mr. Stent. Like to watch me perform tomorrow night? We're putting on the first Berlin production of *Men in White*. I'll leave you a ticket at the box office."

I responded, "I'd be delighted to come, Fräulein U.," but added a typical Berlinesque quip, "especially since I have no other social engagements on my calendar for tomorrow night." On parting, we shook hands, European style.

Helping the driver find his way back to the Gossler, I thought about my lucky find. Quite apart from the ego boost from having managed to make a date with an actress, I liked Hildegard a lot. She had the Berliner's quick and saucy sense of humor, and yet was good-natured. It was easier for me to talk to and laugh with Hildegard than any other girl I'd ever met, including my American ex-fiancée. Not being able to carry on sustained conversations with attractive American women had been the bane of my relations with them. Moreover, it was a tremendous thrill to flirt again in German: I discovered that flirting came easier to me in my teenager's Berlin idiom, developmentally arrested, if not atrophied, by eight years of total disuse, than in my midwestern egghead collegiate English.

There was this troubling thing about Hildegard's father, of course. How could I, a Jew, take up with the daughter of a Nazi, even if he had meanwhile disappeared in Soviet custody, probably never to return, and she was too young to have done Jews any harm? And if her father *did* come back, what would he think of a Jew squiring his daughter? Never mind. Not only couldn't he do anything about it, but as a cowed Kraut he would see the Allied uniform rather than the Jew inside. He would probably start babbling about how he had been forced to join the Nazi Party and how he used his position to save many Jews during the war. Anyhow, I couldn't let Hildegard go, merely because her old man was a Nazi. I needed her for completing the project for which I had come back to Berlin in the first place: liberation from my chronic nightmare by fulfilling the overpowering yearning to return to my hometown in triumph and make love to a blonde Berlin shiksa.

Top of the Line (Second Category)

OPERA IS ONE OF MY HAPPIER CHILDHOOD MEMORIES. I got my first exposure to opera music from gramophone records we had at home. Not knowing what to do with myself most of the day before I started school, I listened to them, squatting on the floor before our hand-cranked phonograph. Our collection included vaudeville comedian patter and musical comedy songs, but I liked the opera records best. My favorite disk was the "Triumphal March" from *Aida*. I played it over and over again. Aida's aria "*Ritorna Vincitor*" was on the flip side. I played it less often, but its title came back to me as the U.S. Army's "Berlin Express" pulled into Wannsee Station. Here I was, the Nazi's intended victim, returned *vincitor*.

I remember my parents going to the State Opera when I was about six, my father in a dark suit, my mother wearing jewelry. My mother had explained to me that my given names, Günter Siegmund, were those of two characters in an opera by Richard Wagner. It was one of her favorites. But she didn't let on that Wagner's Gunther was a jackass and Siegmund the husband of his own sister. Anyhow, I was not named after these characters from the *Ring of the Nibelungen*. Günter had simply been a fashionable name for boys born in the days of the Weimar Republic, while Siegmund was an allusion to the name of my father's father.

Wagner's Gunther is a type of operatic character to whom the Canadian novelist Robertson Davies referred to as "Fifth Business." The standard tragic opera features four main characters: the tenor-hero, the soprano-heroine, the basso-heavy, and the contralto-witch. But these four—in the case of *Götterdämmerung*, Siegfried, Brunhilde, Hagen, and Gutrune (a soprano-witch, as the exception that proves the rule)—cannot make the story go. So another character, the Fifth Business baritone—Gunther—is needed. He is peripheral to the main action, but without him the plot would make no sense. He provides a mainly passive link among the other four.

It so happened that I *was* a Gunther. My wish dreams, from childhood throughout my scientific career, were never to be the star, the Big Number One. Rather, in my fantasies I always cast myself in the role of the Number One Sidekick of the Big Number One, the one who marries not the princess but her lady-in-waiting. (The fate of Wagner's Gunther, who married Brunhilde, no mere princess but the favorite daughter of the God-in-Chief of the Valhalla crowd, exemplifies the danger of overreaching.) My aspirations were embodied by Lanny Budd, the protagonist of the *Dragon's Teeth* novels by Upton Sinclair, which I read at Chicago's Hyde Park High School. Onward from the Versailles peace conference, which he attends as a young aide to President Wilson, Lanny makes the scene wherever the action happens to be—Washington, London, Moscow, Berlin, Rome—as a confidant of President Roosevelt. Although well known to Churchill, Stalin, Hitler, Mussolini, and the Pope, Lanny remains largely unknown to the world at large.

My father, Georg Stensch, the son of Sigismund Stensch and his wife, Caecilie née Salinger, was born in Berlin in 1886. Grandfather Sigismund hailed from Berlinchen ("Little Berlin"), a small town in the deepest Mark of Brandenburg. Like many other young go-getters from the Prussian backwoods of the 1870s, the *Gründerzeit* era of the boom-town capital of Bismarck's new Second Reich, Sigismund had set out to seek his fortune in the city Mark Twain called the "European Chicago." Although he never struck it rich, he did manage to become the owner of a small white-goods shop near the Silesia Gate, then marking the eastern city limits. Grandmother Caecilie had come to Berlin from the Pomeranian hamlet of Kallies, a place even more in the sticks than Berlinchen. She and Sigismund had four children: Aunt Ella (the eldest), my father, Aunt Käthe, and Uncle Fritz.

My mother, Elli Karfunkelstein, the second of five daughters of the glassware wholesale merchant, Richard Karfunkelstein and his first cousin, Clara, was born in Breslau in 1892. The Karfunkelsteins were one of the families of well to-do, assimilated Breslau Jews, who had very little interest in, or connection with, Judaism. My grandfather's main spiritual affiliation was a fancy aesthetes' lodge with Freemasonry-like ritual hocus-pocus, the *Schlaraffia,* which, as its name implied, was concerned more with the *dolce vita* than with the Brotherhood of Man. Grandmother Clara died in 1907, when Elli was still in her teens. The five semiorphaned Karfunkelstein sisters were intelligent, independent women, who wouldn't take any guff from anyone (especially not from their eventual husbands).

Wally (the second-youngest of the sisters) was one of the first women to obtain an M.D., from the University of Breslau. She married a philologist, Hugo Steinthal, from Saarbrücken, where she set up a practice as the only woman gynecologist of the Saar region. One of my earliest memories of the terror of anti-Semitic abuse is an assault on Uncle Hugo by Nazi thugs in 1928. They were lying in wait for Jews coming out of the Berlin Head Office of the CV (for *Central Verein*), or Central Union of German Citizens of the Jewish Faith (on whose national board Hugo served). As Hugo was leaving the building, they beat him up so badly that he had to be hospitalized for several months.

I mulled over this frightful story repeatedly, considering various scenarios of how I would have avoided getting caught by the Nazis. Perhaps by sneaking out of the CV building by the back door, or by going up on the roof, clambering over to the roof of the adjacent house, running down its staircase, and coming out into the street by a different front door.

Katharina, the youngest of the five sisters, who was barely five when their mother died, was my favorite Karfunkelstein aunt. She told amusing stories about her adventures in foreign places, could tap dance, and played the piano—with a seemingly inexhaustible, memorized repertoire of classical as well as popular pieces, including jazz. Always elegantly dressed, she exuded the aroma of exquisite perfumes and spoke several languages fluently. When Katharina talked in Yankee-flavored English to some Americans seated next to her and me at the final track event of the 1936 Berlin Olympic games—at which Jesse Owens won the 100-meter sprint—they wouldn't believe that she was a German who had never been to the States.

Barely out of her teens, Katharina took on one lover after another. When she came down with venereal disease, my outraged grandfather kicked her out, and she thus began her career as an itinerant, trans-European adventuress. A professional matchmaker found her a husband, Bobby van Ameringen, a Dutch businessman, who was dazzled by Katharina's beauty and charms. But within a few weeks of her marriage, Katharina deserted van Ameringen and obtained a divorce from him in Germany that entitled her to a substantial alimony. Van Ameringen felt ill used: Living in Holland beyond the reach of German domestic law, he refused to pay. But when he visited the Leipzig Industrial Fair, Katharina bamboozled my father, whose firm was a regular exhibitor at the fair, to have van Ameringen arrested and put in the cooler until he came across with the delinquent alimony remittances. After his release, van Ameringen stayed out of Germany and made no further payments.

A subject of the Queen of Holland by virtue of her fleeting marriage, my aunt styled herself Katharina v. Ameringen, to give the false impression that the abbreviation "v." referred to the German signifier of nobility "von," rather than to the Dutch "van," which has no aristocratic connotation. She supported herself as a private secretary, but despite her outstanding talents, she could never hold on to a job for very long, because no employer could stand her fiery Karfunkelstein temperament. She seemed to be forever on the move, often visiting the Steinthals in Saarbrücken or us in Berlin.

Grandfather Richard died in 1935, and of all his heirs, only Katharina succeeded in enjoying her inheritance in its entirety. All the others, including eleven-year-old me, lost their patrimony to Göring's confiscation. But Katharina collected her share in cash as soon as Richard's estate was probated and started to spend the money at a dizzy pace. She bought a new wardrobe at top-of-the-line couturiers and a BMW convertible. Because she didn't know how to drive, she hired an instructor (who also doubled as her lover), and toured with him the five-star resort hotels of Central Europe.

I had visited my grandfather a few times in Breslau, in the cavernous apartment in the Goethe Strasse in which my mother had spent her girlhood. By the time of my visits, only Richard and his eldest daughter, Aunt Grete, a spinster high school teacher, were still living in the apartment, so many of its rooms were unused. They had a musty smell; their windows were covered by heavy drapes; and their furniture was shrouded by sheets. It was a lugubrious scene. The object that interested me most in the Goethe Strasse apartment was a large steamer trunk. Richard had taken it with him to America on his visit to the Chicago World's Fair of 1893. I pictured him sitting on this trunk as redskins, galloping along the railroad track on their incredibly fast mustangs, shot their arrows at the train that was carrying him across the wide-open prairie stretching from New York to the Great Lakes. Oh, how I would have loved to have gone with him on that trip to Chicago and shared his adventures in the New World!

When she was nineteen, my mother, Elli, fell in love with a Breslauer of whom her father strongly disapproved (for reasons that I never discovered). My mother's beloved didn't return her love with sufficient ardor to be willing to defy his father-in-law. Deeply disappointed, my mother fled Breslau. Richard arranged for the self-exiled Elli to stay for a while in Berlin with his brother, Karl Karfunkelstein, who owned a small bronze statuary and lighting fitting factory. To pick up some commercial skills, my mother worked as a volunteer in her uncle's office.

Karl's business was in dire need of capital. To get the money, Karl hatched a nefarious plan: He was going to look for a young man who would marry Elli and then invest in Karl's firm the substantial dowry that he knew Richard was going to give his daughter on her marriage. So Karl placed an ad in a Berlin newspaper in which he announced an opening for a traveler, adding the phrase "*Einheirat nicht ausgeschlossen*" ("Marriage into the firm not excluded"), no doubt without my mother or Richard's knowledge.

My father, then a handsome, ambitious twenty-six-year-old salesman, answered the ad. He got the job, the girl, and her dowry. I think it likely that, in view of my mother's beauty and intelligence, he fell in love with her and married her not merely for her dowry. The old saw that my father would later try to impart to his sons, "It's just as easy to fall in love with a rich girl as with a poor girl," was confirmed by his own experience. It was one of the few of my father's pearls of wisdom that made good sense to me. As for my mother, still carrying the torch for the beloved beyond her reach in Breslau, she probably reckoned she would never find another man she would love. So this Berliner, Georg Stensch, seemed presentable enough.

In accord with Karl's expectations, my father did invest my mother's dowry in the firm. It became a partnership between Karl, my father and Friedrich Ballnath, a master metal craftsman whom Karl had recently hired and who brought his technical expertise and personal savings as capital into the enterprise. The partnership was called "Stensch, Ballnath & Co." The absence of "Karfunkelstein" from the firm's name was not attributable to Karl's modesty but to his realization that, in view of his past track record, it would be preferable, credit-ratingwise, not to flaunt his involvement.

Before long, my father and Ballnath realized that Karl was not only a poor businessman but also a crook, trying to cheat them out of their capital. If Karl had picked my father for being a patsy, easily done out of his wife's dowry, he had dialed the wrong number: By the time my father got through with him, Karl was left with little more than the firm's name and office furniture. My father and Ballnath seceded from the partnership, and, having managed to recover most of their original investment, founded a new partnership called "Ballnath & Stensch." The new partnership was a direct competitor of Karl's Stensch, Ballnath & Co. These maneuvers provoked prolonged litigation, but in the end, Ballnath & Stensch prevailed over Stensch, Ballnath & Co., with the latter entity eventually having to declare bankruptcy. My mother fully supported my father in his struggle with Karl, and had no further contact with her uncle throughout the many years both still lived in Berlin.

On the outbreak of war in 1914, the fledgling Ballnath & Stensch partnership became a subcontractor for major munitions' manufacturers. Its bronze-casting foundry was converted to the production of such items as detonators for artillery projectiles and parts of the Fokker mechanical firing synchronizer. This device, which had been invented by Anton Fokker, allowed the lone pilot of a fighter plane to shoot bullets frontward through his own rotating propeller. By the end of the war, the firm was on a solid financial basis and the partners were moderately wealthy men. The factory was reconverted for the production of statues and lighting fittings, and Ballnath & Stensch soon became one of the largest manufacturers in its line of business in Berlin. The hyperinflation of 1923, which financially ruined the German middle class, had little adverse effect on the partnership because of my father's astute structuring of the firm's debits and credits.

After their marriage in 1913, my father and my mother set up household in a small flat in the lower-class, proletarian district of Berlin where my father was born, known now, but not then, as Kreuzberg. My mother gave birth there to my older brother, Rudi, in 1914. With the improving fortunes of Ballnath & Stensch, my upwardly mobile father moved his family to a more spacious apartment in Treptow, an unfashionable suburb in the southeastern part of Berlin, where my older sister, whose name was Clara, but whom everybody called Mausi, was born in 1916. My mother's idea was that, with one boy and one girl, her family was now complete. She declared that she didn't want any more children. Eight years later, on March 28, 1924, I was born to her in Treptow as an unwanted, third child, conceived in Naples. My parents had gone there on their tenth wedding anniversary, by way of a makeup for the honeymoon trip they had never taken.

I don't know what changed my mother's mind about her unwanted third child—perhaps it was merely due to release of postpartum hormones. In any case, she soon began to devote her main attention to me, favoring me egregiously over my elder brother and sister. She spent most of her day with me. When she went for a drive in our Opel, she would usually take me with her. She was one of the first licensed women drivers in Berlin, capable of returning as good as she got any obscene Berlinesque obloquy shouted at her by male chauvinist pig drivers.

My mother spoiled me, granting me almost every wish, and if she did not happen to grant it quickly enough, I enforced her compliance by tantrums and feigned headaches. Some of my parents' friends found my behavior so obnoxious that they no longer wanted to visit our home. Thus for the first six years of my life, my mother was the center of my emotional universe, my constant companion and protectress. Then I entered elementary school, and as my social horizon expanded to include teachers and fellow students, my emotional dependence on her began to wane. With the onset of her fatal illness, when I was eight, I grew more and more emotionally independent. By the time I was ten, my emotions had become well-nigh autonomous, with my eighteen-year-old sister, Mausi, as my only moral suzeraine and object of deep affection. My father could not serve me in this role. He was always making silly jokes and could not be taken seriously, except for what he was teaching me about crookery in the cruel business world

Treptow was home to the petty-bourgeois majority, whose conservative tastes and politics had little in common with the modernist, what-the-hell spirit for which Athens-on-Spree was known during the Weimar Republic. Few Jews lived in Treptow, which was light years removed from the avant-garde action—the Albert Einsteins, the Max Reinhardts, the Marlene Dietrichs, the Bertolt Brechts—to which Berlin of the Weimar Republic owed its cultural glamour.

Treptow could boast merely the AEG electrical plant, the Puhl & Wagner mosaic factory (which was said to be Germany's foremost), a string of beer gardens on the banks of the Spree, a second-run movie theater, and the 400-acre Treptow Park with its little lake, the Carp Pond. Treptow's only cultural attraction was in the park, the Archenhold astronomical observatory, whose 21-meter telescope had once been the world's longest. It was named after Treptow's most renowned Jewish (albeit Christian convert) citizen, the astronomer Friedrich Archenhold, who had raised the funds for the telescope's construction in 1896.

As members of Treptow's upper economic crust, we lived on its finest street, Am Treptower Park, looking out on the park, in the largest apartment in our building. We had a lordly spread of seven rooms, plus kitchen and a tiny maid's chamber. The centerpiece of the dining room was a big, and expandable, table that could seat a dozen people.

Once a month or so, my parents would throw a dinner party. Wine was served at these parties, the only occasions at which alcohol was consumed at our house. We children were excluded from these dinners and had to listen in our bedrooms to the ever-increasing noise made by the diners. The guests usually included some of our relatives, as well as the furniture dealer, Martin Mendelsohn and his wife Bianca, and the traveling salesman, Alfred Reich, and his wife Hannah.

Between meals, Rudi, Mausi, and I used the fully expanded dining table for table tennis. We had to adopt special house rules that considered balls deflected by cracks at the expansion joints or bouncing off the cloth lampshade. Despite these deficiencies of our equipment, we became accomplished players: Mausi was seeded in international Ping-Pong tournaments, and I could beat any other seven-year-old in Treptow.

The salon was lavish, furnished with *Empire*-style gilt-bronze-adorned mahogany tables, chairs, and vitrines displaying porcelain figurines and chinoiserie, a grand piano, and a huge crystal chandelier. It was hardly ever used, except by my mother when she worked the piano. She played mostly Schubert, Chopin, and Liszt, but also popular tunes from shows and operettas, sight-reading them from sheet music. My favorites among her pieces were Schubert's "Impromptus" and Léon Jessel's "Parade of the Wooden Soldiers."

Persian rugs covered the parquet flooring in most rooms. The apartment also featured a choice selection of bronze statues selected from Ballnath & Stensch's line of Graeco-Roman deities, Goethes, Napoleons, Beethovens, Bismarcks, horses, and elephants that were on sale in your better department stores and specialty shops throughout Europe. However, there was no running hot water, no elevator, and no central heating. In winter we were kept warm by floor-to-ceiling tiled stoves burning coal briquettes. To draw hot water for a bath, a coal fire had to be built an hour earlier under an ancient boiler next to the bathtub, an equally ancient four-legged cast-iron-and-enamel contraption. We rarely bathed more than once a week. (Half a century later, I installed a Jacuzzi whirlpool bath in my California house. Since then I have never missed a day of sensuous soaking in the hot eddies.)

That we lived in Treptow in this walk-up, cold-water flat was typical of my father's predilection for choosing top-of-the-line things, but of the second category. The grand piano in the salon was a Schwechten, rather than a Bechstein. Our car was an Opel, rather than a Mercedes, and our chauffeur, who wore a uniform my mother had bought for him second-hand, doubled as a packer at my father's factory. Like the international set, we traveled abroad on holidays, but by third-class train, and we always stayed in the best second-class hotels. My mother played tennis, a sport then reserved to the upper classes, but on fee courts, rather than joining one of the chic clubs. We children skied, but in the backwoods of the Bohemian Riesengebirge, rather than in the fashionable Swiss Alps.

My father's only total extravagance was his sea-going motor boat, the *Titania II*, moored in Treptow near the beer gardens. The riverine jesters on the Spree often pretended to mistake the *Titania II* for the ill-fated White Star liner, and greeted her approach with shouts of "Iceberg Ahoy!" My father sold the boat just before I was born, after it nearly sank, with him on board, in a storm on the Baltic. Although I never saw her, I always believed, and still do, that *Titania II* was a yacht of the first category.

My religious upbringing, in so far as I had any, took place in the context of Germano-Christianized Reform Judaism. There were no observances of Jewish ritual in my home: no Sabbath candles, no Passover seder or matzoth. Instead of lighting a menorah at Chanukah time, on Christmas Eve we switched on one of the first strings of electric candles seen in Treptow, hanging from one of the biggest Christmas trees in the neighborhood.

Although neither my father nor my mother had much interest in, let alone commitment to Judaism, they did belong to a Jewish congregation.

They were members, as had been my father's parents, of the Berlin Reformgemeinde, whose temple stood in the Johannisstrasse in the center of Berlin, not far from the Friedrichstrasse Station. I imagine my parents maintained their membership mainly because of the inertial forces of family tradition. In any case, a cult tax was levied on all Prussian citizens, payable to the established religious confession of their individual choice. Atheists and others opposed in principle to supporting *any* established religion could designate museums and other cultural institutions as recipients of their cult tax. Thus the law saw to it that no one could get out of paying the cult tax by simply renouncing God. If that option had existed, my father might very well have quit the Reformgemeinde.

The Reformgemeinde was founded in Berlin in 1846 to promote the assimilation of Prussian subjects of the Mosaic faith into the German nation, as advocated in mid-eighteenth century by Moses Mendelssohn, Berlin's "Jewish Socrates." Mendelssohn regarded the orthodox Judaic rite, with its Levantine traditions—including Hebrew liturgy, strict dietary laws, and observance of the Sabbath on Saturday rather than on Sunday—an insuperable obstacle to his fervent assimilationist aspirations. To overcome this obstacle, a Reformed Judaic rite was devised, adapted to ambient Lutheran practices. Sabbath services were shifted to Sunday. Almost all prayers were said in German rather than in Hebrew, with all references to an eventual return to the Land of Israel struck from the prayer book. Men did not cover their heads or wear prayer shawls in the temple. (The nonsectarian Latin word *temple* was preferred over the Greek *synagogue*, which implies a Jewish house of worship.) The separation of men and women during services was abolished; and the congregation's chanting was backed by a mixed choir and the organ. Much of the ritual music was based on works by German composers, such as Mozart, Beethoven, and Gluck. The silent prayer was usually accompanied by Händel's *Largo*, rendered by organ and cello.

Few other Jewish congregations would ever follow this out-of-Berlin radical Christianization of the Judaic rite all the way. Yet the Reformgemeinde pioneered many of the practices that now differentiate reformed from orthodox Judaism worldwide, including cremating the dead and extending the bar mitzvah initiation rite to girls. ("Bar mitzvah" was another term that the Reformgemeinde eschewed; we referred to it under the Lutheran name "*Einsegnung*").

The Reformgemeinde enjoyed the support of many prominent and well-to-do Berlin Jews. It provided a religious haven for those assimilationists whose residual self-respect prevented them from taking the ultimate step in the assimilationists' rejection of their origins, namely conversion to the Christian faith. Four of Mendelssohn's six children, including the father of the composer Felix Mendelssohn, did take that step. But my parents would never have even considered conversion. Neither they nor any of my uncles and aunts on either side of the family took any particular pride in their Jewish ancestry, but they didn't ever mention converts in any terms other than distaste. I still share that distaste, even though I would be hard pressed to defend it with rational arguments.

Reform Judaism was distinctly unpopular with the Prussian anti-Semites, who preferred their Jews orthodox and unadulterated, so that no one would be taken in by a Jew masquerading as a real German. Thus, the anti-Semitic Prussian King Friedrich Wilhelm III forbade the holding of Jewish services in the German language, even though in 1812 he himself had decreed the legal emancipation of the Prussian Jews granting them the rights of citizenship. For that reason, the establishment of the Reformgemeinde had to wait until the more liberal Friedrich Wilhelm IV ascended the Prussian throne.

I first attended a Reformgemeinde service when I was four: the *Einsegnung* of my brother Rudi, who was very much in tune with the right-wing, German Nationalist political orientation of the congregation. To celebrate the occasion, my mother had organized a gala affair at the König von Portugal, a top-of-the-line, second-category hotel on the Moritzplatz. There were dozens of guests, including my many uncles, aunts, and cousins, some of whom had come to Berlin for this event from as far away as Breslau and Saarbrücken. The sumptuous banquet was followed by the performance of a little musical, based on a string of popular hit tunes, whose original lyrics had been adapted to the occasion by a librettist hired by my mother. The performers were members of the family and selected friends, including myself, who had a walk-on part as a toadstool. My mother provided the musical accompaniment on the piano. The role of the lead soprano was taken by Bianca Mendelsohn. After the performance, there was dancing to music played by a live band. As a perfect hostess, my mother had also hired a couple of gigolos, of whom there was then no shortage in Berlin, to provide dancing partners for some of the lady guests who, lacking either husbands or sufficient feminine charms, would have otherwise had to sit out the waltzes, fox-trots, and tangos.

The morning after I dropped off Hildegard at her apartment, I checked out a jeep at the military government motor pool and drove out to Treptow, then part of the Russian sector of Berlin. I found it virtually untouched by the war. All the houses were standing, including Am Treptower Park 53 with our apartment. Everything was exactly as I remembered it: here the baker next to the corner pub, there the stationer across the street from the barber —even the large dummy clock still hung in the window of the mezzanine watchmakers at No. 50—except that it all was smaller and shabbier. Only the park had changed dramatically. With almost all the lovely stands of sycamores and chestnuts chopped down for firewood and the meadows overgrown by weeds, the sylvan idyll of my earliest years had been reduced to a sad savannah. The Carp Pond was still there, though.

I rang the bell at our flat. "I was born in this apartment," I said to the astonished woman who opened the door, "and I want to have a look around." What with my being in Ami uniform, she could hardly turn me away. There were two Ivans with side-arms in our former parlor, who barked at me in Russian, in a tone of voice less friendly than would suit words of welcome exchanged among Allied personnel. Were they mad at me because Truman had refused to share our atom bomb secret with Stalin? Were they offended by my talking to the woman in German, which they probably didn't understand? Had I walked in on a black market deal? Did they mistake my American uniform, whose insignia they probably couldn't read, as Wehrmacht issue and think that I was one of the last leftover die-hard Krauts who had just crawled out of a sewer?

I decided not to stick around for a romantico-tragico-ironical scenario of ending my life in the very rooms in which it had begun, not, as in my nightmare, a victim of the Gestapo, but of fellow anti-Nazis. I fled our flat without explaining to the woman my sudden loss of interest in my nativity.

Having rushed down the stairs to get away from the hostile Ivans, I crossed the courtyard, in which the wooden rack on which our maid used to beat the dust out of our carpets was still in place. My first childhood pal, Karl-Heinz Horlitz, used to live on the other side of the courtyard, on the ground floor of the garden house. We had played in the courtyard as preschoolers, doing gymnastics on the carpet rack, and as first-graders we had walked to school together. Sometimes we hitched a surreptitious ride on the white, horse-drawn wagon of the Bolle dairy. Unseen by the driver, we jumped on the back of the moving wagon to ride on the empty rear seats intended for the milk delivery boys, whose calling the progressive child labor laws of the Weimar Republic had recently proscribed. After school, we roamed the Treptow Park. In summer we collected chestnuts, not for eating—they were inedible—but for throwing them from our hiding place in the bushes at the No. 87 streetcar running along the park's edge; and in winter we skated on the frozen Carp Pond. The Horlitzes were people of modest means. Mr. Horlitz worked in a dry-cleaning plant, for which Mrs. Horlitz collected the customers' soiled clothes.

The Horlitz name plate was still under the door bell. I rang, and Mrs. Horlitz opened the door.

"Hello, Frau Horlitz. I'm Karl-Heinz's old friend, Günter Stensch, who used to live in the front house,"

"Oh, my God! I was hoping that you'd come back one of these days. I heard that all of you got away—Herr Stensch, Rudi, Mausi, and you. But poor Frau Stensch! I heard she died before you left."

"Yes, except for my mother, we were lucky. Rudi moved to London in 1936; my father joined him there in 1938. Mausi emigrated to Chicago in 1937; I went to live with her in 1940. We lost only very few relatives. By the way, Rudi and I changed our last name from Stensch to Stent, because 'stench' means '*Gestank*' in English. Rudi also changed his first name from Rudolf to Ronald, but I kept mine."

"We weren't so lucky, Günter. Herr Horlitz was killed in the last days of the war. Karl-Heinz is missing in action on the eastern front. But I'm sure he's alive. He'll turn up. Remember my younger son, Rudi? He was just a tyke when you lived here. Thank God, he's in an Allied POW camp in Algeria. Would you like to have this picture of Karl-Heinz, Günter? Here he is in his Wehrmacht outfit."

The charmingly smiling soldier wearing the eagle-swastika emblem on his breast was instantly recognizable as the dear, good-natured first-grader of whom I had been so fond. I was dumbfounded by the sudden upwelling of affection and sorrow in me for my most-likely-dead friend in the hateful uniform. These unexpected emotions, so soon following brooding about Hildegard and her Nazi father, made me fear that my resolve to gloat over the Germans' terrible comeuppance might not be so easy to keep. Moreover, while contemplating Karl-Heinz's photograph, I also realized how lucky I was to have been a baited Jewish boy in Nazi Germany, rather than one of those cocky, blond Aryan toughs whom I had envied so much. If my fondest wish had come true—to be a Jew no more, free to join the Hitlerjugend, instead of having to flee for my life—I would probably be missing in action on the eastern front by now, like Karl-Heinz.

Frau Horlitz had another piece of news for me. "Günter, remember your parents' friend, Martin Mendelsohn? He survived and is back in Treptow. He's living just a couple of blocks up the street, in the Puderstrasse. It's from him that I heard that you and your family got away in time before the war. Why don't you go and see him?"

Nobody answered the bell at Martin's flat, but a neighbor coming up the staircase told me that he had gone out shopping and would be back soon. When I returned a little later, Martin was at home. I barely recognized his gaunt face; like all Treptow, he was much smaller than I remembered. He didn't recognize me, of course, and stared at me in a daze, practically in tears.

"Who are you? My neighbor told me that a German-speaking Ami came looking for me while I was out. I was sure he was my missing son, Heinz. But you're not him! Do you know what's happened to Heinz?"

"I'm sorry, Uncle Martin, I don't know what happened to Heinz. But I remember him, of course. I'm Georg and Elli Stensch's little Günter."

Once he recovered from his dreadful disappointment, Martin embraced me and told me his story.

"Not long after you left Berlin in 1938, Bianca and I and our two kids fled illegally to Belgium. We were still in Brussels in 1940, when the Nazis came. Heinz and Ursel—remember her too, Günter?—went underground. Bianca and I were arrested and deported to Theresienstadt."

"Of course I remember Ursel. She was a very pretty girl! How'd you manage to survive the Theresienstadt camp, Martin?

"Because of my experience in the furniture business. They assigned me to the carpentry shop. I built coffins, for which there was a big need in the camp. That's how."

"And Bianca?"

"She was sent on to her death in Auschwitz. When the Theresienstadt camp was liberated, I teamed up with a woman fellow prisoner. We made our way back to Treptow together. She and I were recently married. I'd like you to meet her, but she isn't home right now."

"And Ursel?"

"She survived in the Belgian underground and is now living in Brussels. But Heinz disappeared."

I gave Martin my address and phone number at the Hotel Gossler and asked him to call me if there was something I could do for him.

Meeting Martin and hearing his tragic tale impressed on me even more than the encounter with Frau Horlitz how fabulously lucky my immediate family—my father, Rudi, Mausi, and I—had been in escaping virtually unscathed from the horrible fate to which Hitler had consigned millions of people, Jews as well as gentiles. Come to think of it, I survived Hitler better than merely unscathed. I had positively prospered, at twenty-two sent back to the Fatherland as a scientific emissary of Uncle Sam's and heading that very evening for a date with a charming Berlin actress.

The American Presence

HILDEGARD'S SCHIFFBAUERDAMMTHEATER was one of the thirty or so stages on which the plays and musicals of Berlin's Golden Twenties were produced. Bertolt Brecht and Kurt Weill's *Dreigroschenoper* had had its world premiere in that very theater in 1928. Like the Admiralspalast, the Schiffbauerdamm Theater was one of the few halls that had survived the destruction of the city, and was one of the first in which Berlin theater was revived within a few weeks of VE Day.

As Hildegard had promised, there was a ticket for *Men in White* with my name on it at the box office. I took my seat in the middle of the orchestra, noticing that I was the only person in uniform. As we were waiting for the house lights to go down, I heard someone making a racket behind me. I turned around: It was a drunk. Suddenly, the whole audience seemed to be looking and gesticulating at me. Reluctantly, I got the idea. They were expecting the only uniformed representative of law and order to restore peace in the hall. I pretended not to understand what was wanted of me and buried my nose in my program. But then I reckoned I owed it to Hildegard to see to it that her performance came off undisturbed. So I forced myself to go over to the drunk, who fortunately was not a very big man. In what I fancied was a close approximation to Wehrmacht German, using as my model Hollywood's ubiquitous Nazi officer, Helmut Dantine, I yelled at him:

"Shut up and get out of here, you drunken swine!"

The man glared at me in hurt astonishment and shuffled out of the theater. General applause, which I, the American Presence, acknowledged with a smart salute.

Men in White, for which its author, Sydney Kingsley, received the 1933 Pulitzer prize in drama, was a precursor of such TV soap operas as *General Hospital* and *The Doctors*. It dealt with the lives and loves of physicians and nurses working in a large hospital. To my disappointment, Hildegard had only a bit part, which didn't offer much scope for dramatic talent: a nurse who assisted the young surgeon protagonist in the operating room. But she looked so stunning on stage in her white uniform that I could hardly wait for the play to be over. I was anxious to meet her at the stage door, as I had seen stage door Johnnies meet beautiful actresses in many a movie. Anyhow, I couldn't keep my mind on the action, especially when Hildegard was not on stage, because I was mulling over the strategy I had worked out for the remainder of the evening.

I didn't want to risk being thrown out of another military mess. So I had gone to the PX that afternoon and bought some food for a picnic supper in my room at the Hotel Gossler: Spam, sliced American cheese, Ritz crackers, mayonnaise, canned grapefruit juice and pineapple, Hershey bars, and a fifth of gin, as well as paper plates, Dixie cups, and wooden forks and spoons. I had reconnoitered the access to the Hotel Gossler and found an unguarded rear entrance to the servants' back stairs. By this route we ought to be able to get to my room without having to pass the front desk and avoid scrutiny by the receptionist, charged with keeping out Germans, especially Fräuleins taken by transient officers to their rooms for forbidden fraternization activities. Once we were in my room and had finished eating, I would check out the further possibilities the mean mess sergeant had last night prevented me from exploring on the Harnack Haus dance floor. My plan seemed perfect except for one crucial uncertainty: Would Hildegard agree to come to my room? Being nobody's fool, she would divine what I obviously had in mind.

The dark alley into which the stage door opened was deserted: not a single top-hatted, white-scarved, bouquet-holding cavalier in sight waiting for his actress. Only an elderly watchman guarding the stage door entrance. Being an authority figure in uniform myself, I didn't ask for his permission to enter. I merely barked at him, "Where can I find Fräulein U.?" Tipping his right hand to his cap by way of a salute, he used his left hand to point to a door from which Hildegard soon emerged. I found her not quite as stunning as she had looked on stage from the distance of my seat, but still plenty attractive. She was smiling, clearly pleased to see me.

"I had only half-expected that you would turn up tonight, Mr. Stent. You know, so many men don't keep their promises these days! But then I heard backstage about an Ami throwing out a drunk, and I reckoned that that Ami had to be you."

I grabbed the right hand she extended with both of my hands, to show my delight at seeing her again.

"Fräulein U., wild horses couldn't have kept me away from coming to watch you on stage!"

My literal German translation of this trite Americanism made her laugh, because she apparently took it as an original, frightfully clever turn of phrase I had made up on the spot.

"You put on a great performance tonight!"

Unlike an American, she didn't acknowledge my compliment with a simple "thank you." Rather, in accord with the German rule of etiquette that was still in my bones, she belittled the object of my praise:

"Hardly; tonight wasn't really one of my nights."

I proudly took Hildegard's arm, and we left the theater. The time had come to implement my plan.

"How about another try for an *après-theatre* supper with me in Dahlem? But let's avoid the kind of noisy military dining room where we weren't welcome. Why not let me take you to dine in a *chambre séparée*, in the manner of a gentleman entertaining his lady friend in the good old days of

Kaiser Wilhelm? I happen to know of just such a place, namely my room at the Hotel Gossler. And don't worry about getting home after the last U-Bahn train. Tonight I've got my own jeep."

"Terrific idea! Never dined with a gentleman in a *chambre séparée* before."

Maybe Hildegard didn't suspect what I had in mind for us after all? I drove the eight miles out to Dahlem at a breakneck speed through the empty streets. In no mood to make leisurely detours to take in more childhood reminiscence, I went by the most direct route I remembered. I was tense and silent, impatient to get on with my plan, afraid that something might go haywire at the last moment. When we got to the Hotel Gossler, I went through the front door alone, passed the receptionist at the front desk, and opened the rear entrance from the inside to let in Hildegard. In no time, we were in my room.

As I stood behind Hildegard to help her out of her olive-drab overcoat, she turned around to face me, looking at me in a way that even this obtuse twenty-two-year-old doctoral candidate from Illinois could hardly fail to understand. I put my arms around her waist and kissed her. She put hers around my neck and returned my kiss passionately. I noticed that her breath smelled of tooth decay, but, entranced by our embrace, I soon became insensitive to this countererotic antiperfume. With my arms still around her waist, I lifted her off the ground and onto my narrow bed. Her flaxen hair flowed over the pillow. No Gestapo fiend, not even a busybody American duty officer, burst into my room at the last moment to stop that of which I had dreamt so often.

When we began talking again, lying in each other's arms, we spoke sweetly in the familiar, second-person singular *Du*, instead of the formal third-person plural *Sie* we had used hitherto. Then it was still the general custom in Germany, as it is no longer today, for grown-ups to reserve the *Du* only for relatives, close friends, former schoolmates, and children under fourteen. Everybody else was *Sie*. The change from *Sie*- to *Du*-saying was no trivial matter and had to be offered by the person senior in age or standing to the junior. Thus it was only on the twenty-fifth anniversary of their business association that Ballnath, being the elder of the Ballnath & Stensch partners, said to my father, "Isn't it time, Georg, that we called each other *Du*?" There were a few circumstances, however, under which *Du*-saying took effect automatically: Having climbed a mountain higher than 3000 meters together was one of them; the activity Hildegard and I had just shared was another.

I was surprised how much this linguistic symbol of our having become lovers meant to me. With the *Du*, we had given each other a verbal keepsake, a lifelong token of our intimacy and a constant reminder of our mutual belonging. I now realized for the first time how much the absence of "thou" from modern English idiom emotionally impoverishes the language, although it certainly simplifies social intercourse.

I got up to prepare the promised supper, which turned out to be not only *après-theatre* but also *après-amour*. With a towel for a tablecloth on the little table in my sparsely furnished room, I put out two paper-plate-Dixie-cup-wooden-fork-and-spoon place settings, sliced the Spam with my Boy Scout knife, and fixed meat-and-cheese-and-mayonnaise canapés, to be washed down with gin-spiked grapefruit juice, and followed by pineapple-and-chocolate dessert.

"*Du*, Günter mine, you are really a fabulous cook! I can't remember when I had such a stupendous meal."

"No big deal, Hildegard mine. I'm just a professional chef, who happened to have worked his way through high school as a soda jerk and short-order cook, and who made the Soda Fountain Big Time at Liggett's Rexall Drug Store in downtown Chicago's Northwestern Railroad Station."

The first round of canapés was soon gone. Hildegard was wolfing down seconds as fast as I, the professional short-order cook, could fix them. She was obviously hungry and enjoying the exotic and otherwise unobtainable food I was serving. The atmosphere in the room had changed completely, from tense, longing uncertainty to relaxed, blissful certainty. During our meal, while the spiked grapefruit juice was beginning to take effect, we sang songs in the Berlin patois, *sotto voce*. We were also telling Berlin jokes, the kind that crack up the natives but that people not privileged to have been baptized with Spree water tend to find dumb.

"Know this one, Fräulein Actress? 'Daddy, my teacher says our ancestor is a monkey.' 'Baloney, Sonny! Maybe yours is, but not mine.'"

"Ha, ha! I'm dying from laughter, Mister Scientist. Want to hear a *really* funny story? Once upon a time, there was a young actress who *hadn't* been to the State Opera. On her way home from the Schiffbauerdamm Theater, walking toward the U-Bahn, she passed the Admiralspalast just as the people were coming out after the performance.

"Is she anyone I know?"

"No, of course not! This actress saw a lone Ami in the crowd and decided to have a closer look at him. She thought that if he's really an opera fan, he can't be one of your run-of-the-mill GIs."

"Why not? Opera's very popular in the States."

"Don't interrupt! This Ami *wasn't* a run-of-the-mill GI. He spoke to the actress in German and walked with her to the U-Bahn. So she decided that she was going to go with him, wherever he was going, which was probably near American headquarters in Dahlem."

"Did the Ami invite her to come along with him?"

"No. At least, not at first. As they were waiting at the Friedrichstrasse U-Bahn stop, a northbound train pulled in, which was the train she should have taken to get home to Wedding. But she didn't take it, because he didn't. She got on the southbound train with him and changed to the Krumme Lanke line at Stadtmitte, because he did."

Instead of laughing at this funny story, I was flabbergasted. Well, what do you know? Who could have imagined such a thing? A pretty girl like her following a *schlemiel* like me! I went around the table to kiss her and carried her back to bed.

People in the street were staring at us when I brought Hildegard home in the morning. No doubt, they were making a mean-spirited but not altogether counterfactual interpretation of her being dropped off in front of her house at this hour by an Ami in his jeep. I wondered whether her mother might not have been worried about her not coming home after the theater. Was it possible that Hildegard had told her mother not to expect her, anticipating my carefully crafted plan for her seduction? I didn't ask. In view of her funny story, I was afraid of the answer.

Intelligence (So-Called)

I HAD FINAGLED MY TRIUMPHANT RETURN to Berlin in November 1946 by responding to a news item I ran across in a spring 1946 issue of *Chemical and Engineering News*. A Mr. Charles Collins of the Technical Industrial Intelligence Branch (TIIB) of the U.S. Department of Commerce in Washington was reported to be looking for qualified people to screen technical documents in occupied Germany. I was electrified, thrilled, stunned: Jeepers, Creepers! Here's the ticket to Germany that will make my I'm-back-in-Berlin dream come true!

Dear Mr. Collins,

I want to apply for one of the technical document screener positions in Occupied Germany, which, according to a recent article in *Chemical and Engineering News*, you are seeking to fill. I am a graduate student in physical chemistry at the University of Illinois, carrying out research on synthetic rubber under the sponsorship of the War Production Board. I speak fluent German, and have extensive experience in translating scientific and technical German.

Within a few days, Collins had responded on Department of Commerce letterhead:

Your background seems to indicate that you might readily fit into our overseas program. Please come to my office in Washington for a personal interview, if you are still interested in a position as a Technical Industrial Intelligence Document Screener in Germany.

The Washington interview with Collins took less than half an hour. He explained,

"You'll be attached to FIAT, the Field Information Agency, Technical, of the Office of Military Government for Germany. The agency's headquarters are at the I.G. Farben plant in Hoechst, near Frankfurt-on-Main. As a civilian FIAT investigator, you'll have the status of a commissioned officer of the Armed Forces and wear officers' uniforms without rank insignia. Your appointment as a Department of Commerce employee will run for one year, at an annual salary of about $4,000. If you quit before the expiration of your contract, you'll have to reimburse Uncle Sam for what he spent on your overseas travel from the States."

Not long after the interview, a letter arrived from the Department of Commerce. It stated that "actions are now in progress relative to your proposed assignment by the Technical Industrial Intelligence Branch [TIIB] for duty in the British Isles, Germany, and Northwestern Europe." There was an enclosure listing the equipment to be brought from the States by FIAT investigators. The list detailed not only the components of the uniform that I was to bring along—Eisenhower battle jacket, field jacket, overseas cap, olive-drab trousers and shirts, trench coat and field boots—but also specified a flashlight and a Boy Scout knife and the contents of my toilet kit, such as razor, blades, shaving cream, and toothbrush. This list aroused the pleasurable illusion that I was about to go on a long camping trip, something that I had not done, and had been thirsting for, ever since my childhood days in Berlin.

Getting my appointment processed took much longer than Collins had projected, and I began to worry that maybe TIIB had closed down the document-screening project. Finally, in late August 1946, there came a phone call, asking me to report to Washington as soon as possible, for an immediate overseas departure. Most of my basic uniform I patched together from hand-me-downs of my brother-in-law in Chicago, who had been a staff sergeant in the infantry. His Eisenhower battle jacket and khaki shirts were just my size, and my sister shortened his trousers and took in the waistline.

On reporting to the TIIB office at the Department of Commerce, I was officially inducted into government service and handed a letter:

> By your acceptance of this letter, you assure the United States Government that you will report to FIAT all technological, scientific and industrial data that you may acquire in the course of your investigations; that these data will become the property of the United States Government; that, as one of its officers, you will engage only in those activities in Occupied Germany that are necessary, pertinent, and relevant to your assignment, and that you understand that you are expressly forbidden to engage in private commercial transactions.

The TIIB office also presented me with a Special U.S. Passport, issued to people proceeding abroad on official business. Getting this document meant a great deal to me. It certified to all it might concern that I was no longer the stateless nebbish refugee I had been ever since the Nazis stripped me of my German nationality. Hear ye! The bearer of this here passport is a citizen of the world's Number One nation.

I landed at Frankfurt's Rhein-Main Airbase on Saturday, September 21, 1946, stepping onto German soil for the first time since December 31, 1938. Back in the Fatherland at last! The Nazis had lost the war; Hitler was dead and gone; and protected by my uniform, I could safely go wherever I wanted. I would see how the Living Jesus had been bombed out of the Krauts and how they had been punished for their crimes. The more destruction and misery I ran across, the gladder I expected to feel.

FIAT headquarters in Hoechst were closed for the weekend. So a jeep with a German driver took me from the airbase to the Park Sanatorium in Bad Homburg, a spa at the foot of the Taunus Mountains, where FIAT investigators were billeted. To my disappointment, little war damage was manifest in the small Hessian towns and villages, through which the driver —an aggrandizing Kraut evidently feeling entitled to reckless driving at the wheel of a military vehicle marked with U.S. Army insignia—raced at breakneck speed. They looked exactly as I remembered such places: little houses lining the narrow, suddenly curving road, across which chickens were running for dear life; inscriptions in Gothic letters on shops and taverns; and, most powerfully evocative of my childhood memories of the rural Mark of Brandenburg, the smell of manure. It was all so different from the odorless farming communities on the Illinois prairie, along the arrow-straight highway from Chicago to Champaign.

On reporting to FIAT headquarters in Hoechst, I met my future supervisor, Freddy K., only a couple of years my senior. He was a Czech–Jewish engineering student, who had fled to England from Prague just before the war. To avoid being drafted into the London-based Czech Army-in-Exile, he talked his way into the U.S. Army, where he had risen to the rank of sergeant. After his demobilization in occupied Germany, Freddy signed on as a civilian employee of FIAT. A witty, outgoing, intelligent, literate, bon vivant, ladies' man, Freddy was the only one among my FIAT colleagues for whom I would develop any warm feelings of friendship, and not merely because he did me a few big favors during my tour of duty.

As we were served late morning tea by Freddy's secretary in his sparsely furnished office in the requisitioned administration building of the I.G. Farben Hoechst plant, he clued me in about the agency:

"FIAT's mission is to liberate scientific and technical knowledge developed by the Krauts during the war. At the Potsdam Conference, Truman, Churchill, and Stalin agreed that the Allies can help themselves to it. It's part of the war reparations the Krauts'll have to fork over. You'll be one of the field investigators who are supposed to dig up this stuff. And even though you're a civilian, you're subject to military discipline. They can court-martial you for serious violations of Army regulations."

"Who's in charge of military discipline, Freddy?"

"FIAT's chief is Colonel Ralph Osborne. One of the finest commanders I ran across in my years of Army service."

"How long has FIAT been in business?"

"We got started right after VE Day. During the first year of our operations, field investigators were mainly volunteer hot-shot technical experts, dollar-a-year men on paid leave from American industry. They knew exactly what they were looking for and usually found it. For instance, they turned up new methods for making synthetic rubies, or synthetic gasoline, or synthetic rubber. Once one of those industrial hot-shot investigators had found the novel method he was looking for, he took it back Stateside, to have it put into production at savings of millions of bucks in development costs."

"I know all about the synthetic rubber story, Freddy. I'm on leave from the War Production Board's synthetic rubber research program at Illinois. We were amazed when we heard last year that the Krauts beat us to a much better product!"

"OK, then you probably know that these company types passed on the nitty-gritty working details of the German technical breakthroughs only to their own firms. They usually hid them from their Stateside competition. When word got out about this, Washington decided that, from now on, FIAT investigators were going to be paid employees of the Department of Commerce. So this year, they hired a new crew of people's-own investigators, like you. You're supposed to be working for the nation, and not for the moneybags at Du Pont, Firestone, and Upjohn. There's another difference too. You guys won't be sent out to track down the specifics of particular pieces of novel technology. You're supposed to liberate *all* the hot technical stuff you can find."

"How am I supposed to do this? Go from one place to another and simply say to the Kraut in charge, 'Tell me all about the secret technical information you have, Mister!' What if he lies, like 'Frightfully sorry, Sir. I haven't got any secret technical information'? Am I supposed to extract the truth by torture?"

"Unfortunately, torture is not authorized for FIAT investigators. I wish it were, though. With our hands tied, we can carry out our vacuum-cleaner-like intelligence mission only by looking at technical documents. You guys are being assigned a series of targets. At each of them, you are going to screen all available documents. Documents with scientifically hot stuff you mark for recording on microfilm by a camera crew, manned by Kraut FIAT employees. The films are going to be sent here to FIAT headquarters for indexing, translation, and sending on to Washington.

"The first target I'm assigning you to, along with four other new guys, is one of the world's biggest chemical plants, the I.G. Farben-Bayer complex at Leverkusen, on the Rhine, just downstream from Cologne. You're to interview Bayer's new Director-General, Ulrich Haberland, and his staff and then screen their technical documents."

My fellow Leverkusen investigators were a young Austrian–Jewish refugee, who was, like me, a Stateside physical chemistry graduate student; a middle-aged, scientifically and linguistically unqualified Good-Time Charlie from North Dakota; and an elderly, taciturn Austro–Jewish couple from New York with vague scientific credentials. I didn't find these teammates particularly scintillating company, but then the thrill of being back in Germany more than made up for my colleagues' lack of personal oomph.

When we marched into the Director-General's office at the Leverkusen plant, Good-Time Charlie—the only one of us who didn't speak fluent German—acted as our spokesman:

"Mr. Haberland," he said in English, "we're investigators of FIAT, the Interallied Field Information Agency, Technical. We are here to screen and microfilm technical documents that we know you've got at this plant."

Haberland, then and for more than another decade the biggest wheel in postwar German chemical industry, responded in perfect English,

"Yes, Sir. We have lots of technical documents here. They deal with all aspects of chemical research and production, from pharmaceuticals to heavy chemicals. They go back to the middle of the last century. You will have access to all this material. We are going to provide office space in our central administration building for you and your camera teams."

As soon as we had settled down in our spacious, nicely furnished office at Leverkusen, we sat around the big oakwood conference table to plan a strategy for screening the vast number of documents that Haberland said were held at his plant. I asked my colleagues,

"How are we going to tell which documents describe hot technical information? And which are merely old hat, already known all over the world? Take me, for instance, a student of—let's even exaggerate and say, an expert in—the physical chemistry of large molecules. How am I going to decide whether a procedure for making some drug of which I've never even heard, is hot stuff? We'd have to be some kind of universal geniuses to do a real screening job on all the paperwork piled up here!"

The other physical chemistry graduate student chimed in:

"Yeah. And even if we *were* universal geniuses, what with that huge pile of documents we are supposed to screen here, it'd take us forever. We'd all be still sitting here in this comfy office at the turn of the twenty-first century, still slowly turning the pages of loose-leaf binders. Like Emperor Friedrich Barbarossa sitting in his Kyffhaeuser Mountain cave over the centuries, waiting to save the Holy Roman Empire, while his red beard grows through his table!"

Good-Time Charlie also had a good point:

"How are we going to keep our Kraut camera crews busy? It takes a lot longer to read a document than to microfilm it!"

Thus, before even getting started on any screening, we cottoned on to the futility of our mission, and, indeed, to the hare-brained nature of the whole FIAT document-screening program. So we developed an ingenious strategy to carry out our wholly unrealistic assignment. We were going to make random samplings of the untold thousands of documents stored in the Leverkusen archives and mark with an "X" all those shelves holding binders that seemed to contain a preponderance of recent technical or scientific information—valuable or not for the industry of Allied nations. Shelves with documents dating back more than twenty years or relating to office machines, laboratory apparatus, health insurance, or travel expenses would be left unmarked. We would instruct our camera teams to microfilm the contents of all binders on the X-marked shelves. This brilliant rationalization of our screening procedure left us with a lot of spare time for extracurricular activities and resulted in the exposure of enough 35mm film to stretch from Leverkusen to Washington.

Weekends, we usually drove back from Leverkusen to our billet in the Bad Homburg Park Sanatorium, for relief from the boring Leverkusen scene. There were many interesting things one could do to while away the time in the Frankfurt area, including hiking in the Taunus Mountains and taking in an opera, a play, a movie, or a concert. Bad Homburg itself provided a great source of weekend rest and recreation for off-duty FIAT investigators. Just opposite the Park Sanatorium lay the courts of a private Kraut tennis club, which extended guest privileges to the Technical Industrial Intelligence sleuths. The club members expected that at the end of the day, their Ami guests would leave behind some precious, PX-bought tennis balls for their use. Finding partners for playing a set was no problem, because a better class of tennis-playing Kraut women hung out at the club, who were interested in getting to know the better class of university-educated, German-speaking Amis from across the street. Most of the tennis ladies were available also for off-court fraternization.

On my first visit to the club, Hertha B., attractive brunette, wearing an alluring white tennis outfit—tight-fitting, short-sleeved blouse, pleated mini-skirt, and head band restraining her Lorelei-length locks—came up to me and asked in German,

"Like to join my two friends and me? We need a fourth for doubles."

"Sure. Be delighted. But I'd better warn you I'm not a very good player."

"Don't worry. We're no experts either."

At first I was surprised how well I was playing, slamming shots over the net like a pro. Then I realized that the three ladies actually *were* experts: They were feeding me the ball ever so gently and precisely positioned that any clod could hit it back with pizzazz. Obviously, they were buttering me up, just as, once upon a time, Rudi and Mausi had to let their baby brother win at table tennis, according to my mother's house rule, known in the family as "Güntering."

Hertha, whose Luftwaffe flight surgeon husband was still held in an Allied POW camp, offered to take me to a solo ballet performance at the church of Homburg Castle that evening.

"I'm sure you'll enjoy it. The dancer is going to be my friend, Maria Christiana. She does the modern expressive style. Pioneered in the 1920s by Maria's teacher, Mary Wigman."

Maria Christiana danced alone, without musical accompaniment, barefoot, wearing a plain white smock. I had never heard of the expressive style, let alone seen it done. It all seemed weird to me; as far as I could make out, it wasn't ballet at all, which I associated with pantomime plus acrobatics done against a backdrop of musical Schmalz, like Swan Lake. Hertha, herself a gymnastics teacher, enlightened me:

"The expressive style is based on Wigman's theories of natural movement. Wigman thought that classical ballet was a contemptible way of forcing the dancers' esprit into a kinesthetic straitjacket."

That evening, I had my first encounters with modern dance and—after the performance—with fraternization. I preferred the latter, even though I couldn't cast Hertha, being a brunette from Dresden and speaking German with the—to my Berliner's ears, ridiculous—Saxon accent, for the role of the blonde Berlin shiksa of my nightmare.

Being stationed on the Rhine, I was now a lot closer to Berlin than I had been in Champaign, Illinois. But I still hadn't been able to make my dream come true and work my way across the last couple of hundred miles to my native city. And there was no immediate prospect of getting there since, according to a four-power agreement, Berlin was closed to all off-duty travel by Allied personnel. So I begged Freddy,

"Won't you please assign me to an investigative target in Berlin, if not permanently, then at least for a little while? I need some respite from my inane assignment in Dullsville-on-Rhine."

"This isn't easy. Berlin is everybody's dream duty station. There's a FIAT crew working there in the German patent office, screening some 180,000 pending patent applications filed after 1938. It's too bad, but you haven't got enough job seniority to claim that assignment."

Yet within a week good old Freddy had come through.

"I've picked out two small targets for you in Berlin for which I'm sure you're outstandingly qualified. A firm of consultants specializing in the construction of chemical manufacturing plants and a soap and cosmetics factory. You'll be in Berlin for two weeks. First week, you'll interview some of the staff at your targets and start screening their documents. Second week, a Berlin-based FIAT camera crew will come with you to do the microfilming."

Two days later, I had my official travel orders and got on the U.S. Army's "Berlin Express" at Frankfurt Station.

Mother Goes Away

THE REASON WHY EVERYTHING had looked so small to me in Treptow when I revisited it on the day after I met Hildegard was that I hadn't been there since April 1932. That year we moved from Treptow to the fashionable Wilmersdorf district in the West. So I remembered distances and dimensions of objects in Treptow scaled to the body size of the eight-year-old child that I had been, rather than to that of the twenty-two-year-old adult into whom I had meanwhile grown.

Our moving away from Treptow was motivated mainly by the ever-more virulent anti-Semitism fanned by the growth of Nazi Party strength. It seemed wise for us, well known as a Jewish family in our parochial southeastern suburb, to fade into the anonymity of the more urbane Western part of town. Our new apartment was located in the Fasanenstrasse, just around the corner from the Hohenzollernplatz. (The building didn't survive the war. Its vertical walls were standing when I went to look at it in November 1946, but nothing horizontal was left, no floors, no roof.)

I first noticed that there seemed to be something wrong with my mother on the day in April 1932 that we had moved into the new apartment. Since our kitchen was not yet set up, we were having lunch at a pub on the Hohenzollernplatz. Midway through the meal, my mother suddenly jumped up and wailed,

"For God's sake! I forgot to take my jewelry case out of that dresser of the old bedroom set we sold and left behind in Treptow!"

She rushed out of the restaurant, got into the Opel, and raced out to Treptow. Meanwhile, the buyer had the dresser hauled away, but he denied that there was any jewelry box in it when the movers delivered it. My mother made a big fuss. She called the police and demanded that they search the homes of everybody who might have had access to the dresser after we left, including the building superintendent, the furniture mover, and the buyer. These searches didn't turn up the jewelry, but they created a lot of anti-Semitic ill will.

In the following months, my mother often seemed to be very sad and even cried sometimes, which I had not seen her do before. I thought that this downbeat change in my mother's mood was the result of her being still upset over the loss of the jewelry box. Many years later, in Chicago, my sister eventually clued me in on the emotional trauma that had triggered the onset of her depression. Alfred Reich, husband of Lotte Reich and a member of my parents' small social circle, had been my mother's lover for a few years. He broke off their affair at the time of our move to Wilmersdorf. Mausi knew about it all along because my mother had used her teenage daughter as a confidante. My father found out about it as well, by coming across some of the love letters exchanged between my mother and Mr. Reich, which she had left lying around carelessly in her bedroom. According to my sister, my father was very hurt by this discovery, though instantly ready to forgive my mother—but not Mr. Reich. I remember my being surprised at my father's suddenly referring to our long-time family friend as a *Schweinehund*, because I couldn't understand what had caused his sudden fall from grace.

By the fall of 1933, my mother's depression had become so severe—she often said that there was no longer any point in her going on living—that she was put in a psychiatric sanatorium in Strausberg, about 25 miles east of Berlin. My seventeen-year-old sister now took charge of my upbringing and served as my surrogate mother and only source of emotional support for the next three years.

Although it was fun traveling out to Strausberg in a suburban train pulled by a little steam locomotive, the weekend trips I made with my father to visit my mother in the sanatorium were terribly upsetting for me. No one troubled to explain what was really the matter with her. Her dress was disheveled, her hair uncombed, and her face strangely changed. She was usually in tears, and accused my father of having her locked up in this awful place for no good reason. She protested that there was nothing wrong with her that coming home to the Fasanenstrasse apartment wouldn't fix.

The psychiatrists at Strausberg were unable to do anything for her—the development of antidepressant psychoactive drugs, to which she might very well have responded, was still more than three decades into the future. So after a few months, the doctors decided that she might as well go home. Unfortunately, her condition didn't improve on her return to the Fasanenstrasse. She still talked a lot about killing herself and had to be constantly watched. As for me, I began to have nightmares, in which my mother was about to throw herself out of the kitchen window of our fourth-floor apartment, to plunge down to the stone pavement of the courtyard below. I would struggle to keep her from falling, holding on to her for dear life as I leaned against the window sill, and just as she was slipping through my hands and going to plummet to her death, I would wake up, relieved to find that it had only been a bad dream.

Eventually, my mother did make a real suicide attempt. She got out of bed in the middle of the night, went into the kitchen, turned on the gas oven, and put her head into it. Her attempt didn't succeed; in the early morning she was found lying on the kitchen floor, unconscious but not dead, with the smell of gas permeating the apartment. My nightmare had turned into reality.

Now she was taken to another sanatorium, on the Heerstrasse at the western border of the city, not far from Spandau. There I saw her for the last time, when my father and I visited her on Sunday, July 1, 1934. The three of us went for a stroll along the Heerstrasse, my father and me taking care that we had my mother between us so that she couldn't bolt. We walked in the direction of downtown Berlin, to which this broad avenue, designed, as its name implies, for rapid troop deployment, leads in a straight line. After about ten miles and several changes of name, it finally reaches the Brandenburg Gate and turns into the Unter den Linden. Something unusual was obviously happening that Sunday morning, since one Reichswehr truck after another loaded with soldiers in battle dress was speeding downtown along the Heerstrasse. But we didn't have any clue about what it was that was going on. After we had walked for a mile or so, my father said,

"Let's turn around, Elli, and head back to the sanatorium."

My mother acted as if she had not heard him and just kept on marching.

"Elli, we've got to get back to the sanatorium in time for lunch!"

"The damned sanatorium can go to hell, Georg! I'm not going back to prison. I'm going to keep right on walking toward town until I get to the Fasanenstrasse, where I'll stay put from now on."

"Elli, please! You've got to wait until your state of health allows it."

When she just kept on marching, my father became desperate:

"Grab your mother's left arm, Günter. I'll hold on to her right arm, and we'll spin her around."

As we dragged my mother towards the sanatorium, she kept on hollering:

"I'M NOT GOING BACK TO PRISON. I'M NOT GOING BACK TO PRISON."

Fortunately, the Heerstrasse was virtually deserted, except for the passing

Reichswehr trucks. All the same, I was afraid that someone would see us dragging her along the street and alert the Schupos (as police are called in Berlin) that a woman screaming for help was being kidnapped by a couple of thugs. I wished I had nothing to do with that woman. Finally, we got her back to the sanatorium, with each of us in a different frightful state: my mother furiously stymied, my father impotently exhausted, and I sickened and repelled. After this horrid episode, I lived in terror that he would make me visit her again.

Later that evening, we heard on the radio that, on the Führer's personal orders, many traitors had been executed in Berlin and Bavaria. These traitors were not your usual Jew-Bolshevik conspirators, but, incredible as it seemed, prominent Nazis and Nazi sympathizers. Foremost among them was one of Hitler's oldest comrades and dearest friends, Ernst Roehm, the chief of the SA. I couldn't imagine Hitler's reasons for killing so many Nazis. Had he realized the error of his ways and was turning toward decency? Would he give up his awful anti-Semitism and accept me as a loyal German eager to participate in the building of the new Germany? It was announced on the radio that Reichswehr soldiers had been rushed downtown from Spandau to forestall a takeover of government buildings by the traitorous leaders of the Berlin SA.

None of these official announcements made any sense to me. How could the Führer's own SA stalwarts, whose marches through the Berlin streets instilled in me the terror of torture and death, be traitors to their own cause? Wasn't the holy anthem of the Nazi state, the "Horst Wessel Lied," all about how the SA, flag raised high and ranks firmly closed, marches in step for the Third Reich? I also didn't understand the unofficial story put in circulation that the SA traitors had to be killed because they were moral degenerates, *warme Brüder*, who liked to have sex with other men. At the very moment of his arrest, Roehm was reported to have been found in bed with another SA man. Since I couldn't imagine how one man could have sex with another and was too shy to ask my father to explain to me what exactly it is *warme Brüder* are doing, I remained long dumbfounded by the whole affair.

Meanwhile the doctors in the Heerstrasse sanatorium had finally decided that they were unable to help my mother. So she was taken to the last station on her Via Dolorosa, a psychiatric sanatorium in Freudenstadt (meaning, with supreme irony, "City of Joy") in southern Germany's Black Forest. I never went down there to see her, and my father and Mausi made only a few brief visits. Of all our family, it was only her 20-year-old prospective daughter-in-law, Rudi's fiancée, Gabi Teutsch, who kept my mother company in Freudenstadt for as long as a few weeks. Then, in mid-July 1935, Gabi went home to Berlin.

On Thursday morning, August 8, 1935, I was awakened by long, irregularly spaced rings of our telephone. I knew that it had to be a long-distance call, because local calls announced themselves by short periodic rings driven by the automatic dialing system. Long-distance calls, by contrast, were connected by an operator, who rang the bell manually. I thought right away that the call must have something to do with my mother; perhaps it came from Freudenstadt. When I got up for breakfast, I could tell from my father's, Rudi's, and Mausi's behavior that something was wrong, but I didn't dare to ask what the phone call had been about. I was sent off to school without being told anything. But when I came home in the afternoon, Mausi took me aside and told me that our mother was dead. She had disappeared from the sanatorium three days before. They found her decaying body last night, hanging from a tree in the Black Forest, with the cord of the bathrobe she was wearing looped around her neck. My father and Rudi were now on their way to Freudenstadt to bring her body home for burial.

It's most likely that my mother had inherited her psychotic disposition for suicidal depression from her consanguineous parents. Her mother, Clara, had killed herself at about the same age as Elli, and Elli's sisters, Aunts Katharina and Wally, committed suicide a few years after my mother did. In the fall of 1938, Rudi found Katharina's corpse in her London digs, some days after she had taken an overdose of sleeping pills. And Wally took her life in Paris in May, 1940, as the German troops were approaching the city.

The last members of the family who saw my mother before her death were Wally and her nine-year-old daughter, Cousin Eva. I learned about this final encounter only forty years later, when Eva, who had become a French writer, published her autobiography. Eva reports my mother's pathetic flight from Freudenstadt to Saarbrücken in a chapter that is devoted mainly to the hazards to which her parents were exposed as well-known Jewish opponents of the reintegration of the Saar Region into the German Reich. The Saar had recently come under Nazi rule, on the overwhelmingly pro-German vote in the plebiscite of January 1935 provided for by the Treaty of Versailles. In brazen violation of the guarantees of immunity from persecution under the terms of the plebiscite, the Gestapo was arresting Francophile Jews and spiriting them away to German concentration camps. These arrests were made mostly in the early morning hours.

One morning, our bell rings as if it would never stop and wakes me up suddenly. Maman responds without taking the time to put on a dressing gown and whispers a phrase whose precise meaning then eluded me: "It's 7 o'clock already, and 'they' never come that late."

I follow her, too sleepy to feel any fear. She waits a minute before pushing back the bolt. A moment later she cries out:

"Elli! Where are you coming from in this condition?"

I open up my eyes wide. My aunt, always dressed with care and meticulously groomed, is unrecognizable. Her features are puffed up. Her hair is hanging down into her face. She is wearing a rumpled overcoat, which is open in front and reveals the nightgown she is wearing underneath as her only clothing. She has no suitcase. A handbag is all she is carrying, which she is pressing against her body. How could she have traveled in this state from Berlin to Saarbrücken? Maman pushes her gently into the apartment and repeats:

"Elli, where are you coming from?"

Instead of replying, she trembles from head to toe and merely moans in a heart-rending way: "I can't go on. I can't go on."

I don't see her again. Emmi [our maid] is serving me my breakfast in the kitchen, but I'm not hungry. I'm wondering whether Elli is going to be treated in a clinic. When I come home from school, I'm told to tip-toe and to avoid making any noise. She is resting in my father's room and Emmi mutters:

"Your mother has given her a sedative injection. The poor woman is certainly sick."

Later I see a nurse dressed in a blue uniform. Some days thereafter, Maman lets me know that her sister is dead. She didn't supply me with the details until a long time later. Elli hanged herself in a sanatorium, from which she had escaped and to which the nurse had taken her back. They had been treating her there for a serious nervous depression for many months. The unfortunate woman had put it into her head that her three children, including the younger of her two sons, then eleven, didn't really need her any longer. (Eve Dessare, *Mon Enfance d' Avant le Déluge.* Fayard, Paris, 1976, pp. 82–84)

My reaction to learning of my mother's death was mainly a sense of relief. What I had feared for so long, what had been the recurrent subject of my nightmares for the past two years, had happened at last. My mother finally succeeded in killing herself. Although I could tell that Mausi had cried, I had the impression that she, too, was relieved that the hopeless tragedy had finally ended.

I felt guilty for not feeling more greatly bereaved. Having been effectively motherless for the past two years and with my surrogate mother, Mausi, still right there, I didn't sense consciously an immediate loss. Mausi had told me that I better not go to school for the rest of the week, so that I could collect my thoughts and adjust to my new situation in the tranquility of our home. She would telephone the school.

When I returned to school after the weekend, my teacher and some of my classmates expressed their condolences. I thought that I ought to show more grief and not behave as if nothing special had happened, but I couldn't bring myself to act in the way that I imagined was expected of me. All that came to my mind was that from now on I'd be doubly different from my classmates: not only a Jew but also a semiorphan. During the following weeks I often dreamt that I met my mother, sometimes in our apartment, sometimes in a park, and sometimes in one of the distant places we had visited together. I was always happy to realize in my dream that, contrary to what I had mistakenly believed, she wasn't dead after all.

On my mother's death I became a one-third owner of the Essener Strasse property, which my father had bought as an investment in the late 1920s. For business reasons, he had put the title in my mother's name. Thus we three children, rather than our father, inherited the building. In 1935, he had still no intention of quitting the Fatherland merely because some rabble-rousing Nazis led by an Austrian clown had taken over the government. But Rudi had a more realistic view of what lay in store for us in Berlin: The Nazis were here to stay, and they were going to do us in, one way or another. There was no reason to think that Hitler was insincere in proclaiming his mortal hatred of the Jews.

Rudi, the former Prussian patriot, was now eager to get out of Germany, as was Mausi. Under Prussian civil law, both were of age and competent to cede their share of the property to our father, which is what they did at his urging. But I was only eleven. Because of my father's potential conflict of interest regarding my share in our mother's estate (which included also her own fairly substantial inheritance from her recently deceased father, Richard Karfunkelstein) a court-appointed lawyer became trustee of my financial affairs. My trustee refused to cede my equity to my father. So I continued to share ownership with him until the summer of 1938. Then, in conformance with the Nazi-ordained divestiture of Jewish real estate, the property was sold to an Aryan bookmaker, Erich Schmidt. My share of the fire sale price paid by Mr. Schmidt was deposited in one of those blocked Jewish bank accounts soon to be confiscated by Fatso Göring.

German Youth Movement (Second Category)

SOON AFTER THE NAZIS CAME TO POWER in January 1933, I began to hate myself for being a Jew. I had been aware as long I could remember that the goyim hated us, but this had merely instilled the fear of being maimed or killed by the anti-Semites. I had not yet worked out the rational inference that there must be something terribly wrong with us Jews to evoke so much hatred. But I saw my oppressors' point once I became an avid reader of Julius Streicher's anti-Semitic weekly, *Der Stürmer*, whose grotesque and obscene cartoons of my coreligionists were posted in glass-enclosed display boxes on the Hohenzollernplatz for outreach to any benighted Jew-lovers. I became convinced that we Jews really *are* awful people. So I tried to conceal my shameful membership in the Children of Israel and pass as an ordinary anti-Semite. I didn't yet know at the time that in turning toward self-hatred I merely followed in the footsteps of many of my renowned fellow German Jews, such as Heinrich Heine, Karl Marx, Gustav Mahler, Walther Rathenau, Fritz Haber, and Kurt Tucholsky.

There was a creamery on the Hohenzollernplatz in which I heard in the morning of February 28, 1933—a month after Hitler had become chancellor—that the Reichstag was on fire. It had been the maid's day off, and my mother sent me down to fetch our milk, which was ladled out from a big open vat into the pitchers people brought with them. One of the customers shouted,

"The fire's a Jew-Bolshevik conspiracy. The bastards want to start a

Communist revolution to overthrow our new government! There's not gonna be any law and order in Germany until we string up all those filthy Jew-swine!"

It was my first overt anti-Semitic experience after the Nazi takeover and therefore all the more frightening. I bolted from the creamery with my pitcher still empty.

Among the other shops ringing the Hohenzollernplatz there was one that was of special interest to me: a *Braune Laden*, or Brown Shop. It offered all paraphernalia that your politically engaged National Socialist might need: SA and Hitlerjugend uniforms, as well as Nazi books, periodicals, and banners. During the first year of Hitler's rule, the shop did a land office business selling a handy conversion kit for turning worse-than-useless, plain red Communist flags into swastika-on-circular-white-ground-medallioned National Socialist flags. I stared wistfully at the smart uniforms on display in the shop window, wishing I could wear one of them. I ventured in a few times to have a closer look at all that fascinating merchandise, titillated by the peril of being in the lion's den.

I ardently wished that I could join the Hitlerjugend's junior division, the Jungvolk, whose members' ages ranged from eight to fourteen. The Jungvolk boys wore a smart *Kluft* (an untranslatable colloquialism of German youth and student culture, roughly connoting a blend of "uniform" and "togs"). It comprised short black corduroy pants held up by a broad military belt with swastika-emblazoned buckle and attached small dagger, a brown shirt, a black kerchief, a smartly cocked black forage cap, white, knee-length socks, and *Bundschuhe* (the strapped rather than laced boots of medieval peasants). I often watched them march across the Hohenzollernplatz, in step to the beat of their drummers, carrying spears with fluttering pennants displaying the Jungvolk's logo, the pre-Christian Teutonic rune

The Hitlerjugend had its roots in the German youth movement, which arose at the turn of the twentieth century as a romantic reaction against the smug, materialistic, petty-bourgeois society. Its members rejected the

present and sought flight back into an idealized, pre-Enlightenment past, or forward into a utopian, postmodernist future; to favor collectivism over individualism, and nationalism over internationalism. The movement referred to itself as "Wandervögel" (migratory birds) because hiking and camping were its main activities—to escape from the corrupt, ugly city and wallow in the beloved Fatherland's pristine, beautiful countryside. On first sight, the Wandervögel, with their romantic love of nature, may seem to have been an innocuous, or even spiritually uplifting, crowd—a kind of *Sturm und Drang* Sierra Club. But on hindsight, they helped prepare the ideological ground for the rise of National Socialism.

I despaired that, as a non-Aryan, I couldn't join the Hitlerjugend's nifty Jungvolk. But in the spring of 1933 I almost had my wish when I became a member of a Jewish ersatz Jungvolk, the Schwarze Fähnlein, or "Black Squad." The leaders of the SF (the acronym by which the Black Squad was known) asserted the German Jews' integral membership in the German nation, were virulently anti-Zionist, and glorified Prussian military virtues. Their political orientation was close to that of the fascist League of National-German Jews, led by a fellow-parishioner of our Reformgemeinde, Captain Max Naumann, who admired Hitler for his efforts to restore Germany's place in the sun. Rudi had been a member of the SF, but by the spring of 1933 he had better things to do than playing soldier in the backwoods of the Mark of Brandenburg. He had met Gabi, his future wife, matriculated as a law student in the university, and joined the anti-Zionist, nationalist Jewish fraternity, Kartell Convent Jüdischer Corporationen. To keep up the family's engagement on the anti-Zionist front, Rudi sent the SF his Nazi-loving baby brother. At eight, I was the youngest of the SF members, most of whom ranged from twelve to fifteen. Everybody had a nickname; reflecting my status as the SF's Benjamin, mine was Krümel, or "Little Crumb."

The SF's *Kluft* was similar to the Jungvolk's, except that we wore navy-blue forage caps, Jack-Tar sailor's blouses and ultra-short pants (barely reaching down far enough to cover the scrotum and held up by Imperial Army surplus belts with buckles decorated with the Imperial German Eagle perched above the motto "*Gott mit Uns*"). On our marches, we carried spears with fluttering pennants emblazoned with the *Jungenschaftskeil*, the SF's logo:

And every man lugged an Army surplus backpack, called *Affe* (monkey), because its rear flap was covered by a piece of furry horse-hide.

We practiced military drill on our hikes and camping trips (*Fahrten* in youth movement-speak). They involved long marches into the Mark of Brandenburg, toting heavy *Affen* and sleeping on the bare ground in Army-surplus pup tents (such petty-bourgeois comforts as bedding down in sleeping bags were unknown). Going on *Fahrt* was always a torture for me. The long hikes cruelly overtaxed my physical strength and endurance. Although dead tired, I lay awake all night in the pup tent, unable to sleep, with my bones aching from lying on the hard ground, and usually with a migraine headache. Yet quitting the SF, like a *Flasche*, or wimp, just because I suffered, never entered my mind.

After Hitler's appointment as Reich's chancellor in January 1933, the SF joined other right-wing, nationalist Jewish organizations, including the Union of Jewish Frontline Soldiers and Naumann's League of National-German Jews, in trying to convince the Führer that the elite of assimilated Jewish youth was eager to participate in the building of the new German state. They petitioned the Reich's Chancellery to allow the SF's young stalwarts—self-described as "aristocratic in bearing and convictions"—to serve in the Arbeitsdienst. (This was the very unpopular, compulsory, postsecondary school civilian labor corps mounted by the Nazis.) Moreover, if, God willing, it came to pass, they hoped to serve also in the revitalized, expanded new German army. Insofar as is known, none of these groveling overtures was dignified by an answer. Jewish patriots failed to appreciate the centrality for the Nazi canon of the proposition that the only good Jew is a dead Jew. As for me, I knew all about that fatal proposition, which put me in a perpetual state of terror and fear. Yet I was fascinated by and attracted to the Nazi trappings, like a rabbit facing a boa constrictor, and I wanted to join the self-professed murderers in the worst way.

The military drill we did on our *Fahrten*, especially practicing guard duty, was not a wholly academic exercise. We were in constant danger of being attacked by a Hitlerjugend detachment. In the one such attack I experienced, we were luckily (and most unusually) camping indoors, in a barn in a small village in the outer reaches of the Mark. The Nazi gang came from their camp to storm the barn at about midnight, and our guards managed to get inside just in time and lock the door. Our troop leader assigned defensive positions, with the oldest boys deployed at the places most vulnerable to a break-in, such as the doors and windows, and with me, the youngest, at the center of the defensive circle. We made a lot of noise, shouting commands like "Ten men to secure the southeast window!" to give the impression that there were many of us inside, ready to do battle. But we were too cowed to make verbal responses to the obscene anti-Semitic taunts and chanting of "*Juda verrecke!*" ("Jews, drop dead!") by the howling mob out in the street. They were pounding the side of the barn with their fists and kicking it with their boots. We could not see them because it was a pitch-dark, moonless night. The whole village must have been roused by the ruckus. All I could think of was what the bloodthirsty beasts were going to do to me once they got inside.

The shouting and kicking the barn side stopped after about an hour. Our leader warned us that this was probably just a ruse, to lull us into a false sense of security before their attack: "Remember the Greeks at Troy!" Finally, one of our toughest guys crept outside to have a look and found that they had left. The farmers had probably persuaded them to slake their thirst for Jew blood in some way that didn't entail the destruction of an Aryan barn.

I relived the terror of that beleaguered barn in many a nightmare. Usually the Hitlerjugend did break in. As one of the Nazi beasts was about to plunge his dagger into my chest, I took the only possible escape route, namely forcing myself to wake up. It was at about that age that I first began to develop this handy mental skill for neurotics, which I later used routinely to stop my canonical, postemigration "I'm back in Nazi Berlin" nightmare.

Between weekend and holiday *Fahrten*, we met for *Heimabende* at one another's homes. Accompanied by our troop leader on his lute, we sang songs laden with political or ideological meaning. Favorite themes included the sixteenth-century German peasant revolt and the Dutch liberation of the Lowlands from the Spaniards. But they never mentioned romance. Boy–girl love didn't seem to have a place in the youth movement. In so far as women were mentioned at all in the lyrics, they were usually being said farewell to, as one is marching, battle bound, out of the town, or sailing from the home port to face the enemy at sea. Indifference to women was one of the movement's few shibboleths whose thralldom I managed to avoid totally.

Cossack songs also had a magical fascination for us, despite (and not, as I would now like to believe, because of) the Cossacks' penchant for staging brutal pogroms in the Ukraine. A performance of Serge Jaroff's White Army Don Cossack Choir at Berlin's Sportpalast was a major musical event for the youth movement crowd (not excepting the Jungvolk and the Hitlerjugend). From time to time, I still play the Don Cossacks' "Evening Bell" on a scratchy Electrola record and get goosepimples.

Grotesque as it must seem, I still sing along whenever I hear not only one of our old SF songs but even one of the Nazis' I learned when I was one of their closet admirers. A few years ago, I mortified a friend of mine who went with me to a Washington showing of the reconstituted 1937 Nazi exhibition "Degenerate Art." *Alta voce*, I sang along with the tape-recorded Nazi songs to which I was listening on earphones at the museum. Thus I gave her and the other people who had come to revile Nazi barbarism the impression that I had come to wallow in nostalgia.

When I was in the fourth grade, I wore my SF *Kluft* to a special showing of the movie *Hitlerjunge Quex* in the Atrium cinema, about a mile from my primary school in Wilmersdorf. Made by the *Universum Film Aktiengesellschaft*, or UFA movie studio and released in the fall of 1933, the film was one of the first feature-length Nazi propaganda movies. Attendance was obligatory for all students and welcomed as a relief from classroom tedium. I suppose that, as a Jew, I could have been let off from going to the movie. But I would have sooner died than asked my father to write a note requesting my being excused from this golden opportunity to masquerade as a Nazi. I didn't tell him about it, and I don't think he would have cared one way or the other had he known. My mother, who would have been horrified, was too ill by then to pay much attention to what I was doing.

We all marched in formation up the Kaiser Allee to the Atrium, with the party-member teachers in their Nazi uniforms and the students mostly in Jungvolk outfits. The navy blue of my SF *Kluft* stood out in the sea of brown shirts and caused great interest among my fellow marchers. They wondered whether I happened to belong to an elite marine division of the Hitlerjugend. As we were moving up the Kaiser Allee, we sang the Hitlerjugend anthem "Our flag is flutt'ring in the van; we're marching for Hitler, man for man." Its lyrics were written by the Reich's youth leader, Baldur von Schirach, who drew a twenty-year prison sentence at the Nuremberg trials.

Hitlerjunge Quex dramatized the life and death of the young anti-Communist martyr, Herbert Norkus, killed in 1932 in a street brawl between Nazis and Communists in Berlin's Plötzensee district. It was the first (and maybe only) grand propaganda movie in the history of the cinema specifically targeted at children, as a long-range investment in the producers' political future. The film's objective was to inspire a sense of responsibility in the kids for fighting the nefarious enemies of the New Germany. It certainly inspired nine-year-old me, who would have marched to the front then and there.

In the fall of 1934, the Nazis noticed to their regret that the number of Jews fleeing Germany was declining dramatically, after their most welcome mass exodus in the wake of the anti-Jewish boycott of April 1933. Overt anti-Semitic violence had abated, and, except for their exclusion from government service and institutions of higher learning, German Jews could still lead a fairly normal life. Some emigrés were even coming home. Jewish businessmen, like my father, were making good money under the improved economic conditions of the Third Reich, and feeling secure. There seemed to be no pressing need for them to decamp.

This unwelcome development of restored Jewish complacence led the Gestapo to take a dim view of Jewish nationalist organizations that advocated the absorption of their members into the new German state. Groups such as the SF, with its thesis that there are some Jews who are a breed of Super-Germans, came to be perceived not only as an ideological challenge to Nazi racist doctrine but also as an impediment to a speedy cleansing of Germany of Jews.

Paradoxical as it must seem to people unfamiliar with Nazi mentality, the Gestapo now came to favor Zionist organizations, which advocated immediate Jewish emigration to a remote Levantine backwater. And so toward the end of 1934, the Gestapo dissolved the SF, precisely because it stressed so steadfastly its close ties to the Fatherland and was asking for a role for its members in building the Third Reich. The SF's forlorn efforts to qualify as young Germans just as good, just as brave, and just as patriotic as the Hitlerjugend stalwarts had gone *kaputt*. Many veterans of the nationalist German–Jewish youth movement did get their chance eventually to become soldiers, but not as defenders of the New Germany. They fought to bring it down, as members of the Allied armed forces.

On the fourth day of my return to Berlin in November 1946, I finally visited one of the two targets that Freddy had cooked up for me, the KLEINOL soap and cosmetics plant in the Neukölln district. The firm turned out to be owned by Anglo–Dutch UNILEVER, and hence, as property of an Allied nation, probably exempt from FIAT's scrutiny. Yet the chief chemist seemed happy to have the opportunity to chat with an American colleague when I called on him in his office at the plant:

"I would've been glad to be of assistance to you, Sir, in your search for technical documents, but there's nothing here for me to show you. We've just got a little quality control laboratory. All research and development is done at our main laboratory in Mannheim. Actually, I've got *some* technical documents, namely a few mimeographed patent literature surveys distributed by UNILEVER and a correspondence file with letters regarding patent disputes over hair-dyeing compounds. But they won't tell you anything about soap and cosmetics production, in which, I take it, you are mainly interested. Best place for you to find technical documents on soap and cosmetics is our main office in Hamburg. Why don't you pay 'em a visit? If you've got any further questions, don't hesitate to come back."

Obviously there was no point in ordering a local crew of FIAT photographers to come to this boring place to microfilm anything.

Feeling that I had done my investigative duty, I went to have a look at my old *Penne* (the affectionately sardonic Berlin argot for "secondary school"), the Bismarck Gymnasium in the Pfalzburger Strasse in Wilmersdorf. I had entered the Bismarck's *Sexta*, the lowest of the nine forms leading to the university entrance certificate, or *Abitur*, in the spring of 1934. The school was a *humanistisches Gymnasium*, meaning that its curriculum emphasized Latin, Greek, literature, and history. Mathematics, science, and modern languages, such as English and French, were relegated to minor roles.

It was late afternoon when I got to the school, and a dense fog had settled over the dark city. The ground floor and main portal of the red-brick Gothic Revival building seemed undamaged, but I couldn't make out the rest of it, which was hidden in the fog. Whatever I could see of the school seemed smaller than I remembered.

On my pounding, the front door was opened by an old man bundled up in a heavy winter coat.

"I'm a former student. Could I have a look at the place?"

"Welcome back! I'm the new *Rektor*. Please come in. You're by no means the first Old Boy to return in Allied uniform."

This surprised me, since, in my days, the Bismarck Gymnasium had very few Jewish students. And who else but a Jew would be coming back after the war in Allied uniform? Because of the school's conservative nationalist tradition, Jewish parents preferred to send their sons to the more liberal Goethe Gymnasium, a few blocks up the street. We Bismarckianer tough guys held the spineless Goethianer wimps in contempt.

The new *Rektor* took me to the former *sanctum sanctorum*, the teachers' common room on the ground floor, in which, in my days, no student had ever set foot. The dark, unheated and nearly barren common room hardly matched my boyhood fantasy of it as a plush, wood-paneled, thick-carpeted parlor with leather-upholstered easy chairs and sofas, with whiffs of manly cigar smoke in the air, where our oppressors were *unter sich* (among themselves) and conspired in the victimization of their students.

"The school was set on fire by Soviet troops during the fall of Berlin. The burnt-out upper stories haven't been repaired yet. So the building can't be used for instruction. I'm the only one to live here. I live in the basement apartment, formerly the stoker's. But the school hasn't given up the ghost. For the time being, our classes are being held in the building of the Cäcilien Schule, the former girls' Lyceum just off the Hohenzollernplatz. But the formerly unthinkable *has* happened: Can you believe it? The Bismarck Gymnasium has gone coed. We've got mixed classes."

"What? Gone coed? I *can't* believe it!"

"Remember the Super-Nazi who was *Rektor* in your days? He died in the last days of the war, fighting as a Volkssturm reservist. I took over from him when the school reopened after Hour Zero. Your geography teacher became a general in the Luftwaffe. He's in prison now, awaiting trial as a war criminal."

"What about my home room teacher, Dr. Frisch?"

"He's still on our staff, teaching Latin in the Cäcilien Schule building."

The new *Rektor* gave me a tour of the semiruin. The fog, which penetrated the building through its paneless windows, scattered the beam of the *Rektor*'s flashlight as we climbed the stone staircase, its steps worn by half a century's trampling of boys' boots. He led me along the high-ceilinged, Gothic-arched second-floor hallway to my former classroom.

I couldn't summon a single happy memory about the Bismarck Gymnasium. I had been pleased that I was the only Jew in my class, but that very fact had kept me in a near-total social isolation. I hadn't been subjected to much overt anti-Semitism by my classmates, and yet I had no friends among them. Verbal Jew-baiting had come mainly from the older boys in the upper classes, by way of an escalation of the general harassment to which they were wont to treat all their juniors. My teachers had been distant, strict, and sarcastic. That they treated me with even less than average friendliness was probably more attributable to my being an undistinguished, solid 3 student (on a grade scale of 1 to 5) than to their anti-Semitism. One of my report cards bore Dr. Frisch's notation: "He ought to make better use of his talents and intensify his diligence. He participates in a lively manner in general classroom discussion but tends to create disturbances."

The new *Rektor* took me down into the school yard, where we used to spend our two daily recesses, rain or shine. I remembered the noon recess of Monday, March 18, 1935, when all classes were called out to line up in the yard in closed formation. The old Nazi *Rektor* stepped forward:

"Boys, this was an important weekend for us and for Germany, one that will make us all very proud. On Saturday, the Führer announced that Germany has thrown off one more of the shameful shackles of the Jew-inspired Dictate of Versailles. As of this weekend, Germany has a Luftwaffe again, which can defend the Fatherland against the aggressor enemies that encircle us. And now, boys, I've got a great surprise in store for you."

Out in front of the assembly stood our geography teacher, dressed as a colonel in the smart, new Luftwaffe uniform, with silver braid all over. A silver-sheathed dagger was dangling from his belt, which Luftwaffe officers wore instead of a saber, and his blue tunic was bedecked with medals, which, in irreverent Berlin argot, were referred to as "*Lametta*" (tinsel). As everybody in the school knew, our geography teacher had been an ace fighter pilot during the war.

"Boys, I'm proud to've been recalled to active duty, to help the Führer make Germany invulnerable. I trust that when the opportunity arises, all of you will do your duty too."

The old *Rektor* commanded: "Let's affirm our unwavering commitment to serving the Fatherland, by raising our right arm for the German salute and join in a triple *Sieg Heil! Sieg Heil! Sieg Heil!*"

I would have liked to raise my right arm and join in with the *Sieg Heils*. I was indeed proud of our geography teacher in his gorgeous blue outfit having become such a big wheel in the new German air force. But I was afraid that my classmates would think it presumptuous of me, as a Jew, to give the German salute and shout "*Sieg Heil!*" And so I just stood there in silence without raising my arm, an intruder in my colleagues' happy celebration.

Actually, I needn't have been so squeamish about giving the German salute. More than a year later, the Jewish fencer, Helene Mayer, who was still allowed to represent the German Reich at the 1936 Berlin Olympics, raised her right arm to *Sieg Heil!* at the ceremony in which she was awarded her Silver Medal. The Gold and Bronze medalists in the women's fencing category, Ilona Elek-Schacherer of Hungary and Ellen Preis of Austria, also happened to be Jewish, but they, of course, didn't *Sieg Heil!*

Hitler's remilitarization of Germany was received so enthusiastically by the Bismarck Gymnasium student body—myself included—not merely because it would make our country strong again. We believed also that war was a good thing. Contrary to your pacifist weak sisters, we thought that war brings out the best in people, and we were desolate to have missed out on the glorious times of 1914–1918. So I was disgusted when I heard that a couple my parents knew wouldn't let their boy play with any military toys, such as wooden guns and tin soldiers. I felt sorry for the poor kid, so mistreated by his crazy parents.

My classmates and I were happy when Mussolini invaded Ethiopia in October 1935. At last, there was war again! On our own initiative, we put up a map of the Horn of Africa on our classroom wall. By pinning little flags on the map, we kept track of the shifting positions of the Italo–Abyssinian front. Our sympathies were pro-Abyssinian and anti-Italian, because the nasty Fascist dictator Mussolini was persecuting our fellow Germans in the Alto Adige. On formation of the Berlin–Rome Axis a few months later, however, sympathies for the two sides in the Abyssinian War flip-flopped, virtually overnight. I did not take part in my classmates' celebration of the capture of Addis Ababa by the Italian forces in May 1936. By then I had left the Bismarck Gymnasium for another school.

I thanked the new *Rektor* for his guided tour of the old *Penne*.
"Please greet Dr. Frisch from me. Perhaps he'd be interested to hear that one of his less-than-diligent *Sextaner* rowdies of the 1934 entering class, who tended to create disturbances, expects to get his doctorate in chemistry from the University of Illinois within a year."

The Proud Dead

ON THE SIXTH NIGHT after my return to Berlin, *Men in White* wasn't on. So I arranged to pick up Hildegard in Wedding, to take her to dinner at the Femina. This was a famous nightclub in the Nürnberger Strasse, which had survived from the Golden Twenties. According to Hildegard, we would be able to dine and dance at the Femina for lots of marks and a few cigarettes. I would have to wear mufti, because the place had been posted off limits for Allied personnel. I put on the one suit I owned—a gray, double-breasted, worsted wool outfit—and a college man's soft, button-down-collar Oxford cloth shirt with a Windsor-knotted tie.

Hildegard's face was luminous when she opened the door of her apartment in Wedding. I gave her a small bouquet I had bought at the PX, and kissed her demurely in front of her mother and little brother. She seemed relieved that I had shown up, with flowers yet. Perhaps she had considered the possibility that I had regarded our *après theatre* lovemaking at the Hotel Gossler as a one-night stand and that the other morning she had seen the last of me.

The haggard Frau U. and Hildegard's emaciated little brother stared at me as if I were an alien creature from outer space, especially after I took off my GI wool-lined trench coat and they beheld a civilian dandy the likes of which had probably not been seen in Wedding for some while. Hildegard complimented me on my dapper attire. Maybe this was the first time in her life that she was going to go out with a man in mufti. All her beaux no doubt wore Nazi uniforms before VE Day, and Allied uniforms thereafter.

I gave Frau U. a can of coffee and Little Brother some Hershey bars. They thanked me politely, but were otherwise tongue-tied. Frau U. asked me no questions, and I didn't know what to say to her. I didn't talk to Little Brother either, even though I was usually quick to kid around with boys his age. My silence was due, not to the specter of the vanished Nazi father, Albert U., but to the awful sadness of mother and son. I felt sorry for this certifiably Nazi family, despite my original resolve to gloat over the Germans' terrible comeuppance. This resolve seemed to be eroding fast, as I had first noticed a couple of days earlier, when I saw the photograph of my missing childhood pal Karl-Heinz in his Wehrmacht outfit. Having become Hildegard's lover had, no doubt, something to do with my pitying her folks, but I also asked myself how could her little brother have helped it that his old man was a Nazi? I felt ashamed of what I felt was my lack of moral fiber in being unable to maintain the righteous contempt for Krauts that, as a Jew, I was not at liberty to abandon.

I took care to address Hildegard in the formal *Sie* in front of her mother. Calling her *Du* would have rudely flaunted the fact (no doubt known to Frau U. anyhow) that her daughter and I had become lovers within twenty-four hours of our first meeting. A few days ago, I had wondered how Hildegard's Nazi father *might* have reacted to a hypothetical meeting with the Jew who was squiring his daughter. Now there was an actual confrontation of Hildegard's mother with the Jewish seducer. It wasn't hard for me to picture a grim-faced Frau U. as an archetypal Nazi harpy, with her hair tied in a bun, a swastika brooch fastened to her collar and wearing the uniform of the Nazi Women's Auxiliary, ranting and raving about Hildegard's having stayed out all night with a Jew. I could imagine her calling Hildegard a *Nutte*, or strumpet, and threatening to throw her out if she ever did such a thing again. But here was this cowed, speechless woman, who could thank her lucky stars that she had a daughter attractive enough that she could hustle a little coffee and a few candy bars. As Hildegard didn't seem to be interested in prolonging my interview with her family, we soon left for the Femina.

I parked my jeep around the corner, so as not to draw attention to the illicit presence of an Ami on the premises. We walked in, ignoring the sign posted by the British military police at the night club's entrance that read

OFF LIMITS TO ALLIED PERSONNEL

This sign didn't trouble me because I interpreted it as a benevolent, paternalistic measure to protect me, a member of the new *Herrenvolk*, from coming to harm in this den of Kraut underclass thieves. We passed scrutiny by a bouncer at the door. Most likely, he let in *only* Allied personnel in civvies and their guests, because Germans were unlikely to be able to afford the astronomical prices charged here. In view of the "U.S. Scientific Consultant" shoulder patch sewn on my trench coat, he couldn't have had much trouble identifying me as an Ami.

The club was dimly lit, but even in the darkness one could tell from their haircuts that most of the male patrons were out-of-uniform Amis and Brits, with German women in tow. I was proud to find that Hildegard was the fairest of them all. The menu was very limited: cabbage soup as a starter, goulash with boiled potatoes for the main course, and vanilla pudding under raspberry syrup for dessert, all washed down with a bottle of Rhenish white. But it was certainly a better meal than the picnic supper to which I had treated Hildegard in my room and probably much better than anything she had eaten in a long while. She obviously enjoyed the food, especially the wine. Glasses in hand, we locked arms to drink *Brüderschaft*, which confers the right to say *Du*. In our case we had already established that right via another procedure, of course, but there seemed no harm in solemnizing our liaison by more than one rite. Hildegard's eyes sparkled, and my spirits soared as I looked into them and saw that this extra-special girl was mine.

"*Du*, Günter, I've got exciting news. I've been given a part in the first German production of a Soviet play, *A Day of Rest (Den' otdycha)*. It's going to be premièred at the Schiffbauerdamm on New Year's Eve. My part isn't major-major, but bigger than what they gave me in *Men in White*. Rehearsals are going to start within a couple of weeks."

We clicked glasses:

"*Prost*, Hildegard mine! You're on your way to the Big Time. I know you're going to wow them. Me, I've only got boring news. I finally went to check out the first target on my assignment list for Berlin, a soap and cosmetics plant in the Neukölln. They didn't have one single interesting document worth screening and photographing. So I left empty handed. But even if I don't discover any technical secrets of interest for the Allied soap and cosmetics trade, my trip to Berlin is certainly a tremendous success. After all, I discovered you! You're a three-star attraction, which, in my tourist guide book, means 'worth the trip.'"

Instead of laughing over what I had meant to be funny hyperbole, Hildegard blushed deeply. It shamed me when I realized that she had taken my callous humor seriously. So I quickly changed the subject, and told her about the visit to my old *Penne*, the Bismarck Gymnasium.

There was dance music by a piano, saxophone, clarinet, and bass fiddle. Unlike the band at the Harnack Haus, this combo didn't go in for Austro–German Schmalz. They were into American jazz, including boogie-woogie. My limbs began to twitch in phase with the syncopated rhythm, and I pulled Hildegard to the dance floor. She was not a particularly good dancer, but it didn't take her long to catch on to the basic double-toddle of Chicago-style jitterbugging. When they played a slow number, we danced cheek-to-cheek. I pressed her body against mine, put my mouth close to her ear, and softly sang the lyrics of an Evergreen tango from the Golden Twenties. Aunt Katharina had taught it to me when I was twelve:

> *Je ne suis pas curieux, mais je voudrais savoir,*
> *Pourquoi j'aime les femmes blondes, et pas les femmes noires.*
> I'm not especially curious, but I would like to know,
> Why I don't care for the dark-haired, yet love the blondes so.

Hildegard was entranced by my singing and virtually melted in my arms. Probably, she took my words literally again and believed that I really meant I could love only her with her flaxen locks and not the brunette down the block. But if these lyrics had any deeper meaning for me, it would have been nothing personal: *Les femmes blondes* would simply have meant "shiksas," and *les femmes noires* "Jewish girls." She made me sing it again and again that night, and on every possible occasion thereafter. It was to become *our* song.

Fortunately, as it turned out, Katharina had taught me her own version of this stanza, whose authentic third and fourth lines, which wouldn't have been of any relevance for our romance, actually are

Pourquoi les femmes blondes, ont des sourcils noirs.
Why blonde women have dark eyebrows.

As we were lying in my bed at the Hotel Gossler after our dinner at the Femina, I felt the aura of coming down with a cold, and soon fell asleep. After a few hours, I woke up with a sore throat, and, as Hildegard determined by holding her hand against my forehead, I was running a temperature.

"You'll stay in bed, Mister, take lots of aspirin and drink lots of water. And I'll stick around to take care of you."

"Thanks, but you better leave. When the maid comes to clean up, she's going to tell the duty officer that there is a German woman in my room, and there'll be hell to pay. I'll take you home in my jeep and then go back to bed."

"Nothing doing, Mister! When I say stay in bed, I mean *stay* in bed!! No back-talk! Got it? I'll take the U-Bahn back to Wedding."

As she left, I promised to come and see her either at the theater or at home, as soon as I got over my cold. I was touched by Hildegard's concern for my welfare and liked being bossed around by her.

I slept most of the day, falling in and out of feverish dreams. Successive episodes of my febrile fantasies always deal with some seemingly urgent theme, their action being repeated compulsively over and over again. That day, I kept on hallucinating about my dead mother, of whom I had rarely thought or dreamed. I suppose that being back in Berlin at last, ill, and with a good woman willing to look after me, had summoned childhood remembrances from the deeper recesses of my memory of her, the protectress of my earliest years. By late evening, my fever had reached its crisis, after which it went away during the night. Next morning, I still had a sore throat, but I felt well enough to disobey Hildegard's orders. I had so little time in Berlin left that I didn't want to waste a second precious day of my two-week stay cooped up at the Gossler.

The fever dreams about my mother made me think that I ought to visit her grave. This wasn't one of the sites featured in my I'm-back-in-Berlin nightmare. But now that I *had* come back, didn't filial piety demand that I pay this obeisance to my late mother? It was a cold but sunny day, and so, sore throat or not, I drove out to Weissensee, where my mother is buried.

The Weissensee cemetery is one of the largest Jewish graveyards in Europe. Opened in 1880, at a site then far beyond the Berlin city limits, it extends over an area of more than two square miles. By the time the Nazis had exterminated Berlin Jewry, Weissensee had provided places of eternal rest for more than 100,000 people. The Jewish community had been split into a couple of dozen congregations, all with their own synagogues and idiosyncratic rites and liturgies. Yet Weissensee was the ultimate destination for almost all Berlin Jews (including agnostics, atheists, and even some Christianized apostates). Only the ultra-orthodox Addas-Jisroel had its own place of burial.

When I was seven, my father began to take me to the Weissensee cemetery on what seemed to me to be almost every other Sunday, to visit the double grave of his parents. Neither my mother, who had been anything but close to my father's family, nor my teenage brother and sister, who had better things to do with their Sundays, came along on these outings to Weissensee. They remain in my memory as one of the main father-and-son togetherness activities of my childhood. They also provided me with a lifelong lesson about how piety demands serving the dead, even if one believes that the idea of an afterlife is nonsense and that the dead have no way of knowing what we might be doing on their behalf. I think I was able to intuit at that tender age, without even understanding the terms "piety" and "self-respect," that piety is a form of self-respect.

I liked walking with my father along the elm-tree-shaded, promenade-like paths crisscrossing the cemetery in a geometrical pattern. They divided it up into alphabetically designated "fields," with each field comprising thousands of graves. The paths were lined by ornate family mausoleums and monuments erected by Berlin's richest Jews for self-commemoration. They represented a fascinating hodgepodge of styles—Egyptian, Assyrian, Hellenist, Romanesque, Art Deco, modernist—which I found appealing. Only a few had any Hebrew inscriptions; most bore legends typical of German tombstones, such as

> WE WILL NEVER FORGET OUR DEVOTED, KIND-HEARTED, LOVING WIFE,
> MOTHER, SISTER, GRANDMOTHER AND AUNT,
> SIEGLINDE KOHN

Some tombs even displayed portraits of the deceased, a goyish practice forbidden by Hebrew tradition. I didn't know, and maybe neither did my father, that this camp collection of vulgar heathen tombs contravened the ancient, nobly egalitarian Jewish funerary custom of marking graves of rich and poor alike with only the simplest of headstones. Weissensee is a colossal testimonial to the advanced state of assimilation of Berlin Jewry in Wilhelmine and Weimar eras. Like Angkor Wat, it eventually became an abandoned funerary city that allows the latter-day visitor to recapture the values of a lost civilization.

My father would explain to me who many of these rich people resting in the mausoleums were: the Mosses of the *Berliner Tageblatt* newspaper empire, the Jahndorfs of the KaDeWe, the Wertheims of the Wertheim department stores, the Kempinskis of the Kempinski restaurants. Sometimes he fibbed and made up explanations, such as when he told me that one of the most striking of the mausoleums, bearing the name SALINGER, belonged to relatives of ours on his mother's side. It gave me a sense of being personally acquainted with all these powerful people.

I wasn't all that interested in the plain grave of my grandparents. Since Sigismund died before I was born, and Caecilie when I was four, my paternal grandparents meant almost nothing to me. But I was proud to visit the special section called the *Ehrenfeld*, which honored Berlin Jews who had given their lives in defense of the Fatherland during World War I. These heroes had proved—hadn't they?—that in spite of professing the Mosaic faith, we German Jews were loyal citizens of the Reich.

My father stopped taking me to the cemetery when I was eight, after we moved from Treptow to Wilmersdorf, maybe because our new apartment in the Fasanenstrasse was too far from Weissensee. In any case, once my mother had been confined in a series of sanatoriums on the outskirts of Berlin, we spent many of our Sundays visiting her. When my mother died, my father did not take me to the cemetery for her funeral. By the time I finally left Berlin in 1938, I had visited her grave only two or three times.

On that November morning in 1946, there was no need for me to consult a map for the drive to Weissensee from Dahlem, diagonally across the whole city from its southwestern to its northeastern border. I remembered that once you got to the Alexanderplatz, you just follow the Greifswalder Strasse out for several miles, until you come to the Lothringer Strasse, where you turn right toward the cemetery's main gate.

As I crossed the Alex, I looked in vain for my favorite Berlin statue, the 25-foot-tall bronze Berolina perched atop an equally tall marble pedestal, protectress of the city and icon of her women. According to the turn-of-the-century Silesian dramatist and novelist Gerhart Hauptmann, they are "the most lively, merry, lovable, clever, faithful, piquant, noble, understanding, beautiful and charming creatures in all the inhabited Earth." Berolina was gone, melted down in 1944 to recycle the tons of precious copper she had under her corset. No wonder that with its protectress gone, Berlin was demolished.

I had expected that, as the most extensive physical vestige of the former Jewish presence in Berlin, the cemetery would be in total ruin, desecrated and devastated by the Nazis. But the gate's finely sculptured ornamental wrought-iron work was still in place, as were most of the yellow-brick-faced structures in the entrance area just behind the gate, such as the administration building, the reception hall, the House of Mourning, and a cute little house with the visitors' toilets, styled to match the ceremonial buildings. They had suffered no more than minor damage and had evidently not been set on fire on Kristallnacht.

How was I going to find my mother's grave among the 100,000 burial plots? I couldn't recall where it was, except that it was at the opposite end of the cemetery from the grandparents'. But I did recollect that the field in which she was buried had been opened up only recently and that it therefore would hold only graves of people who died in the 1930s. So I planned to look for a field with modern headstones bearing appropriate dates of death. This stratagem turned out to be unnecessary. I simply sonambulated toward my mother's grave, never troubling to stop and wonder which way to go. Left turn at the reception hall, right turn into the path that runs parallel to a cemetery wall. The path was lined on both sides with mausoleums and monuments I remembered. Some of them were intact and others in various states of collapse, yet still recognizable.

Weeds were growing all over the graves, and the paths were covered by dead leaves, although many of the trees had been cut down, probably for firewood. But there was little of the wanton destruction I had expected to find. Probably the Nazis had decided to defer the major effort needed to raze hundreds of mausoleums and a hundred thousand gravestones until after the Final Victory. As plans found after the war disclosed, they meant to build an autobahn through the cemetery. I seemed to be the only visitor in the vast graveyard that November day, which gave me the feeling of being the lone survivor of my tribe.

As I came to an opening in the wall, I knew instinctively that to get to my mother's field, I had to turn left, go through the opening, and cross the nursery on the other side of the wall. Where formerly had been rows and rows of pretty flower beds, there now was only a fallow field. Remains of horticultural equipment still lay about, as an archaeological clue to the former function of the area. On the far side of the nursery, I found the field with burials dating from the mid-thirties. All the graves were overgrown with ivy, which also covered the narrow trails separating their successive rows within the field. The burial sites turned out to have been filled in strict chronological order, and so I traipsed across the leafy ground, moving closer and closer to August 1935, the month of my mother's death.

At last I spied the seven-foot-tall rectangular slab of highly polished black marble that I recognized as my mother's. It bore as its only inscription, in modern, sans serif bronze letters cast in Ballnath & Stensch's foundry:

> ELIZABETH ELLI STENSCH
> Geb. 4.3.1892
> Gest. 6.8.1935

The little bench my father had bought from the cemetery administration eleven years ago still stood next to my mother's ivy-covered grave. I sat down, and a Berlin song that my father used to sing came to my mind:

> Den schönsten Platz, den ick uff Erden hab,
> is die Rasenbank am Elternjrab.
> The most beautiful place on Earth I have,
> is the garden bench at my parents' grave.

It seemed little short of sacrilegious that a risible piece of *Kitsch* from the music halls of Wilhelmine Berlin was my first thought on this momentous encounter with my mother's grave. But the lyrics actually came close to expressing my feelings. There was something uniquely precious about sitting here on this bench, with the remains of my long-dead, lost mother within a few feet of me. I felt that her grave—with its headstone and bench—tied me to Berlin like no other place that I had revisited, that it had been, and will always be, truly mine. Tears came to my eyes.

That coming to my mother's grave should have aroused in me a depth of feeling for her of which I had not been aware for all those years surprised me. How could it have been, I now wondered, that making this visit had not been one of the major motivation for getting myself back to Berlin in the first place?

In trying to evoke my mother's face, I came up with the countenance of the woman on the snapshots taken on our family trips—to the Baltic seashore, to Switzerland, to the French Riviera—that Mausi had managed to bring with her to Chicago. Throughout the war years, I had looked again and again at these pictures from my childhood, and they, rather than direct remembrances of her, now formed the basis of my image of my mother.

The School for Self-Esteem

THE VISIT TO MY MOTHER'S GRAVE reminded me that, in anticipation of her eventual release from the Heerstrasse Sanatorium, my father had decided to abandon our upscale Fasanenstrasse flat. Her psychiatrists had advised him that morbid remembrances might connect her to the Fasanenstrasse. So at the end of 1934, about six months before her death, we moved into a new apartment, a few blocks away, up the Kaiser Allee. I had wondered at the time of our move what my mother would be like when she did come home, afraid that she might not be her old self, keep on behaving strangely, and try to kill herself again. As it turned out, the move was all for nothing. My mother never did come home.

After leaving my mother's grave at the Weissensee cemetery, I went to see whether the building into which we had moved in vain anticipation of her coming home was still standing. Driving along the Kaiser Allee, I passed the ruin of the Atrium—the cinema, to which I once had marched in formation as a little would-be Nazi to the showing of the *Hitlerjunge Quex* movie. Its massive, curved colonnaded front aspect, styled to match the theater's Latin name, still stood, but the rest of the house was gone. The facade had the patina of an antique monument, and small shrubs were growing out of its cornices. Allied bombs had transformed this piece of architectural kitsch into a romantic ruin evocative of the splendor of the Roman Forum.

The building at Kaiser Allee 31A had survived the war intact, except for its artillery-pockmarked stucco facade, while most of the other houses that once lined the broad Kaiser Allee had been reduced to rubble. The fat retired acrobat who had reigned as superintendent in our days was gone, and a sign in the entrance hallway announced that our former apartment on the first floor had been converted into a boardinghouse. As I had done in Treptow, I rang the doorbell and asked the woman who answered it to let me in and have a look around. Here, in the American sector, there were no hostile, armed Ivans, so I could make a leisurely tour of my childhood abode. The flat—a twelve-room spread, featuring a glass-enclosed winter garden, a blue-tiled bathroom with a king-size sunken tub, and a hallway large enough to accommodate a regulation-size Ping-Pong table—looked just as we had left it. The brown plush carpeting we had installed in 1934 still lay in the hallway and the corridors.

Six months after my mother's death, the banquet celebrating the double wedding of Rudi to Gabi Teutsch, and of Mausi to Rudi's KC fraternity brother, Heinz Münsterberger, was held at the Kaiser Allee apartment. A year before the wedding, Heinz had quit his medical studies at the University of Berlin, because Jews would no longer be licensed for the practice of medicine. On Mausi's pleading, my father took in her ex-student fiancé as a salesman at Ballnath & Stensch.

Both couples had a civil wedding at the municipal registry office on February 9, 1936, Rudi's twenty-second birthday. After the civil ceremony, Gabi (in a short-sleeved, white satin wedding gown, elbow-length white gloves, and veil) and Rudi (in tails and white tie) were wed before God in the Reformgemeinde temple by the Reverend Dr. Rosenthal. The Reformed ritual was Christianized and wholly in German, lacking such features of the traditional Jewish chassene as holding the chuppah over the couple or stomping the wine glass from which the couple drank after the rabbi's final seven blessings. Mausi and Heinz were in the audience but did not take part in the ceremony, because Heinz was rabidly opposed to any religious ritual, not excepting the utterly secularized liturgy of Reform Judaism. After the ceremony, both couples received a throng of well-wishers in the lounge of the temple. And then we returned to the Kaiser Allee, where the superefficient, nineteen-year-old Mausi, with the aid of an au pair girl and a hired helper, had prepared her own and her brother's wedding feast.

There were about thirty guests, which included only relatives of the two couples. This banquet was our final extended family shindig before the lights went out. It was the last time until after the war that I saw my aunts and cousins on my mother's side, who had come to the wedding from Breslau and Saarbrücken.

Mausi and Heinz defined my notion of the perfect couple: spiritual and romantic togetherness, epitomized for me by their weekend trips in Heinz's folding kayak on Berlin's vast interlocking network of rivers, lakes, and canals. My sister was a good-looking, altruistic, artistically gifted sportswoman, totally devoted to her man. And Heinz was a tall, handsome, athletic intellectual, who was the smartest and best-informed person I knew until, in my mid-twenties, I fell in with a crowd of world-class academics. Heinz seemed to have an inexhaustible store of historical, geographical, political, and scientific knowledge, as well as being totally *au courant* on current affairs, able to answer almost any question I might put to him. His speech was like a professor's, deliberate and carefully articulated.

I believe that it was my failed quest for the ideal conjugal life modeled for me in my childhood by Mausi and Heinz—kayak and all—that brought me a lot of frustration when I began to take out American girls at high school and college. The American ritual of dating seemed to me more like a crass adversary proceeding than the spiritual and romantic togetherness that I was looking for. That is not to say, of course, that it wasn't mainly my own emotional deficits that prevented me from finding my ideal model of the boy–girl relation.

After the Gestapo-ordered dissolution of the SF in the fall of 1934, I joined another Jewish Wandervögel-like outfit, the Bund Deutsch–Jüdischer Jugend, or BDJJ. It had been founded in the previous year as a coalition of all remaining non-Zionist Jewish youth groups who, like the SF, considered themselves primarily Germans. But they were less rabidly Fascist or militarist and did not demand their natural right to take part in the building of the New Germany. They just wanted to be allowed to continue living in the Fatherland, with full entitlements as citizens. The BDJJ's ideological orientation was close to that of Uncle Hugo's CV, namely that we are not members of any so-called Jewish people but German citizens who happen to believe in the Hebrew religion. The Gestapo had not dissolved the BDJJ, along with the SF, in 1934, because they were probably unaware of the BDJJ's professed aim of achieving a synthesis between a German and Jewish way of life, rather than of preparing us for quitting the Fatherland.

The *Fahrten* of the BDJJ were much more relaxed than those of the SF. Instead of the SF's forced marches that tested every member to the limits of his endurance as a preparation for military service, they went in for easygoing nature tourism. We wore no specific *Kluft*, because meanwhile the Gestapo had forbidden Jews to wear uniforms in public, including even the old Imperial Army uniform in which the patriotic veterans of the Union of Jewish Front-line Soldiers loved to parade. Outwardly, I complained that I missed the good old SF and that I didn't like associating with a bunch of BDJJ wimps, who seemed only one step removed from petty-bourgeoisie. But inwardly I was happy to be released from the suffering I endured on the SF *Fahrten*.

In the fall of 1936, the Gestapo dissolved the BDJJ as well. The Nazis had decided to tolerate none but Zionist youth organizations, who, in their view, were the only ones making any serious efforts to encourage their members to clear out. Thus ended my four-year, preteen youth movement career.

Diplom Ingenieur Harry Pauling was the other of the two investigative targets that Freddy had cooked up to legitimize my spurious FIAT mission to Berlin. The firm specialized in the design of plants for the production of industrial acids, of which it had erected many inside and outside Germany. I wondered whether my target's eponymous Harry Pauling might not be a relative of Linus Pauling, the famous physical chemist and chairman of Caltech's chemistry division.

In my junior year at college, I had been assigned Linus Pauling's *The Nature of the Chemical Bond*. It was the first obligatory textbook I enjoyed in all my previous thirteen years of doing time in German and American schools. Thanks to the brilliance of Pauling's exposition, I was able to grasp the principles of quantum mechanics and their application to chemistry, which, so I had been warned, you've got to be a genius to savvy. Grateful that he made me a genius, I made Linus Pauling my first academic hero.

Harry Pauling was not at his office when I called on him in my investigative mission in the second week of my mission, and Mr. Stahl, his business manager, had never heard of Linus Pauling.

"There are only a few technical documents on hand here, Sir. Many of the records of our Berlin were destroyed when our former head office near the Kaiser Wilhelm Memorial Church went up in flames during an air raid toward the end of the war. Most of the surviving documents were then evacuated to Bavaria, but the railroad wagon containing them was looted on the way. The small fraction of the documents that finally reached Bavaria is still there. Why don't you go to Bavaria and have a look at them, if you're so interested in the production of acids? I'd be happy to give you the address where the documents are stored."

I wasn't going to be fobbed off so easily.

"Please show me whatever technical documents you do have in your office."

He produced a few cartons containing some ring binders, and I was thrilled to find that there were a couple of dozen files stamped "SECRET." All I could make out was that they contained documents pertaining to a secret procedure for manufacturing sulfuric acid. I had no idea, of course, whether this procedure would be of interest to anyone in the Allied countries. But at least I had found something SECRET worth photographing, to justify my two-week junket to Berlin. I officiously instructed Mr. Stahl:

"You will make these files available to my microfilm team, which is going to come here in a few days,"

With my mission successfully accomplished, I headed back for the Hotel Gossler. On the way, I stopped to have a look at the grounds of the Private Waldschule Kaliski, the last school I attended before leaving Berlin in 1938. The PRIWAKI, as we called it, was a fancy private Jewish school, housed in a millionaire's mansion in the poshest part of Dahlem. I found that the PRIWAKI building, unlike the Bismarck Gymnasium, had survived the war intact. But the romantically landscaped garden, with its swimming pool and tennis court, had been disfigured by the addition of two Nazi-style office buildings. After the PRIWAKI was closed in the spring of 1939, the German foreign ministry had taken over the property and built the two satellite structures on the grounds. Throughout the war, the mansion had been used

The School for Self-Esteem

by Ribbentrop as a suburban *pied à terre,* a refuge from the bombs falling on his downtown office in the Wilhelmstrasse. Now the place was being used by the city as a home for the physically handicapped.

I had switched from the Bismarck Gymnasium to the PRIWAKI in the fall of 1935, after the Prussian minister of education had declared that all non-Aryan students ought to leave the public schools and that the Jews should look after the education of their children in segregated facilities. It was clear that life was going to be very unpleasant for any Jewish kids whose parents didn't take the hint, as my father wouldn't. So Mausi decided for him that I had better transfer to a Jewish school. I found this idea highly repulsive, because in anti-Semitic German idiom, the very word *Judenschule* epitomizes a filthy, noisy, and chaotic pigsty. In calling us to order, our Bismarck Gymnasium teachers sometimes yelled at us to remember that the gymnasium was not a *Judenschule.*

Friends had recommended the PRIWAKI to Mausi. To humor my beloved sister, I graciously consented to be interviewed by the school's eponymous founder-owner, Lotte Kaliski, then not yet thirty, and by its thirty-five-year-old director, Dr. Heinrich Selver. Fräulein Kaliski had a pretty face and curly hair, and Dr. Selver looked debonair.

It was love at first sight. Nothing could have been further from my image of a *Judenschule* than the PRIWAKI. The mansion and grounds were well kept; they exuded an elegant, yet informal, homelike ambiance, in stark contrast to the grim Bismarck Gymnasium scene. The Jewish teachers were as young and friendly as Fräulein Kaliski and Dr. Selver, and the well-dressed children obviously came from upper-class families. None looked as though they hailed from the Scheunenviertel near the Alexanderplatz, where the Yiddish-speaking Eastern Jews plied their rag peddler's trade. Best of all, the school was coeducational.

The public elementary school to which I went in Wilmersdorf had a girls' division, physically separated from the boys' by the schoolyard, across the middle of which ran a high cyclone fence. During recess, I could see the girls on the other side of the fence, as distant objects of desire. Once the girls' parallel class came over to our room for a joint lantern slide show. We boys acted up so much in our agitation that the event was never repeated. I had been amazed to learn by reading a German translation of *Tom Sawyer* that in the American paradise, teenage boys and girls attend school together, so Tom could conduct his romance with Becky Thatcher in the classroom. And in the PRIWAKI I too would have girls as my everyday classmates.

Dr. Selver agreed to accept me, despite my undistinguished academic record at the Bismarck Gymnasium. The three years that I would spend in the PRIWAKI were going to be the happiest (or the least unhappy) of my Berlin childhood.

My admission to the school was contingent on my taking private lessons in French and enrolling in bonehead modern Hebrew, to catch up with my class, which had been studying these languages, rather than Latin and Greek. I didn't mind taking lessons in French, of which I had some conversational smattering, but I was unenthusiastic about having to learn Hebrew. I never did make much progress with Hebrew beyond mastering the alephbeth. But in French, I soon caught up with the class, because Mausi arranged for me to be tutored by Gabi's elder sister, Steffi Teutsch, who had recently completed her teacher's training. Steffi also taught me chess, and we played a game or two after each lesson.

Lotte Kaliski's styling her school as a *Waldschule* was not to signify that it was tucked away in a *Wald*, or forest, like the witch's gingerbread house. Rather, she meant to indicate that it, like the Walden schools now popular in the United States, was inspired by the pedagogic ideals derived from Rudolf Steiner's "anthroposophy." The *Waldschule* movement arose in Germany in the early part of the twentieth century. Its aim was to spare innately hale children the affective disorders caused by serving a twelve-year sentence in authoritarian *Pennen* such as the Bismarck Gymnasium. The *Waldschule's* aim was to develop the child as an all-around human being, by replacing tyranny with friendship as the foundational teacher-pupil relation. In the case of the PRIWAKI, the concept of the child's wholeness was meant to include a proud awareness of being Jewish.

The PRIWAKI with its elementary and secondary divisions had a total enrollment of about 350 students. It was run as a *Tagesinternat,* or daytime boarding school, with formal instruction from 8 A.M. to noon. A communal full midday meal (*Mittagessen*) was served in the dining hall, followed by informal instruction in sports, gardening, and French and English conversation until 6 P.M. Thus the school dominated the daily life of its students, with their family contacts limited to the *Abendbrot* evening snack at home and to weekends. For many, the school *became* their family, as it became for me. The family sentiment was undoubtedly fostered also by a feeling of being sheltered within the precincts of the school from the cruel, Jew-baiting Nazi world on the outside. Instead of looking forward to school holidays and vacations (and wishdreaming that the school had burned down), as I had done at the Bismarck Gymnasium, I hated to miss even a single day at the PRIWAKI. I could hardly wait for vacations to be over.

Within a year of the Nazi takeover in 1933, it had become obvious to Fräulein Kaliski and Dr. Selver (as it had not to my father and many other loyal German citizens of the Jewish faith) that there would be no long-range future for us in the Fatherland. Consequently, the curriculum of the PRIWAKI was intended to prepare its students for their emigration, rather than for the German *Abitur*. French, Hebrew, and English were emphasized in the curriculum, because these were the languages spoken in the countries where most of the students were likely to end up. Instruction in English was especially intensive, which provided me with a fluent command of my future primary language, albeit with a lifelong indelible German accent. On the other hand, I remained woefully ignorant of the very academic subjects that would turn out to be relevant for my future career, such as mathematics and science, and even of history and world literature. Nevertheless, the PRIWAKI faculty, which consisted of youngish Jewish intellectuals interested in all aspects of Western and Near-Eastern culture, did teach me what it means to be an intellectual.

The French teacher, Paul Jacob, also known as "Monsieur Jacquot," had great histrionic talents (I think he was a failed actor) and accompanied his beautifully accented French elocution with evocative facial expressions and bodily movements. When I lived in Paris in the early fifties, I tried to ape Monsieur Jacquot's Francophone persona.

Erwin Jospe, our music teacher, didn't just teach us how to play the recorder and sing German, French, English, and Hebrew songs. He also introduced us to musicology, especially to the history of music and the analysis of opera. He worked us through Bizet's *Carmen*, playing the opera's instrumental music on the piano and singing all the parts himself. I owe to him the understanding that music is more than entertainment and that it has an emotional meaning.

Zionism was taken very seriously at the PRIWAKI. My family had always rejected Zionism as a ridiculous idea. Who would want to live in a place where there are nothing but Jews? They defined a Zionist as a Jew who tells another Jew that he ought to move to Palestine. But our Jewish history teacher, Ludwig Kuttner, made us read Lev Pinsker's late-nineteenth century, proto-Zionist tract, *The Auto-Emancipation of the Jews*. Pinsker declared that, in view of anti-Semitic persecution being as virulent as ever, post-Enlightenment emancipation of European Jewry had been a failure. Eighteenth-century advocates of assimilation, such as Moses Mendelssohn, had been whistling in the dark: The status of the assimilated Jew demands an impossible compromise. Pinsker insisted that our salvation lay not in assimilation, but in vigorous assertion of Jewish nationality. We must settle in some land of our own. (Pinsker did not think that that land was necessarily Palestine.) Although most of Pinsker's tract was beyond my grasp, I could see that what the Nazis were about to do to us justified his prophecy that we'd all be dead in the long run unless we could take refuge in a sovereign Jewish state. Yet I clung to my anti-Zionist, pro-assimilationist *idées reçues,* regardless.

Thanks to the influence of Arthur Kohn, the German teacher, I first became aware that there was such a thing as moral philosophy, as distinct from morality. While discussing Schiller's *Wilhelm Tell,* Dr. Kohn asked whether we thought that Tell was justified in ambushing and killing Gessler, the tyrannical Austrian governor. The general opinion of my classmates, informed, no doubt, by Old Testament values, was that Tell was not only justified in killing Gessler, but also was rightly celebrated by the freedom-loving Swiss people for his courageous assassination of the fiend. However, I, the little moral-philosopher-in-the-bud, disagreed:

"I think nobody has the right to take the law into his own hands. I think Tell was a murderer."

Dr. Kohn was astonished. "You know, Stensch, this is the first intelligent thing I've ever heard you say in this class."

One of the most important reasons for my liking the PRIWAKI so much was the friends I made there. With three of them—Ralph Koltai, Günter Steinberg, and Gerd Rawitscher—I formed a quadrumvirate of bullies at the top of the class pecking order. We were the strongest and could beat up even some upperclassmen. Occasionally we fought with one another, but on the whole we preferred terrorizing the other boys.

There was a girl in our class, Ursula A., on whom my two buddies, Rawitscher and Koltai, as well as I (and some other, no-account boys), had a crush. But none of us knew what to do about it. Dating, in the American sense, did not exist, and we were too shy even to pass—Tom-Sawyer-to-Becky-Thatcher-like—mash notes to her in class. My main form of courtship was to tease the object of my adoration. I constantly ridiculed Ursula, a native of Dresden, for her Saxon-accented German, and I would also let out the air from her bicycle tires. More to the point, while we were taking our afternoon swims in the school pool, I would sometimes grab her from behind and, feeling her already well-developed bosom, push her head under water. Rawitscher tried a different approach and succeeded in going for a bicycle ride with her in the Grunewald, which gave me fits of jealousy.

Ursula had tremendous sex appeal for the boys, even though (as later I ascertained from scrutiny of old photographs) there were several other girls in our class who were much better-looking, more amusing, and smarter than Ursula. But none of us was erotically attracted to them. Ursula's paramount oomph for us inexperienced *nudniks* must have arisen from her self-confident demeanor as queen of the ball, rather than from her natural endowments. I had often gone out of my way to pass Ursula's house on my bike, gazing up longingly at her window. I was hoping that, maybe, she would look out and see me and ask me to come up. But that never happened.

A decade later, on my way back to the Hotel Gossler after my postwar visit to the PRIWAKI grounds, I made a little detour via the street on which Ursula had lived. Looking up from my parked jeep to what used to be her window still aroused some yearning for her. Too bad she can't see me now, the U.S. Scientific Consultant, back on an intelligence mission in Berlin, doing the town with his blonde starlet! Come to think of it, I thought as I was looking up at Ursula's onetime window, what has my blonde starlet been up to for the past few days? Maybe she is worried about me? Shouldn't I have let her know that I had recovered from my fever?

When I got back to the Gossler, the receptionist handed me a letter that, he said, a young woman had left for me in the afternoon. It was from my blonde starlet.

19 November 1946

For You, Günter!

It's possible, of course, that you will come and visit me tonight. I want to leave at least a few lines here for you, since I am someone who reckons with all possibilities. This evening I have an all-night rehearsal for a radio play, from 6 P.M. until 5 A.M. I am sorry that I won't have any time for you. How are you otherwise? Probably better than last week!!

Since I know that you are neither disappointed nor unhappy that I'm not at home, I would like to ask you to fetch me from the theater tomorrow. Will that be possible? If not, leave me at least a little message here. Or is that asking for too much? Is there any news with you? None?! Only that your lack of interest in women has intensified since you met me!!

I hope you haven't forgotten where my theater is!!

Hildegard!

She really *is* something special, I thought. So she *had* noticed my gauche remark on the night of our first meeting that my main interest in life is science rather than women after all! But having no way of knowing whether I meant to insult or flatter her, she showed the good manners to ignore it. And I was moved—no, shamed—by the way in which she preserved her dignity in the face of my not having kept the promise I made when she left my bedside of getting in touch with her as soon as I got over my cold. To act with all the dignity I could muster to try to match hers, I resolved that those wild horses wouldn't keep me from fetching her at the theater tomorrow night.

Why hadn't I kept my promise to Hildegard? Hadn't I been happier keeping company with her than with any other woman I met in the two years since my beloved college fiancée broke our engagement? It wasn't that I was indifferent to Hildegard—considering her no more than an easy lay or just another conquest, like Don Giovanni's *duecento trent'uno [donne] in Al'magna*. No, I cared for her a lot. Could it really be that I didn't get in touch with her as promised, merely because of a temporary satiation of my erotic needs—like not being hungry after a good meal?

I didn't ask myself these questions at the time, because I had no insights into—indeed, was not interested in analyzing—my own feelings. But as I reflect half a century later on this first instance of my incoherent and eventually horrid treatment of Hildegard, I think I now know the answer: Because I had lived under a self-imposed emotional isolation since my early teens, it simply never occurred to me that Hildegard might have an inner life of her own.

Although my emotional self-isolation probably began with the onset of my mother's depression, it came into full bloom after Rudi and Mausi's double wedding in 1936, when my father and I were left alone in our grand Kaiser Allee spread. My mother had died; Rudi and Gabi had moved to London; and Mausi and Heinz had taken a little flat of their own. On this dissolution of our family household when I was twelve, I lost all live-in adult formative guidance. Separated from Mausi, my surrogate mother, whose primary familial obligations were now owed to her husband, and my being unable to take my father seriously, I had to find my way to deal with my mother's suicide and the Nazi threat.

The self-protective strategy I worked out at the onset of puberty for handling this emotional turmoil was to barricade myself against unpleasant or threatening input from the outside world. Already deficient in understanding the emotional foundations of interpersonal relations, I walled myself off from the thoughts, experiences, and feelings of others. Thus, besides having turned into a self-hater, I also developed into an empathy-deficient loner.

Practicing Escaping

WITH MY MOTHER DEAD AND RUDI AND MAUSI GONE, our grand spread in the Kaiser Allee had become too big for my father and me So in the spring of 1936, we moved to a smaller apartment in the Schöneberg district, at Badensche Strasse 53. This was to be the last place in which I lived before fleeing from Berlin in 1938.

To while away the time before picking up Hildegard at the Schiffbauerdamm theater on the day after my postwar visit to the PRIWAKI mansion and finding the letter she left for me at the Gossler, I drove to Schöneberg to have a look at our last apartment. The building at Badensche Strasse 53, like that at Kaiser Allee 31A, had survived the war. The elevator being out of order, I climbed the stairs to the fifth floor and found that our old lodgings had no front door. They were burned out, without window panes, and uninhabited. But the internal walls were standing, which made the floor plan still recognizable.

One of the rooms had housed the virtually unused salon, with its lavish furnishings, including the huge crystal chandelier, the *Empire* mahogany vitrines with their chinoiserie, and the grand piano. We had dragged this swank mobiliary along from one apartment to the next, as a pious token, I imagine, of our once intact family. I had had the master bedroom with a panoramic view over Schöneberg, in line with the never-abandoned tradition established by my mother of spoiling me at the expense of all other members of the household. By now, the panoramic view from my room over Schöneberg was gone, blocked by a new building whose foundations were being laid at the time we had abandoned the apartment.

There had been a large, glass-enclosed bookcase in my room, which held the remnants of the family library, as well as my own books. The family books included an abbreviated edition of the Great Brockhaus Encyclopedia (which I consulted for information about sex, such as the female reproductive organs, shown as gross—in both the technical as well as aesthetic sense—anatomical sections), and the first erotic novel I had read, *Jahrgang 1902*, dealing with the affair of a fifteen-year-old boy with a mature woman during World War I. My own books included the collected works of my favorite author, Erich Kästner, including *Emil and the Detectives* (whose setting was the Hohenzollernplatz neighborhood I had once roamed) and *The Thirty-Fifth of May*, as well as many volumes of Karl May's adventure stories.

The nameplate over the doorbell of the building superintendent's apartment on the ground floor still bore the name "Jascowski." I rang the bell and identified myself to Frau Jascowski as the son of the former tenant of the bured out fifth-floor apartment.

"Why, you've got to be Günter Stensch! How's your father and your beautiful sister, Frau Münsterberger? I hope they're alive and well. They were such nice people. We thought of you often during the war."

"They are alive and well, Frau Jascowski. My father lives in London, and my sister in Chicago. I joined her there in 1940. That's how I became an Ami."

"What a pity you had to leave your home because of those damned

Nazis! My husband was killed in Berlin in the very last days of the war. Drafted into the Volkssturm. He always said Hitler would be the ruin of Germany. How right he was! Back in Berlin for long?"

"No. Just for a couple of weeks. On special mission for the U.S. military government."

"We still got some of your books and a few of the other things that you left behind in the apartment when you suddenly disappeared on Christmas Day, 1938. Have a look at them!"

There it stood: my glass-enclosed, dark-oakwood bookcase containing the remnants of my library. The Brockhaus Encyclopedia was there, as were *Jahrgang 1902* and the Karl May volumes. The works of Erich Kästner, however, were gone. The Nazis had put them on their *index librium prohibitorum,* because Kästner had joined other intellectuals in signing an anti-Hitler manifesto just before the Führer became chancellor. Noticing the upwelling of my emotions on beholding these things that once were mine, Frau Jascowski asked me whether I wouldn't like to take some of the books with me. Embarrassed by her offer, and much to my later regret, I said, "No, thank you, I wouldn't know what to do with them."

It took me years to make up for this rash refusal and to reconstruct my lost library, searching second-hand bookstores all over for copies of the books that would have been mine for the asking.

Not long after our move to the Badensche Strasse apartment in 1936, my father engaged the services of a matchmaker, Frau Cohn, to find him a wife. As a middle-aged widower with few social contacts and absorbed by his business affairs, this was, no doubt, the easiest way for him to locate a suitable candidate. Instead of looking for one, Frau Cohn made up her mind that she wanted my father for herself.

They must have reached some level of intimacy because, as it soon transpired, my father had blabbed to her about how he and Heinz had recently tried to smuggle out money to London, hidden in a Ballnath & Stensch table lamp. (It was the cash my father had given to Mausi for ceding him her one-third interest in the Essener Strasse property. The lamp arrived in London without the money, which had been stolen by one of the firm's packers before the crate left the factory.) Worse yet, he had bragged to Frau Cohn how, in violation of the 1935 Nuremberg Law for the Protection of German

Blood and German Honor, he continued to have sexual relations with shiksas. And worst of all, he had entertained Frau Cohn with Hitler jokes, of which he had a large repertoire.

In the end, my father decided he didn't want Frau Cohn, and so she blackmailed him: If he wasn't going to marry her, she would denounce him to the police for having smuggled money abroad in cahoots with his son-in-law (*Devisenschiebung*), fornicated with Aryan women (*Rassenschande*), and denigrated the Führer (*Führerbeleidigung*). Even though all three of her threatened disclosures were true, my father told her to go to hell.

One morning in December 1936, detectives called at the Ballnath & Stensch office, asking for my father and Heinz to come along to the police station with them for questioning. Heinz happened to be out of the office. The police told my father that an anonymous letter had come in denouncing him and Heinz for *Devisenschiebung,* and my father alone for *Rassenschande* and *Führerbeleidigung*. My father asked to see the letter. It was in Frau Cohn's hand. He identified her to the police, categorized her allegations as a pack of lies, and accused her of blackmail. Despite the gravity of these charges, my father was released after the interview.

Mausi managed to tip off Heinz, who not only stayed away from the office, but never went back home. He met Mausi downtown, and they withdrew some money from their bank account, mainly to buy tickets and winter clothes for their escape. That same evening, we saw them off at Stettiner Station, on their way to Denmark. They had no intention of coming back to Berlin and were planning to emigrate to Chicago, where distant relatives of Heinz's, Alice and Jules Weil, were living. At the outbreak of World War I, the Weils happened to be touring American vaudeville stages as a husband–wife dancing act, and, rather than going back to wartime Germany, they settled in Chicago. The Weils had sent Heinz and Mausi the affidavit of support needed for applying for the U.S. immigration visa.

My father was making plans for his own escape, but in contrast to Mausi and Heinz, he intended to come back eventually. Using his brother, Uncle Fritz, as an intermediary, he hired an Aryan lawyer to fight Frau Cohn's denunciation, confident that, despite the truth of her allegations, the lawyer would get him off the hook. After all, there was no evidence other than the word of one lying Jew against that of another. Meanwhile, he and I would lie low in Italy's most fashionable, world-class skiing resort in the Dolomite

Alps, Cortina d'Ampezzo, in the fabulously expensive, five-star Hotel Cristallo. Normally, my father would have hardly picked a top-of-the-line place such as Cortina for our stay, but it turned out to be the only refuge abroad to which we could flee on short notice and pay for our keep in German money. For the sake of promoting people-to-people contacts among the newly betrothed Axis partners, the Nazis allowed German citizens visiting Italy to prepay their hotel bills in nonconvertible Reichsmarks.

As soon as my father, not without some pain, had shelled out a small fortune for a month's room and full board at the Cristallo, we were off to the Dolomites, with prepaid, round-trip, third-class railway tickets. We had to take the long way round to Cortina, via Zurich, Milan and Venice, rather than going by the much shorter, direct route to the Dolomites, via Innsbruck and the Brenner Pass, because we couldn't travel across Austria. In that winter of 1936–1937, the Austrian chancellor, Kurt von Schuschnigg, had annoyed Hitler with his pig-headed obstruction of the Nazification of Austria. To show his displeasure with the Austrian government's opposition of the popular will, the Führer forbade German citizens to cross the Austro–German border without a special permit stamped in their passports. This permit was not given to Jews. We had very little cash, except for the few marks that my father had managed to smuggle out.

We broke the two-day train ride with a twenty-four-hour stopover in Venice. This was my first encounter with *la Serenissima,* and, blasé little twerp that I was, I thought she was overrated. My architectural tastes were corrupted by the modernist penchant for clean lines permeating the Berlin aesthetic and by the Nazi/Fascist proclivity for monumentalism. On the single ride we took on a vaporetto along the Grand Canal, I thought, a true transalpine barbarian to my bones, that the palaces fronting it were just a bunch of decrepit dumps for which a Bauhaus-style facelift would do a world of good.

At the Cortina train station, we were met by a horse-drawn sleigh from the Cristallo, driven by a top-hatted coachman. When we arrived at the hotel, there began the embarrassing routine that would be repeated throughout our month-long stay. After depositing our luggage at the reception desk, the coachman indicated to my father that a tip would now be in order. At first, my father pretended that he didn't understand what was wanted of him, but finally, after the coachman had made his request clear in unmistakable terms, my father told him that he had no money for tips. As the coachman shuffled off, it was evident from his demeanor that my father had made a negative contribution to people-to-people contacts among the Axis partner nations.

The Cristallo was palatial: elegantly furnished, crystal-chandeliered lounges, bars, and dining rooms downstairs and large, comfortable bedrooms upstairs with views over the mountainsides. The members of the staff of most interest to me were the pint-sized, Call-for-Philip-Morris bellhops, or *piccolos,* in red livery topped by a brimless cylindrical hat, worn at a rakish angle. They couldn't have been much older than me. I thought that I would have loved to be one of them and strut about the hotel in their gorgeous outfit. The *piccolos* would probably have been only too happy to take *my* place, exchanging their poor cowherd fathers for rich daddies like mine, with whom they would stay as guests in the hotel and piss on the mean concierge.

When we went down for our first evening meal, my father was wearing a business suit. The headwaiter wouldn't let us into the dining room.

"I'm very sorry, Signor, but formal evening dress—dinner jacket with black tie for gentlemen—is required here for dinner."

To my embarrassment, my father made a big fuss.

"This is ridiculous! The Cristallo representative in Berlin didn't tell us anything about this foolishness. So I didn't bring my dinner jacket along. We've prepaid a month's stay in full pension at this hotel, with three meals per day, and I insist that we get them."

"No problem, Signor. I'll arrange to have your and the Signorino's dinner served in your room. You're welcome to take your breakfast and lunch in the dining room, for which your present attire is perfectly adequate."

My father went to complain to the hotel manager, who was adamant.

"I deeply regret this misunderstanding, Herr Stensch, and the failure of our Berlin representative to acquaint you with the standards of dress prevailing in hotels of our category. But we cannot change our house rules on your behalf. You are free, of course, to move to one of the more modest *albergos* in town, where formal attire is not required at dinner. And I will personally see to it that, on your return to Berlin, you will receive a full refund for any unused portion of your prepayment."

Since our cashless condition made moving to another hotel in Cortina out of the question, there was no option but for us to stay at the Cristallo and eat our dinners cooped up in our room. The room service was surly, because the waiters knew that my father was a deadbeat who wouldn't give them a tip for lugging our food upstairs. Except for the morning coffee, we only drank tap water, because all beverages were supplementary. It was all very humiliating for me—not exactly the grandiose life at a world-class five-star hotel to which I had been looking forward when we set out for Cortina.

These shameful childhood experiences with the untipped coachman and waiters at the Cristallo are, I suppose, responsible for my lifelong aversion to requesting any personal services from hotel staff. Even though for some time now I have had enough spare change for handing out a few tips, I still try to avoid being shown to my room by a bellman on arrival or having my luggage taken down to the lobby on departure, and I never call room service to bring up breakfast.

It well-nigh broke my heart that I could only watch the skiers cavort on the slopes, since our penury didn't allow me to rent skis, ski pants, and boots. I couldn't even go for hikes on the snow-covered mountain trails, because I had only brought street shoes. So all I could do all day long was to hang around in the splendiferous lounges of the hotel, watching the *beau monde*, which seemed to consist mostly of elderly Italian, English, and American couples. I never ran into any kids my own age in the Cristallo with whom I could strike up a friendship, and so, except for my father's not exactly thrilling company, I was left all alone.

The hotel had a reading room, where an international collection of newspapers and magazines was available to the guests. There I discovered a stack of back numbers of the *National Geographic*, and my main project for killing time became poring over them, cover-to-cover. This not only improved my English but also taught me about aspects of God's Own Country that I hadn't picked up from Karl May's adventure stories about Redskins and Trappers, and also boosted my yearning for living in the States. I was especially taken by the ads, which revealed that in America ordinary folk, and not just millionaire plutocrats, can go on steamer trips to Hawaii, own fast motor boats, drive fabulous automobiles (there was a picture of a Plymouth sedan priced at $510), attach house trailers to them (to take along one's own hotel rooms), send sons to private military schools, and take a "Pause That Refreshes" by drinking delicious Coca-Cola (which I had never tasted) in the company of beautiful girls. The military schools and the beautiful Coca-Cola girls fascinated me most of all. No one would have suspected by scrutiny of the *National Geographic* that the United States was then in the depths of a paralyzing economic depression.

As the month-long prepaid exile in our Golden Cage was ending, Uncle Fritz phoned my father to let him know that the lawyer thought the prospects for his beating the denunciation rap looked good. Frau Cohn had agreed to declare that she had misunderstood my father, in return for his not pressing the blackmail complaint. But the police had not yet decided what to do with his case. So he wasn't yet fully in the clear, and it would be unwise for him to show up in Berlin at this time.

Since I was tired of hanging around and wanted to get back to the PRI-WAKI and my school friends, my father agreed that I would return to Berlin alone. He would cool his heels in London, where Mausi and Heinz were now waiting for their U.S. immigration visa. So the whole family was going to be reunited in England, except for little Günter, who would be heading home to face the Nazis all by himself.

As we checked out from the Cristallo, the concierge informed us, "Unfortunately, *Signori*, the coachman is occupied." So we had to lug our suitcases to the train station ourselves.

On coming back to Berlin from Cortina, I moved in with Uncle Fritz, Aunt Hannah, and my Cousins Eva (age fifteen)and Walter (age fourteen). The few weeks I spent with my relatives were happy ones. I felt more sheltered and less lonely than at any time since the onset of my mother's illness. I shared Walter's bedroom, and although Aunt Hannah was strict about putting the light out at what seemed an unreasonably early hour, every evening we kept on talking from our beds, long into the night. There was a floor-to-ceiling tiled stove in Walter's room, which we used for rock-climbing exercises. And on weekends we roamed the Berlin streets and parks. This was my first (and only) experience of a live-in companionship with a boy my age, for which I was particularly hungry after my lonely weeks at the *Cristallo*.

It was also thanks to Walter that I became an opera buff. His gymnasium class had been taken to a matinee of *Der Freischütz,* and he had liked it so much that he started going to the opera on his own. To let his cousin in on a good thing, he took me along to a performance of *The Merry Wives of*

Windsor. A would-be man-about-town stepping out into grown-up high life, I loved it, from the overture in the darkened house and the rise of the curtain on the brilliantly lit set of the garden of Mistresses Ford and Page, to the grand finale in Windsor Forest. Our next opera was *Madame Butterfly*, at which I first heard my future national anthem, as Pinkerton's leitmotif.

Sunday mornings, Walter and I queued at the box office for cheap seats. It wasn't hard to get tickets for the second-string City Opera in Charlottenburg, but to swing an affordable ticket for the State Opera one needed the great luck of getting to the box office just after the tickets had first gone on sale or some of them had been returned for a refund. So we managed to make it into the State Opera only a few times.

At carts of second-hand book vendors in the streets, Walter and I bought the little orange Reclam paperback librettos, sometimes "in the English way" (as streetwise Berlin kids euphemized shoplifting). But when we stole books, we did it as a sport rather than for lack of money or kleptomaniac sublimation of frustrated desires.

Once, as we were waiting in the ticket line at the State Opera, a big man came into the lobby, yelling, "*Juden raus!*" He shouted that he was a Sudeten German, exploited and oppressed at home by the dirty Jews who run the *Tschechei* (the derogatory German name for Czechoslovakia).

"It's a bloody outrage that here in Berlin, where the Führer is putting things right at last, crooked Jews are still allowed to take opera seats away from honest Germans."

Walter and I were terrified. We didn't stir, staring down at the floor, hoping to become invisible. Nobody left the line, and the man eventually disappeared. We never went back to the State Opera.

By the Easter school holidays, my father still had not come back to Berlin, and Mausi and Heinz were about to leave for America. So I traveled to London alone to see my family. I was dreadfully upset by the prospect of not seeing Mausi again for God knows how long. More upset, I think, than I had been two years earlier when she informed me that our mother was dead. My fondest wish was that I would be able to join Mausi and Heinz in America before too long. But for now, I had to go back alone to Berlin.

Within a week or so after my visit to London, my father got a message that his lawyer had managed to persuade the police to drop the case against him. In retrospect, it seems amazing that at this late stage in the progressively more brutal Nazi persecution of the Jews—little more than a year before the Kristallnacht pogrom—it was still possible for a Jew to hire a lawyer to beat down a denunciation to the police for having committed the crimes of *Devisenschiebung, Rassenschande,* and *Führerbeleidigung*. Countless people were in concentration camps on much flimsier charges, with nobody caring whether these charges were true (as they *were* in my father's case). So, on Uncle Fritz's advice, my father came back to Berlin from London, despite desperate efforts by Mausi, Rudi, Heinz, and Gabi to restrain him, even at the last moment, as he was boarding the boat train for the Harwich-Hook Ferry at Liverpool Street Station. My father and I moved back into our apartment, and for him it was business as usual.

With Mausi gone for good, my father hired a live-in housekeeper, Fräulein K. She was a plain, roly-poly Jewish spinster in her early thirties, who wore her hair in a bun. Our housekeeper was Jewish, not by my father's choice, but in conformance with the same Nuremberg racial law that he had already violated, which sought to protect German honor also by proscribing domestic service by Aryan females in households with adult Jewish males.

I liked Fräulein K. She was a good cook—I especially savored the cakes she baked—and she didn't try to boss me gratuitously. She soon noticed, as my father hadn't, that I sometimes labored to breathe, and wheezed. Under her grilling, I admitted that those were not the first times I had trouble getting air into my lungs—in fact, it usually happened when I had been on *Fahrt* with the Schwarze Fähnlein.

Paying no attention to my strenuous objections, she dragged me to her doctor, who diagnosed my wheezing as bronchial asthma. With antihistamine drugs then still undiscovered, the only treatment he could prescribe was taking a cold shower every night before going to bed. The idea was to force hyperventilation of my bronchial tract by my gasping for air induced by the cold shock. I scrupulously followed doctor's orders, feeling proud of myself as one tough guy. The treatment didn't do a thing for my asthma, but it made me feel better just to know that what ailed me had a name.

The small room in which Fräulein K. lived was connected directly to mine by a (locked) door. The furniture in her room included a vanity with a large mirror placed kitty-corner opposite the connecting door. One day I noticed that I could view the mirror from my room through the keyhole.

When Fräulein K. had gone out on an errand, I adjusted the position of the mirror, so that it would produce a reflection orthogonal to the line of sight through the keyhole, providing me with a view over most of her room, including her bed. By that indirect optical pathway, I watched her undress nearly every evening, with the light turned off in my room, which, as I discovered by experimentation, improved my visual image of Fräulein K.

Night after night, I was in a state of well-nigh unbearable erotic agitation. I had never seen a live, nude woman before, only statues. So I was unprepared for Fräulein K.'s ample breasts and pubic hair. Sometimes she didn't undress fully within my field of view, and I lost sight of her while she was still in her underwear, which excited me even more. I fantasized getting into bed with our housekeeper-Venus, begging her to instruct me in doing whatever it is a man and a woman do together and which I was aching so badly to learn. A few times, while she was undressing and scratching herself under her arms, I resolved to turn my fantasy into reality and enter her room through the unlocked door opening into the hallway. But lacking the nerve to implement my resolve, I never opened her door.

In May 1938, my father gave notice to Fräulein K. because he was about to marry my future stepmother.

The Surprise Stepmother

AFTER MY VISIT TO OUR OLD APARTMENT in Schöneberg, I drove to the Schiffbauerdamm theater to pick up Hildegard. When I got to the stage door, the elderly porter saluted me and said: "Fräulein U. is waiting for you inside, Sir." Hildegard flew into my arms, and we kissed and kissed. It didn't seem to bother her that some of her colleagues were watching us carrying on. Though ecstatic to hold her in my arms again, I felt embarrassed at making a spectacle of myself and gently pushed her away.

As soon as I had dissolved our embrace, she started to bawl me out.

"Why didn't you let me know that you were still alive? Admittedly, your clever actress managed to guess correctly that you weren't on the critical list any longer when she came to the Gossler yesterday and found that you were up and out. But weren't you clever enough, Mister Scientist, to realize that she was worried about her Wunderkind and that you ought to pass on to her the good news of your recovery? After all, Amis like you don't grow on trees!"

"I'm sorry I gave you cause for distress, Fräulein Actress, but I thought that coming down with a little cold wasn't such a big deal."

Ever the clod, I didn't indignantly deny Hildegard's ironic conjecture in her note that, since I hadn't been heard from, my earlier professed lack of interest in women had only intensified after I met her. All I managed to produce by way of sweet talk was "Your suggestion, Fräulein U., that I call for you at the theater tonight made me very happy. It relieved me of the anxiety of wondering how and when we would meet again."

"*Du*, Günter, I hadn't realized that you're such a big *Dummkopf*. What was there to wonder about?"

She hooked her arm under mine and we walked out of the theater, pressing our flanks together. We went to a restaurant, the Gambrinus, around the corner from the theater, where they knew Hildegard and would accept cigarettes instead of ration cards for serving us supper. During the meal, I gave Hildegard an account of what I had been up to after leaving my sickbed. I told her that I had gone to visit the grave of my mother.

"I didn't know that there was a big Jewish cemetery in Weissensee. For God's sake, Günter, why didn't you tell me that you were going to go out there? I would've loved to come along and see it, especially to visit your mother's grave with you!"

"I would've been glad to take you, Hildegard-mine. But the patient hadn't even thought of going to Weissensee when you left him on his sickbed."

I couldn't bring myself to confess that her offer to take care of me in my illness was probably the main cause for bringing on my febrile hallucinations about my dead mother and for visiting her grave. Instead of advancing to that level of verbal intimacy with her, I merely kissed her.

Then I told her about my visit to the PRIWAKI. It turned out that Hildegard also didn't know that there had been any Jewish schools in Berlin. In fact, her notions about the Jews seemed to be highly abstract. Apparently it had not occurred to her that they are born as babies, go to school, fall in love, may be unemployed, get sick, and are buried in cemeteries, just as everybody else. During her childhood she must have thought of them simply as dangerous devils, world poisoners of all nations. After Germany's Zero Hour in May 1945, the Jews suddenly turned into the great contributors to German arts and science, martyred victims of the Nazi butchers. No doubt, it was easier for her to think of me as a freak Ami who happened to have been born in Berlin than as a real, live specimen of that mysterious, vanished Jewish race, which was just as well, as far as I was concerned.

After supper at the Gambrinus, we went to spend the night at the Hotel Gossler. There I told Hildegard more about my family background, especially how, after the onset of my mother's fatal illness, my elder sister took care of me until she got married and emigrated with her husband to Chicago and how I acquired a surprise stepmother.

"After my sister left for the States, my father hired a housekeeper. She took good care of us for a year, until he suddenly fired her, without telling me why. It all came out, though, when, to my surprise, he proposed a week or so later that we visit the zoo."

"Why were you surprised? Seems normal to me for Daddy to take Sonny to the zoo."

"Not this Sonny! Daddy knew that I was interested in trains and motor cars and bored by stupid lions and crocodiles. The only thing I liked about the zoo were the open-air military band concerts, and there wasn't one that day. After we took a quick look at the dumb beasts, my father made another unusual proposal, namely that we stroll down the Kurfürstendamm."

"What was so unusual about *that*?"

"He'd always badmouthed the Kufürstendamm as a street where suckers are overcharged for stuff you can get elsewhere at half the price. When we got to the Café Möhring at the corner of the Uhlandstrasse and he said, 'Let's go inside for some ice cream,' I was beginning to wonder whether he'd lost his wits."

"Why? What's so crazy about offering to treat Sonny to some ice cream?"

"Because he'd explained to me time and again that all those fancy cafés are into nothing but *Nepp* (as Berliners call price gouging). Their ice cream at two marks a scoop is no better than what you can get from any street vendor for twenty pfennig."

But in we went, with him heading straight for a table at which a fashionably dressed, heavily made-up, dark-haired woman was seated. She looked up and smiled at us.

"'Günter,' my father said, 'meet Friedl, your future mother.'"

"You're making this up, Günterlein!"

"Like hell I am. A few days later, they were married at Schöneberg Town Hall, and Friedl simply moved into our apartment. She replaced the fired housekeeper and *eins, zwei, drei,* took charge of our ménage."

Hildegard laughed: "Funniest story I ever heard, Günter! Was she a good mother?"

"I got on pretty well with her. She obviously sought my approval and spoiled me. She also bought new clothes for her ill-dressed men, including my first pair of long trousers. By Friedl's dress code, every garment had to be *uni,* or of a single, plain color. Only gray or navy-blue suits were allowed, with red *uni* ties to be worn with gray suits and gray ties with Navy blue. Multicolored ties with patterns of any kind were *streng verboten* for her gentlemen. Far from resenting these restrictions, I admired her taste."

I could tell that Hildegard wasn't satisfied with my answer to her question. Obviously she wanted to hear something about my feelings for Friedl, but maybe she already knew me well enough not to press me for more details, and I didn't volunteer any.

When I brought Hildegard home in the morning, we arranged that I would pick her up at the theater three nights later. She was going to have more late-night rehearsals during the next two evenings.

I didn't feel like telling Hildegard the whole, somewhat sordid story of my father's remarriage. In fact, his surprise introduction of Friedl hadn't totally bowled me over. Cousin Walter had overheard his parents saying that my father had engaged the services of another matchmaker. I'd been hoping that my future stepmother would turn out not only to be an elegant and beautiful young woman, with whom I would be proud to be seen, but also one who might be available to me sexually. Friedl was certainly elegant, though neither beautiful nor young. She wasn't ugly either, and, no doubt, a little younger than my father, maybe in her early forties. He could have chosen worse, I reckoned. But she didn't arouse any sexual desire in me. Eventually, after she and I escaped together from Germany, there would be many opportunities for a quasi-incestuous relation—we often slept in the same room and sometimes even shared the same bed. Nevertheless, it never occurred to me to approach her. Nor did she try to seduce me, as, in view of her ill repute, some of my relatives had expected she would.

Gabi's mother, Käthe Teutsch, who, as did my mother, came from Breslau, had made some inquiries there about Friedl, whose full name was Elfriede Gomma, née Roth. She discovered that Friedl, ex-wife of one of the Gomma brothers who owned Breslau's premier kosher butcher shop, had been divorced under scandalous circumstances. It came out at the divorce hearing that she robbed her husband and brother-in-law by helping herself to part of the daily receipts.

The shop's clientele also deemed Friedl a *ganef,* accusing her of giving her customers short weight. After her divorce, Friedl reportedly lived in Vienna as the kept woman of a rich man; and when that relation broke up, she came to Berlin to look for a well-to-do husband.

Cousin Walter told me that his parents and Aunt Ella were trying to talk my father out of wedding this wicked woman. He married her anyway, in June 1938. Since he would not have lightly ignored the advice of Fritz and Ella (to whom he was devoted), he must have fallen under the spell of Friedl's powerful persona.

Those stories about her filching in the Breslau butcher shop didn't bother me much. The Gomma brothers were probably *schnooks,* and her *kosherdik* customers just a bunch of orthodox fuddy-duddies. As far as I was concerned, what really counted was her buying me fancy things that my father never would have, such as a Zeiss Ikonta camera and a portable Olympia typewriter.

Friedl pretended to be a recent transplant to barbaric, dreary Berlin from urbane, gay Vienna. To savor in exile the ambiance of her favorite city, she often went, dressed to the nines, to the *thé dansant* at the Café Wien on the Kurfürstendamm, where music was provided by her favorite artiste, the Hungarian fiddler, Barnabas von Gézy. In truth, she was a Berliner. All of her family lived in Berlin: her widowed mother, Frau Roth; her younger sister Betty, married to Ernst Haase, owner of a small printing firm, and their thirteen-year-old son, Manfred. There was also a younger brother with whom neither his mother nor his sisters had much contact, since he had married a shiksa and converted to Christianity.

I never did discover what Friedl's true feelings were toward me during the year and a half we lived together, six months of which the two of us spent alone, without my father, under the harrowing circumstances of our escape. I don't remember that any angry words ever passed between us or that I ever felt any animosity toward her. Yet I didn't like her especially, let alone have any filial love.

I took part in a Reformgemeinde service for the last time in July 1938, at my joint bar-mitzvoid *Einsegnung* with Cousins Walter and Eva. Uncle Fritz and Aunt Hannah also belonged to the congregation, but they were even less engaged in its affairs than my father. Fritz was exactly the kind of member of whom the congregation was most proud, more proud even than of its well-known academics, professionals, and merchants. Fritz had fought for the Fatherland in the First World War as a noncommissioned officer in the infantry. He was a real *Frontkämpfer*, a bearer of the Iron Cross (Second Class), and not one of your run-of-the-mill, Jewish medical officer veterans.

Except for the Reverend Dr. Rosenthal's sermon, the confirmation ceremony was brief. I had to step up to the altar and read a passage from the Scriptures. However, my reading was not, as tradition demands, in Hebrew from the unfurled Torah, but in German from a leather-bound Bible. Only our parents and our Aunts Ella and Käthe had come to witness our initiation into the Community of Ultra Reformed Jews. And by way of a celebration, there was merely a small family dinner in Fritz and Hannah's apartment.

It was all in dramatic contrast with Rudi's confirmation gala in the salad days of 1928. Times had changed. Then, for Berlin Jewry, the Nazis were no more than a distant menace on the horizon, and the idea that Hitler would come to power too horrible to contemplate. Now Hitler had been master of our destiny for five years. Our situation had gone from bad to worse and was likely to continue getting ever worse. So it was not a propitious time for throwing a big party, even if Jews had still been allowed to rent the ballroom at the König von Portugal on the Moritzplatz, which they were not.

After the ceremony, Dr. Rosenthal presented each of us with a copy of a German version of the Old Testament devoid of any Hebrew characters. He had pasted a card bearing a personal dedication and a passport-picture-size portrait of himself on the flyleaf. The dedication included an admonition, not taken—God forbid!—from the Scriptures, but from Goethe:

> *Ein Jeglicher soll sich seinen Helden wählen,*
> *dem er sich den Weg auf den Olymp nacharbeitet.*
> Everybody must choose a hero,
> in whose tracks one works one's way up Olympus.

All my life, I remembered and tried to abide by these words of Goethe. Dr. Rosenthal would have been disappointed, however, if, by any chance, he had expected that I would elect to follow *him* up Mount Olympus. My main hero-designate for the climb would be my lifelong scientific mentor, Max Delbrück, whom I met in the States when I was twenty-four.

Not long after my father's remarriage in the summer of 1938, Mausi wrote that she and Heinz wanted me to come and live with them in Chicago. They thought it was high time that I got out of Germany. I was wildly enthusiastic about this project. First, I longed for Mausi, my real surrogate mother, and her Heinz, whom I admired tremendously for his intelligence and encyclopedic knowledge. Second, for me, as for the typical German boy nurtured on Karl May's stories, America with its Red Indians seemed like a wonderfully exciting place. Moreover, as I knew from my intensive study of the *National Geographic* in Cortina, America had the prettiest girls, the tallest buildings, and the most cars: It was the Country of the Future! At first, my father was reluctant to let me go—it was bad enough that Mausi was so far away. But then, in view of our ever-more repressive treatment by the Nazis, it finally dawned even on him that there was no future for us in Germany. He could see that my physical safety should be his paramount concern. So he agreed that I would join Mausi in Chicago, or in "Schickergoy" ("*beschickert*" meaning "drunken" in Yiddish), as my father liked to call my future home town.

Heinz and Mausi sent me the affidavit of support required of applicants for U.S. immigration visas. It arrived in late September 1938, just at the time of the Anglo–French sellout of Czechoslovakia at Munich. Chamberlain and Daladier's peace-in-our-time knuckling under to Hitler convinced many German Jews that he would be unstoppable in Europe, which caused a rush on U.S. consulates by would-be emigrants to the States.

The hitherto underused annual quota of 27,000 German-born immigrants established by the Immigration Act of 1924 would have allowed immigration visas to be issued immediately to most of the German Jews who had affidavits of support and were otherwise qualified for admission to the States, especially if, in case the 1938 quota *were* exceeded, additional visas would be issued by placing a lien against the 1939 quota. But the U.S. State Department claimed—reportedly under orders from the White House—that its consular staff was too small to process in a timely manner the large number of immigrants provided for under the German quota. Consequently, U.S. consulates issued waiting numbers to German–Jewish visa applicants, in the order in which their applications had been filed.

Fortunately, I put in my application at the U.S. consulate in Berlin at an early stage of the Great Visa Rush. I was told that it would be a few months until I would be called. People who filed a month later—because they had still been waiting for their affidavits—were given to understand that it would be six or seven *years* until their numbers would be called up, round about 1944 or 1945. The waiting numbers were transferable between U.S. consulates. Thus one wouldn't lose one's place in line if one moved from Berlin to, say, London. I was thrilled by the prospect that within a few months I could turn my back on that rotten Old World, which was evidently selling out to Hitler. I was going to be on my way to a new life in the glamorous New World, which was led by the great man to whom Joseph Goebbels habitually referred as "Crypto-Jew Fränkel Rosenfeld."

As Dr. Selver and Fräulein Kaliski had foreseen, the PRIWAKI began to lose more and more students to emigration, giving those of us who were still left behind the feeling of awaiting rescue from a sinking ship. By the end of the summer of 1938, they themselves had emigrated to the States, and Monsieur Jacquot became the school's director. Steinberg had moved to New York, and Ursula was getting ready to leave for Palestine. Her businessman father had been one of the few PRIWAKI parents with Zionist sympathies.

My own career at the PRIWAKI ended in September 1938, when my and Rawitscher's fractious behavior finally led to our expulsion. As far as Monsieur Jacquot was concerned, the last straw was that we had brought a young substitute teacher to tears. We interrupted her lecture by shouting that we couldn't hear what she was saying because the sun coming in through the window was blinding our eyes. We jumped up from our seats to pull the curtains. The class laughed, while the teacher asked Rawitscher angrily for his name. He refused to tell. She then asked me for his name. I said he's a stranger whom I've never laid eyes on in all my living days. More class laughter. So she ran out of the classroom crying, to tell Monsieur Jacquot that she was quitting her job. She didn't want to have to deal with such unruly ruffians, in a school that evidently knew no discipline. Rawitscher was expelled outright, while I was given the *consilium abeundi*, meaning that my father was informed by Monsieur Jacquot that it would be in the interest of all parties concerned that he take me out of the PRIWAKI.

My father had little choice. I was remorseful over having been kicked out of the school at which I had been so happy. But I took solace from the belief that Rawitscher and I had become folk heroes, because punishment for open defiance of a teacher ought to have earned us the admiration of our peers. (If only Ursula had still been there!) Thinking wishfully that my departure for my new life in Chicago was drawing nigh, I reckoned that, before long, I would have had to quit that stupid *Judenschule* anyhow.

By the time the PRIWAKI closed its doors in the spring of 1939, the emigration of its student body was nearly complete. Of the five hundred or so children who attended it in its seven-year history, only thirty-nine are known to have perished in the Holocaust. Most of these Holocaust victims were not caught in Germany, but in Nazi-occupied countries, such as France, Holland, and Belgium, to which they emigrated before the war. Almost all our teachers managed to escape. Only two disappeared in Auschwitz; one was our Jewish history and Hebrew instructor, Ludwig Kuttner, a hero. My class treated him very badly, because, being the bunch of self-haters we were, we resented the subjects he taught. Herr Kuttner did not avail himself of several opportunities he had to escape to Palestine. He chose to remain on a farm near Paderborn, to take charge of the agricultural retraining of a couple of dozen Jewish boys, preparing them for kibbutz life in the Palestine they never lived to see.

Fräulein Kaliski settled in New York (where she founded a school for retarded children), Dr. Selver and Herr Jospe in Chicago, and Monsieur Jacquot and Dr. Kohn in Israel (where the former headed a children's village and the latter taught school).

Dispersed over the five continents for more than half a century, most of us remained bound by powerful emotional ties to the school community. Fräulein Kaliski, Dr. Selver, and their staff had managed to turn the children they had had in their charge for a mere two or three years into relatives, whom one is free to love or hate, but never free to feel indifferent toward. When a PRIWAKI reunion was held in New York City in November 1989, more than eighty of the erstwhile children, now well into middle age, turned up. Rawitscher, Koltai, Steinberg, and Ursula A. had all come.

The most prominent alumnus at the reunion was Michael Blumenthal, from the class below mine, who had been President Carter's secretary of the treasury. By way of an uncanny coincidence, three people from my class had wound up on the faculty of the University of California at Berkeley, one a professor of physics, another of botany, and the third, namely me, of molecular biology. Dr. Selver, Monsieur Jacquot, Herr Jospe, and Dr. Kohn had all died meanwhile. Lotte Kaliski *was* present, and when she made much of us, Rawitscher and I tactfully refrained from reminding her that Monsieur Jacquot had kicked us out.

Kristallnacht

IN THE EARLY MORNING OF THURSDAY, NOVEMBER 10, 1938, I got out of bed in our apartment in the Badensche Strasse and went to my window to have a look at the men down below, who were already at work excavating a huge hole next door. The hole was to accommodate the foundation of a new building fronting the Badensche Strasse. It seemed a pity that the new building was going to block the panoramic view from my room. But what am I worrying about? By the time the building is up, I'll be long gone. I'll be in Schickergoy, safely out of the Nazis' reach, starting a new life with Mausi and Heinz in the New World.

As my gaze shifted from the construction scene below to the southwestern horizon, I noticed a smoke plume rising skyward. There seemed to be a big fire in the neighborhood. Why, the smoke was coming out of the cupola of the Prinzregentenstrasse synagogue! The synagogue was on fire! I ran into the kitchen, where Friedl was making breakfast, to let her in on this amazing news. Just then, the phone rang. It was a man from the Berlin police. Because my father had recently done him a favor in a business deal, he warned my father, without identifying himself by name, that during the next few days, they would be rounding up as many Jewish men as they could lay their hands on, especially professionals and well-to-do businessmen. It would be wise for my father to disappear from his apartment and not to show up at his office either. Without further explanation, the policeman hung up.

Friedl suggested that her mother, Frau Roth, had a flat in the center of town, near the Mark of Brandenburg Museum, that would be a safe place for my father to hide. It was unlikely that during their roundup of Jewish men the police would be ringing the doorbell of a lone, elderly widow with an Aryan-sounding surname. Soon after this first call, the phone rang again. This time it was Uncle Max Salomon, the ex-husband of my father's younger sister Käthe Salomon.

"Awful things are going on, Georg! During the night, all the Jewish-owned stores had their windows smashed and all the synagogues were set on fire. Lots of Jews were dragged from their homes and beaten in the streets. Mobs are still roaming through town, looking for shops to trash that had been overlooked during the night and for Jews with whom to have a little fun. It was all because of the shooting of the German consul in Paris by that Polish Jew three days ago. So don't go out into the streets, for God's sake!"

"So *that's* what's happening, Max! I was just warned that the cops are going to round up as many Jewish men as they can find. I'm going to lie low at the place of Friedl's mother, near the Mark of Brandenburg Museum. It might be a good place to hide for you, too."

I wasn't going to heed Uncle Max's advice to stay at home. I was dying to see what was happening in town. When my father had left and ignoring Friedl's desperate pleas, I took off for the Prinzregentenstrasse synagogue on my bike. A lot of people were gawking at the burning building. They seemed to be more awed than gladdened by the dramatic scene. Heavy smoke was pouring out of the synagogue, which was ringed by fire engines. The firemen were scurrying about with their hoses, and it looked to me as though they were trying to put out the fire. As it turned out, they were merely trying to prevent the fire from spreading to the neighboring Aryan structures. For the first time (as far as I can remember) I felt anger, rather than merely fear, over what the Nazis were doing to us. Why were those sons-of-bitches burning down this beautiful modern building?

I went to find out whether the more fashionable Fasanenstrasse synagogue—ministered by Berlin's most distinguished rabbi, Leo Baeck—was burning as well. The street scene along my bicycle route seemed fairly normal. People were going about their Thursday morning business. But I did pass several shops whose windows had been smashed. It was easy to tell that they were Jewish owned, because some months earlier, their proprietors' names had to be painted in 10-inch-high white letters on the window panes.

This was to give notice to Aryan would-be customers that these places belonged to enemies of the German people.

As I got to the corner of the Fasanenstrasse and the Kurfürstendamm, I heard shouting. A dozen goons, some in Hitlerjugend uniforms, were smashing the windows of a perfume shop, yelling anti-Semitic obscenities. A small crowd was watching passively. No Schupos were in sight. I stopped at the curb to take in the spectacle from some distance. So that I could make a fast getaway in case any of the bastards spotted me as a Jew, I stayed on my bike. Two elderly women were passing near me, and I heard one of them say to the other: "I hope those assholes are going to cut their hands on the glass." I wasn't sure whether their scorn was evoked by pity for the persecuted Jews, or simply by outrage over this disturbance of the public order.

I had a Walter-Mitty-esque fantasy: *Moi, je suis le Caporal Günter Stensch de l'Armée Française, membre de la Force Interalliée d'Occupation à Berlin, chargée de la suppriration des Nazis.* Invulnerable in my French uniform, wearing my *képi* at a smart angle, I sit astride my motorcycle with a machine gun mounted on the handlebars. I pull the trigger and spray the howling mob with bullets. Everybody falls dead; blood is all over the Kurfürstendamm. *Chouette, alors!* That'll teach those *cochons* a lesson they'll never forget! (*I* certainly never forgot it. It was the first thing I thought of when, eight years later, I came back in my Ami uniform to that corner and its still existing perfume shop.)

The Fasanenstrasse synagogue was on fire as well. Here the damage to the building was much more obvious than in the Prinzregentenstrasse; part of the roof had collapsed. The firemen were spraying water, not on God's burning house but on the walls of the adjacent apartment buildings. I figured that maybe I'd better cycle home.

While I was out, Frau Roth had telephoned to let Friedl know that "her cousin had arrived from Breslau" and that he needed some clothes. Friedl said she would see what she could do about it. She packed a small travel bag with one of my father's suits, some shirts, some underwear and socks, and his toilet things, and asked me to take it to him at Frau Roth's. This seemed like a fun assignment, an undercover mission for Secret Agent Günter Stensch. I looked out of the window furtively to check whether there were any Schupos down in the Badensche Strasse staking out our building. The coast seemed clear, so I sneaked out of the front entrance and, travel bag in hand, putting on a carefree mien, strolled down to the U-Bahn stop on the Bayrische Platz, turning around casually from time to time to make sure I wasn't being followed. I went down to the U-Bahn platform, hid behind a pillar, and watched the stairs to make sure no one was on my tail. Having thus cleverly evaded the close police surveillance to which I imagined our house was subject, I took the train to the Mark of Brandenburg Museum stop.

My father was not the only man hiding out at Frau Roth's. Friedl's brother-in-law, Ernst, and Uncle Max were there as well. Max had learned meanwhile that the roundup of well-heeled Jewish men was indeed under way. They were being taken to the Sachsenhausen concentration camp just north of Berlin, modeled after Dachau, the infamous archetype of the genre.

I had heard all about what was going on in those camps. The guards were members of the black-uniformed *Schutzstaffel*, or SS, organized in 1929 by Heinrich Himmler as Hitler's personal bodyguard. The SS monsters kick and beat you all the time; they make you stand at attention for hours and hours, night and day, rain or shine, in freezing cold or blazing heat; they shave your head, starve you, and for the slightest disobedience or talking back, they hang you from your arms, which are tied behind your back. Maybe, even though I'm only fourteen, they'll arrest me too, what with my being the owner of a substantial bank account, and I'll wind up in Sachsenhausen as well. How would I be able to stand it? I've got to get away before I too wind up in Sachsenhausen!

My father happened to have a valid passport, albeit stamped with the big, red *J*, for "Jew," by means of which he could leave Germany legally. Unfortunately, almost none of the countries with which Germany had reciprocal agreements for visa-free entry, such as Britain, France, and Switzerland, would any longer admit German citizens with *J*-stamped passports without special visas. Contrary to what became widely believed, the idea of marking the passports of German Jews with a *J* did not originate with the Nazis. It was done in the fall of 1938 at the request of Dr. Heinrich Rothmund, chief of the Swiss federal police, as a paradoxical result of the then recently convened international Evian Conference on facilitating the emigration of German Jews. While the participating national governments urged one another to let in substantial numbers of German–Jewish refugees from Nazi persecution, all put forward good reasons for unfortunately being unable to do so themselves.

Since Holland was one of the few countries that still admitted bearers of valid German passports without special visas, even if they were *J*-stamped, my father intended to try to escape to Amsterdam. There he hoped to collect money owed by his Dutch customers to Ballnath & Stensch and to get a British visa that would allow him to join Rudi and Gabi in London. My father asked me to head back home from Mrs. Roth's and tell Friedl that he would go to his bank that afternoon, to withdraw as much cash as he could. Once he had the money, he'd get on the first available Lufthansa flight to Amsterdam.

At suppertime, our doorbell was rung very energetically. Friedl and I both went to open the door, trying to hide our fright. Two Schupos were standing outside.

"We want to speak to Georg Stensch."

"Unfortunately, Herr Stensch isn't in."

"You're Frau Stensch?"

"Yes, and this is our little son."

"When do you expect Herrn Stensch?"

"He should've been home by now. We're wondering what's happened to him. We're getting a bit worried."

"Was he at his office today?"

"Of course. He's always at his office, except when he goes on business trips."

"No, he *wasn't* at his office today. We looked for him there."

"My goodness! He didn't say anything to us about going on a trip when he left home this morning."

"May we come in and have a look around, Frau Stensch?"

"Of course. Please make yourselves at home."

I admired Friedl's cool aplomb. She was flirting with the cops! Talking to them in her fake Viennese accent, which she thought was terribly attractive to men but which gave me the creeps. The Schupos inspected our apartment perfunctorily; they didn't look under the beds or open closets to check whether my father was hidden somewhere. Probably they were eager to get home for supper. It was a good thing they didn't notice that our dining table was set for only two, giving the lie to Friedl's claim that we were expecting my father to show up momentarily.

"We'll be back tomorrow, about this time. Tell Georg Stensch to be at home then."

"I certainly hope he'll be back by then."

By Friday, the synagogue and department store fires had been put out.

The Jewish shop owners were under orders to clean up the mess caused by the spontaneous popular expression of outrage over the murder of a German diplomat by one of their Jew boys. They were to clear the sidewalks of glass splinters and replace the jagged broken window panes.

In the evening, one of the two Schupos was back ringing our doorbell.

"Is Georg Stensch home now?"

"No. We've haven't heard a thing from him since he left yesterday morning. We're getting desperate. It's not like him to disappear just like that! We've heard rumors that Jews have been attacked in the streets, and we're worried that something awful has happened to him. Can you help us find him?"

"We'll try. In any case, tell your husband to report to the Schöneberg police station as soon as he turns up."

He made an entry into his notebook and left without asking to be let in.

Saturday afternoon, my father went to Tempelhof Airport, but the Amsterdam flight was sold out. So he was going to get on the night express to Holland. We saw him off on the crowded Zoo Station platform. There were lots of Schupos about, but it didn't seem likely that they were there to look specifically for Herrn Georg Stensch. No one paid any attention to us. He gave Friedl all his cash, except for the ten marks that Nazi currency regulations allowed one to take out of the country. The only thing of value my father had on him was his gold pocket watch with a solid gold chain. We kissed good-bye and wished him good luck for the frontier crossing at Bentheim on the Dutch border. He promised to phone us as soon as he had safely reached Amsterdam. God knew when we would see each other again; never in Berlin, that's for sure!

With my father gone for good, I was now left behind in Berlin as the last of our original family household. Who would have thought that it would be up to little Krümel to close up shop? I wasn't afraid that with my father gone, I would be in greater danger from the Nazis. On the contrary, after the Frau Cohn extortion episode and our having to run off to Italy, I had come to think of my father as a source of hazard rather than security. His departure didn't seem to make much of a difference, because I felt that I had been on my own anyhow in those two years since Mausi left for America. Of course, Friedl was still around, but I hardly thought of her as a protectress. All my hopes were now pinned on getting my American visa and making my escape from Berlin to join my sister in Chicago.

Sunday morning, there was no phone call from my father. On the *Deutschlandsender* newscast they announced that an Amsterdam-bound Lufthansa plane from Berlin had crashed in Holland on Saturday. Everybody on board was killed. It was the flight my father hadn't been able to get on at Tempelhof! But why wasn't he calling us? We were afraid that he had been arrested at the border and that he might already be on his way to a concentration camp.

Finally we did hear from him early Monday morning. He was calling from Amsterdam, where he had just arrived. Thank God! When his train pulled into Bentheim Station, the German frontier police took all the *J*-stamped passports and made their bearers get off with their luggage. My father and his two dozen or so Jewish fellow travelers had to line up on the platform to watch their train pull out for Oldenzaal on the Dutch side of the border. After standing on the platform for an hour or two, they were interrogated, one by one, in the police office at the station, presumably to identify anyone who was on one or another of the Nazis' wanted lists. Customs then strip-searched them and scrutinized their luggage to look for hidden money and other valuables. My father's gold watch, which, fortunately for him, he wore openly on his vest, was confiscated. Finally, my father, like most but not all of the *J*-passported passengers, was put on the evening train for Amsterdam. There, he managed to collect some of the money owed his firm by its Dutch customers and obtain the British visa that he needed to travel on to London, where he lived for the next twenty years.

Escaping

I PICKED UP HILDEGARD AT HER THEATER three days after our last date, to take her to a party thrown by a FIAT document screener in his Dahlem billet. He had been quartered in the Bad Homburg Park Sanatorium when I had first arrived from the States the previous September. In the meanwhile, he had managed to get himself assigned permanently to everybody's dream duty station, the Berlin patent office. I ran into him at the FIAT field office, and he invited me.

"Bring along some booze and a girl, if you know one."

"It so happens that I do know one, but she's a Kraut."

"Doesn't matter. Just bring her along."

My FIAT colleague was living in a fancy apartment in a requisitioned villa not far from the Gossler. The place contained massive pieces of German furniture, an upright piano, and shelves full of books left behind by the evicted occupants. The hardwood floors were bare, however, presumably because the oriental rugs that had surely covered them once upon a time, had been liberated long ago. A couple of dozen people had come to the party. All the men and a few of the women were American or British civilians. The rest were German girls. The party was well under way by the time we arrived, as indicated by the advanced inebriation of some of the revelers. Couples were dancing cheek-to-cheek to music coming from an old portable acoustic phonograph.

The Allied personnel women were generally unattractive, as they had been at the Harnack Haus dining room, and while some of the German girls were not unsexy, there was something cheap and vulgar about the lot, as there had been about the women at the Femina nightclub. Here, as at the Femina, I judged Hildegard the most desirable woman present, an opinion that, as I could tell from the stares of the other swains (mostly older men, well into their thirties), I was not alone in holding. We kept mostly to ourselves, and while we were dancing ultra-close, Hildegard made me sing *"Je ne suis pas curieux"* over and over again.

After I had sung our song often enough, Hildegard asked me to tell her about the Kristallnacht pogrom I had mentioned on the night we first met.

"What happened to you that night, Günter?"

"Nothing really bad happened to me personally or to my family, but many Jews suffered a lot, dragged through the streets and beaten, especially in Austria. Next morning, I biked through the Wilmersdorf and Schöneberg streets, to watch the burning of the Jewish synagogues, and the smashing of the Jewish shop windows by the howling Nazi mob. My father barely avoided being sent to the Sachsenhausen concentration camp and escaped to England, leaving Friedl and me behind in Berlin."

"I'm terribly sorry for all the awful things we Germans did to the Jews. But I just can't believe that my own father—my wonderful, kind Daddy—had anything to do with any of this, even though he was in the Party."

It didn't seem very likely to me that, as a Nazi *Bonze*, her old man didn't take part in hounding Jews, however wonderful and kind a daddy he might have been to his kids. But not wanting to disillusion Hildegard about her father, I kept my conjecture to myself.

"Now, I'll tell you something about myself, Günterlein. In 1943, when I was only sixteen, I joined a drama ensemble of the Wehrmacht Truppenbetreuung, our equivalent of your USO. My unit was putting on plays and musicals for the troops on the eastern front. They were still fighting in Russia, but soon started retreating west, keeping just one step ahead of the Ivans. When the Ivans crossed the Vistula south of Warsaw in January 1945, everything fell to pieces. The final curtain went down on our Truppenbetreuung theater. My unit was disbanded. I had to fend for myself, under enemy fire.

"I ran into a young Waffen SS officer. He became my protector and lover, and we continued on our westward flight together. By April 1945, we had got close to Frankfurt-an-der-Oder. There we became surrounded by Ivans. They didn't take SS men prisoner. Any they nabbed, they rubbed out on the spot. So my friend switched his black SS outfit to an ordinary, gray-green Army uniform he had taken off a soldier's corpse. A Soviet patrol caught us while we were hiding in a Mark of Brandenburg forest, not far from Berlin. The Ivans ripped off his green jacket, to look under his armpit for tattooed letters of his blood type. Himmler had had every SS soldier tattooed, to make sure that in an emergency his guys got transfused with the right blood type. The Ivans found his tattoo, and,—piff, paff, pouff—put a bullet through his brain."

Hildegard didn't tell me what the Ivans did to her, and, in view of what I imagined it was, didn't ask.

"They finally let me go, and I made my way back to Berlin a month or so later, after the fighting was over."

So it turned out that Hildegard was not only the daughter of a Nazi *apparatchik* but also the ex-mistress of an SS officer. How could I reconcile my self-respect as a Jew with having any further dealings with this girl? Since, despite this revelation, I still didn't want to give her up, I realized that—not surprisingly for a self-hater—I couldn't have much self-respect.

As we were lying in bed at the Hotel Gossler after the Ami party, Hildegard said,

"*Du*, Günterlein, tell me how you and your stepmother got out of Germany before the War, after your father escaped at the time of Kristallnacht."

"Well, it's a long story, and if I go into the details, it'll keep us up all night."

"Never mind!"

"After my father got away to England, Friedl and I reckoned that we had better join him in London as fast as we could. Obviously, the Kristallnacht pogrom had been just a rehearsal of things much worse to come. For starters, the Nazis soon issued a string of new regulations that would take us Jews completely out of public life, besides robbing us of our money through Fatso's collective billion-mark fine. From now on, we couldn't any longer attend German schools or universities, go to theaters, operas, concerts, or movies, or enter parks, swimming pools, and beaches. We were also forbidden to enter the Unter den Linden area and—cruelest blow of all—not allowed to keep carrier pigeons. To brand us officially on every document as enemies of the German people, all Jewish men and women had to add "Israel" or "Sara" to whatever first name their parents had given them at birth.

"My dear Israel! Why didn't you tell me you had another middle name, besides Siegmund?"

"Don't you ever call me that! I hate it! Call me Sigi, if you don't like Günter! Anyhow, Friedl and I took a chance and didn't register for that humiliating name change. We figured we'd be gone by the deadline for the renaming of all Jews. Mind you, we didn't have our passports, and we figured that the Schöneberg police wouldn't give 'em to us until my father showed up. What we had in mind was to try the only way of fleeing the coop available to us. Getting out illegally, *bei Nacht und Nebel* [by night and fog]."

"How?"

I explained that ways of escaping from the Fatherland illegally were a favorite topic of conversation among Berlin Jews and that there were regular bulletins about who had made it and who hadn't by this or that route. The most promising seemed to be sneaking over the border in open country—the "Green Frontier"—between Germany and Holland or Belgium. To make it across, one had to keep out of the way of not only the German but also the Dutch or Belgian frontier patrols. They were handing over to the Germans all Jewish refugees they caught. But any refugee who managed to get beyond the border zone could report to the local police in the interior of Holland or Belgium and get a temporary residence permit. To avoid getting caught on the frontier, one had to hire a guide. And that took lots of cash, because those guides didn't come cheap.

Hildegard interrupted: "Where were you going to get lots of cash from?"

"My Dad had buried a wad of 500-mark bills wrapped in old newspapers in the soil of one of the flower boxes on the balcony of our apartment. When we dug up the buried treasure to pay for the tour guide, we found that the package had become soggy in the wet dirt and that the banknotes had started to rot."

"Like father, like son! He's a genius too?"

"Right! Fortunately, your future Ami-scientist boyfriend managed to dry and smooth out the rotting bills with an electric iron."

I went on with my story. Friedl's brother-in-law, Ernst, put us in touch with a Mr. Levin, who was in touch with the ideal tour guide: Mr. X, a coal miner living in Herzogenrath, a German village on the Dutch border, near Aachen. The border ran right behind X's house, separating his garden from his neighbor's on the Dutch side. X met his clients at Aachen Station and brought them to his house after dusk. In the dead of night, he took them for a little stroll through his garden, to cross the border and leave them at his neighbor's house. Next morning, the neighbor would take his nighttime visitors to Maastricht Station and put them on the train to Amsterdam. X would also took care of smuggling his clients' money and valuables to Holland via the same route. It was a package tour, all fees, tips, and transfer costs from Aachen to Amsterdam included, at a few thousand marks per head.

We asked Levin to sign us up, including smuggling out our money and Friedl's jewelry. We figured it'd take us about three weeks to settle our affairs in Berlin. Told him that we'd like to leave just before Christmas. Once in Amsterdam, we were going to register with the Dutch police and apply for a temporary travel document issued to refugees, for which you had to bring along some proof of identity, such as a birth certificate and a photo ID card. My father was going to apply for British entry visas for us in London, and once we had all the necessary papers, we'd be on our way to England to join him. It seemed all wonderfully simple.

Even after Kristallnacht, it was still possible for legally emigrating Jews to pack all their household belongings—including furniture, chandeliers, and grand pianos—into a single, giant crate, called a "Lift," and have the whole caboodle shipped abroad. But we weren't going to emigrate legally. We were going to vanish furtively from our apartment and sneak across the Green Frontier, by *Nacht und Nebel*, and couldn't very well risk drawing police attention to our planning an illegal getaway by the spectacle of movers packing a huge Lift out in the street. So the only things of ours we could hope to save were items small enough to fit into standard crates packed in the privacy of our apartment and discretely carted away by a forwarding agent. The rest of our stuff we would have to try to sell or simply leave behind in the apartment.

We put a want ad in a local newspaper and managed to sell the grand piano to a musician (throwing in a Ballnath & Stensch bronze bust of Beethoven as a bonus) and the *Empire* furniture of the salon to an antique dealer, for hard cash, at a fraction of their actual worth. No buyer could be found at any price for our jumbo crystal chandelier with its hundreds of wonderfully cut, sparkling glass pendants. They told us that crystal chandeliers had gone out of style. The rest of the furniture and lighting fittings wasn't worth bothering about. We called a forwarding agent to pack our silver, linen, porcelain, oriental rugs, and, most importantly for me, my precious Olympia portable and my Zeiss camera. We asked him to put the crates into temporary storage in his warehouse until we could instruct him where to send them abroad.

There was another surprising loophole in the Nazi restrictions limiting the export of Jews' personal belongings. We could still ship as many suitcases and steamer trunks as we wanted abroad as unaccompanied railway baggage. And by greasing the palms of the baggage handlers at Zoo Station, you could avoid having your luggage searched by German customs for forbidden stuff. After Kristallnacht, the baggage counter at Zoo Station was forever crowded with Jews shipping their bags to foreign destinations.

We sent off a dozen suitcases to London, containing Friedl's enormous wardrobe, including some expensive furs and a vast number of fancy shoes and handbags, as well as the new togs that Friedl had selected for my dad and myself. My brother Rudi cleared our suitcases through customs at Victoria Station. According to him, hundreds of unclaimed suitcases from Germany were piling up, completely congesting the customs shed.

Meanwhile Levin told us that Monday, December 26, 1938, had been set for the date of our crossing the Green Frontier in X's garden. X was going to be in Berlin on other business in a couple of days and call on us to explain the details of our little excursion. X showed up in ordinary street clothes rather than, as I had expected, in a miner's outfit—black overalls, lamp-topped cap, and pick-ax in hand. He told us to take the night express to Cologne on the evening of Christmas Day, December 25, and buy one-way tickets. Next morning, in Cologne, we were to buy round-trip tickets Cologne-Aachen-Cologne and get on a local train to Aachen. He asked us to go through all this rigmarole, just to get to Aachen, because the German frontier police were making spot checks on the international express trains between Cologne and Aachen, ahead of the final control at the border. They were less interested in passengers traveling on local trains terminating in Aachen. But in case we *were* checked by a patrol on the local train, we were going to claim that we were making an excursion from Cologne to visit historic Aachen. To prove that we were tourists for real, we'd show them our round-trip tickets from and to Cologne. Also, we weren't supposed to have any luggage with us other than a small case containing only a few items typically needed on short trips, such as pajamas, spare underwear, and toilet articles. In Aachen, we were supposed to go to the station restaurant and wait there for X, who'd show up at about four in the afternoon. From then on, we'd be taken care of until we got on the train to Amsterdam at Maastricht Station on December 27.

About a week before we were going to decamp, Levin came to our apartment to collect the tour fee and the valuables to be smuggled out by X for us, like our leftover cash in partly rotted German banknotes and all of Friedl's jewelry—except for the few pieces that she might reasonably wear on an excursion to historic Aachen. Levin would get the stuff to X ahead of us, because it was dangerous for Jews to travel anywhere near the border with a lot of valuables on their persons. Levin told us that we'd have a Mr. Y as a fellow Green Frontier tourist. He was going to bring Mr. Y to the train.

"On Christmas Day, Hildegard mine, I woke up in a mixed mood of happiness and sadness. I was happy that we'd soon be out of Hitler's reach and sad over having to leave so many of my things behind: my books, my bicycle, my electric trains, our phonograph records. I knew that this, my last day in Berlin, would be a watershed in my life: I was going to count time as before and after December 25, 1938. I didn't think I'd ever see Berlin again, not to speak of imagining meeting you at the doors of the Admiralpalast eight years later, my dear Fräulein U."

"Thank you, Mister Stent. I was only eleven at the time!"

I didn't think I'd ever see Berlin again, because I believed Hitler's claim that the Third Reich would last a thousand years. So I was sure that whatever I was doing, I was doing it for the last time in Berlin: having my last look at the panoramic view over Schöneberg, taking my last morning bath, eating my last lunch in our dining room, listening to the last playing of the record with the "Triumphal March" from *Aida*.

We did our final packing, putting all our remaining clothes into a couple of suitcases for a last shipment of unaccompanied baggage to London, except for the things we were going to wear on the trip or carry in our little overnight cases. When we were done, the apartment still looked as if it were occupied. There was furniture in all the rooms, except the salon, which contained nothing but the magnificent crystal chandelier. The bookcase full of books was still in my room, and the kitchen was stocked with pots and pans and other kitchenware. We hadn't notified the landlord that we were going to skip out, and hadn't asked to have the telephone, gas, or electricity shut off.

Ernst came to the apartment in the afternoon to help us bring the suitcases to Zoo Station. He carried them into the elevator while Friedl, dressed in a smart two-piece suit and wearing high-heeled, alligator-skin shoes suitable for the *thé dansant* at the Café Wien, put on her fancy Persian lamb coat. I had one final look around the apartment. For the last time, I turned out the lights and shut the door behind us."

Our fellow traveler, Mr. Y, was a well-dressed, middle-aged Jewish lawyer. We had a second-class compartment to ourselves. Friedl wouldn't put up with my Dad's habit of traveling third class. I stood at the window as the train was pulling out of Zoo Station. The brightly lit tower of the Kaiser Wilhelm Memorial Church passed by in the distance. That passing picture of the church stayed with me for the rest of my life. It stood for my final separation from my home town.

We hardly slept during the night trip to Cologne. Friedl was crying from time to time. Not out of sorrow for leaving the Berlin she had always maligned. She was scared of getting caught on the Green Frontier. Y was very nervous. He probably shared Friedl's fears. Although I wasn't totally unworried either, I was looking forward to the kind of exciting adventure I had read about in lots of books and seen in lots of movies.

Next morning in Cologne, we bought the round-trip tickets to Aachen and changed to the local train. We got to Aachen Station at about half-past three in the afternoon, half an hour before we were supposed to meet X, our guide. There hadn't been any spot checks by the police on the train. Nobody had paid any attention to us. It was the second day of Christmas, and the station restaurant was deserted. X wasn't there. So we ordered coffee and waited for him to turn up.

He hadn't shown at four, not at four-thirty, not at five. Friedl and Y were getting more and more nervous. By six, they were in a panic. What were we going to do? The restaurant people were beginning to look at us suspiciously. The dapper Y and Friedl in her Persian lamb and alligator shoes didn't look exactly like their usual customers. Anyhow, few of them would have lingered over a cup of coffee for more than two hours. Not to speak of the frontier policemen working at the station dropping in for a quick schnapps. We were afraid the waiter would call the cops to tell them that there were some suspicious characters hanging around the restaurant. So we decided we had better clear out in a hurry. But where to? And how was X going to find us if we left the place where he was supposed to meet us?

One of us just had to go to Herzogenrath and find X. Both Friedl and Y were too jittery to carry out that mission. So that left only me to try contacting X. Aha! Here is another mission for Secret Agent Günter Stensch. I got into a taxi waiting outside the station.

"Take me to Herzogenrath, please. I've got to pick up something there and then come back here to Aachen Station."

"Got enough dough to pay for the trip, buddy?"

I showed him the 50-mark bill Friedl gave me, and he took off. When we got to Herzogenrath, I made the driver stop at the village tavern. I went in and asked, "Can you tell me, please, where Mr. X, the miner, is living?"

"Yes, I can, but he ain't home."

"Why isn't he home?"

"Because he was arrested yesterday and taken to jail. The cops are still all over his place."

I tried to hide my terror and just mumbled: "Well, there doesn't seem to be much point in my going to Mr. X's house, does there? Thanks, anyhow."

Instead of rushing out to the waiting taxi, I managed to stroll out of the tavern casually, as slowly as I could, and asked the driver to take me back to Aachen Station.

"How old were you then, Günter? Only fourteen? *Mein Gott!* When it comes to guts, man, you sure *were* a Wunderkind!"

"Friedl and Y were totally crushed by my news. Flipped their lid. Became nearly hysterical. I told 'em there was only one thing we *could* do, namely to get out of the restaurant before the cops showed up, and hop on the next train back to Cologne, using the unused return part of our round-trip tickets."

Friedl and Y calmed down a bit once we were safely on the Cologne train. Our situation wasn't good. Friedl and I had less than fifty marks in cash between us, which wouldn't last very long if we were going to stay in Cologne. The rest of our money was gone; we had given it all to Levin, and by now the cops probably had it. Going back to our apartment in Berlin seemed out of the question, since X had no doubt given the cops the names of his customers.

It was nearly midnight when we found two rooms in a cheap hotel in Cologne, not far from the station, one room for Friedl and me and the other for Y. Friedl phoned Ernst in Berlin to tell him about our disaster and how we were totally broke. Ernst told her to get some sleep; he'd see what he could do for us and call us back at the hotel tomorrow.

When Ernst phoned next afternoon, he had good news: He'd been able to withdraw a few thousand marks from his bank account. What's more, through the grapevine, he got the name of a woman who owned a cigarette shop in Cologne and who was booking guided tours across the Green Frontier. He was going to take the night express to Cologne this evening and come to our hotel tomorrow morning. Ernst was our savior!

The cigarette lady's guided tour led across the Green Frontier to Belgium and was just as expensive as X's garden stroll to Holland. Under the new package deal, we were going to be taken by taxi from Cologne to a farmhouse on the outskirts of Aachen. We'd wait there until late evening, when another taxi was going to take us on a back country road to a lonely spot in the forest close to the frontier. We'd get out of the taxi and meet our guide, who'd take us for a short walk across the frontier to a farmhouse on the Belgian side. There we'd rest, have a bite to eat, and then be taken by yet another taxi to either Brussels or Antwerp, whichever city we wanted to go to.

Friedl had lost her nerve for crossing the Green Frontier and was sorry we hadn't tried first tried to get our passports from the Schöneberg police, so that we could have left Germany legally. But it was too late for that now, with X in jail and our names surely on the wanted list for attempted illegal flight and currency smuggling. Ernst advised us to choose Antwerp rather than Brussels as our destination, because he had heard that life was less expensive and housing easier to find in Antwerp. The date for our second attempt to cross the Green Frontier was set for New Year's Eve. Ernst went back to Berlin; meanwhile we were cooling our heels in Cologne for another couple of days, wondering all the while whether maybe we'll be stood up again and stuck in this hotel, as we were stuck in the Aachen station restaurant. Friedl was too upset to leave our room and stayed in bed while waiting for the taxi. Y had been able to make his own arrangements. We wished each other good luck, and he left the hotel.

"Your gutsy Wunderkind boyfriend reckoned, Hildelein, that he might as well go sightseeing. He'd never been in Cologne before. Although I had traveled a lot abroad, the only big German cities I knew were Breslau and Munich. I went to have a look at famous Cologne cathedral. I was keen on seeing it. Not for the Gothic splendor, but for the 160-meter-high twin bell towers. Did you know that they are still the tallest structures in Germany, taller even than our own Berlin *Funkturm*? Father Rhine was a disappointment, though; his flat shoreline at Cologne didn't look at all as romantic as in the pictures I'd seen of the mighty river winding its way past the Rock of the Lorelei."

In the morning of December 31, a taxi did pull up in front of our hotel to pick us up. There were already two passengers aboard, Mr. and Mrs. Z, our new fellow Green Frontier tourists. There was very little conversation during the two-hour drive to the Aachen outskirts. Everyone was preoccupied with his own thoughts, and we didn't mention to the Z's that this was our second attempt. As advertised, the taxi took us to a farmhouse, where we were given coffee. The Z's and Friedl were very nervous and fidgety. At about nine in the evening, another taxi came for us. It had started to snow, and as the car made its way toward the Ardennes Forest, the snow began to cover the ground.

Suddenly I saw a light moving in circles in the middle of the road ahead, which, as I remembered from the war games we used to play in the Schwarze Fähnlein, was the military signal "Halt!" So I realized well before the others what was about to happen. The taxi came to a stop in front of a German frontier cop standing in the snow with a lantern in his hand and a carbine slung around his shoulder. We were mortified: Disaster had struck again! During the taxi ride, I had horror visions of being chased by a German patrol with Alsatian dogs, as we were hightailing it through the snow. I hadn't imagined, though, that our escape would end in the taxi before it had even begun. Friedl and the Z's had turned ashen-faced, and I was shivering from fear. Only the driver didn't seem fazed. The cop asked no questions; he just ordered us to stay in the car. He got on the running board, and the driver drove us to a nearby German frontier police post, without having received any further instructions. According to a road sign, the post was near the town of Rötgen.

Two more cops with carbines came up to the taxi and ordered us into the post barrack. To my surprise, the taxi, whose driver I had expected to be arrested along with us, simply drove away. The cops were frosty but not abusive. They made us sit on a bench, and then took us, one by one, to be interviewed by the post's chief in his office. When my turn came, he asked me for my passport. The sangfroid of Secret Agent Günter Stensch was gone: I was so afraid that I could hardly speak. When I managed to mumble that I hadn't any passport, he demanded some other kind of identification. My hands were trembling so much that I had trouble pulling out my birth certificate and my Makkabi Sport Club membership card from my jacket pocket. When I finally did get them out and handed them to him, he examined them carefully, and when he was done, he didn't give them back to me. Instead he asked,

"What did you think you're doing? Didn't you know that it's forbidden to leave Germany without a valid passport?"

"Yes, Sir, I did know that. My stepmother and me, we were only trying to join my father, who's in London. To get to England through Belgium."

"Didn't you know that you'd be put in prison for what you are trying to do?"

"Yes, Sir, I did know that."

He told me to leave his office and go back to the bench. When the interrogations were complete, Friedl and Mrs. Z, who had recovered sufficiently from their initial state of shock to be able to cry and were in tears, were led into one room, and the equally terrified Herr Z and me into another, to be strip-searched. Luckily none of us had any contraband or money in excess of ten marks.

After we had been in the barrack for a couple of hours, the chief called us in again. I was sure that he was going to tell us that the paddy wagon had arrived that would take us to a concentration camp. But no, he handed us our identity papers and told us that he'd let us go on to Belgium. In fact, our guide was already waiting outside. But he warned us not to get caught by the Belgian patrols. Because if they brought us back, we would be taken to Aachen headquarters, to be dealt with as we deserved.

We were bowled over by this totally unexpected turn of fate—on the bench I had already pictured myself hanging from my arms in the concentration camp. Tremendously relieved, we hugged each other, and the women stopped crying.

"Well, Günterlein, I'm glad to hear that not *all* Germans were mean to the Jews! The Rötgen police post chief was obviously trying to help you escape."

"No, Hildelein. He hardly looked to me like Mr. Nice Guy. At first, these strange happenings at the frontier police post didn't seem to make any sense. But eventually it dawned on me that the cops must have been expecting us. Otherwise they would have surely arrested, or at least questioned, the taxi driver. And otherwise our Belgian guide would have hardly gone to the frontier police post and said to the chief, 'Hello, Captain. I've come to pick up some Jews whom I'm going to sneak across the Green Frontier.' No, my dear, the cops must have been in on the deal! Most likely, they got a cut of the cigarette lady's fee. Besides, the deal might have given the chief a few merit points with headquarters if he arrested a Jew whose name happened to be listed on the Gestapo's all-points bulletin of wanted people. But no-account Jews like us who weren't on the wanted list, these he could simply let go and pocket his cut of the payoff money. We were really in luck that whatever wanted list our names were on after X's arrest had not yet made it to the Rötgen police post!"

But the women's happiness was short-lived. The Belgian guide told us that our walk would take about two hours, through woods and across meadows, which unfortunately were covered with snow. Fortunately, it would be mostly downhill. Friedl was still dressed in the Persian lamb coat and the *thé dansant* alligator shoes she was wearing when we left Berlin. It was hard to imagine an outfit less suitable for wandering through the snow-covered Ardennes Forest. And what's more, Friedl had never in her life gone on a hike, let alone in the snow. I carried both of our overnight cases, but couldn't be of much help to her otherwise in her ordeal. Adding to Friedl's distress, our guide made us lie down in the snow whenever he thought he heard the voices or saw the lights of a Belgian patrol.

In contrast to Friedl's suffering, I enjoyed this great New Year's Eve hike, which, as far as physical exertion was concerned, was nothing compared to what I had regularly endured on the Fahrten of the SF. While trudging through the snow, I was already mentally composing the detailed account of our escape that I was going to put on paper and send to my sister and my brother once we had gotten to Antwerp.

Because of Friedl's painfully slow progress through the snow, it took us three, rather than two hours to reach the farmhouse, not far from Eupen. Once indoors, Friedl collapsed onto the farmer's bed and just moaned. I pulled off her Persian lamb and her wet shoes and stockings, and covered her with a blanket. The Z's and I joined Friedl in the large double bed, and we all slept side by side for a few hours. It was the morning of January 1, 1939. We had made it.

After a breakfast of coffee and rolls covered by the tour fee, the third and last taxi came to take us to our destination. The Z's were going to Brussels and were dropped off first. When we got to Antwerp, we asked the driver to take us to the office of the Jewish Refugee Emergency Aid Committee. We were broke and had no place to stay. Fortunately, the office was open on New Year's Day. We registered and were allowed to use the office telephone to call my father in London to let him know that we made it safely across the Green Frontier. We asked him to mail us some money c/o General Delivery at the Antwerp central post office and to send the suitcases with our clothes to Antwerp Central Station.

The committee's people gave us a few francs spending money and told us to take a room for the time being at the Max, a mini-hotel in Antwerp's red light district, near Central Station, renting rooms mostly by the hour. Friedl was in a state of total mental collapse. She undressed, got into the bed that we were going to share in our tiny room, and didn't get up for the next three days. I went downstairs to the Max's restaurant with the Committee's francs jingling in my pocket and ordered a large plate of Belgian *pommes frites*.

"*Du*, Hildelein, they tasted better than any other French fries your Günterlein had ever eaten!"

It was early morning by the time I had finished my tale of how we sneaked across the Green Frontier. Hildegard wanted to talk about it, but I was too exhausted, and I took her home to Wedding. We made a date for evening-after-next, to go to the Schlosspark Theater in Steglitz and take in Franz Molnar's play, *Liliom*.

In Hitler's Lair

IN THE LATE MORNING, after having caught up on the sleep I lost while telling Hildegard my Green Frontier story, I took the U-Bahn into town from Dahlem. I wanted to have a look at what was left of Hitler's lair, the New Chancellery, on the Wilhelmplatz in the Russian sector (then recently renamed "Ernst-Thälmann-Platz," in honor of the martyred prewar leader of the German Communist Party). As I stood in front of the Chancellery's ruin, I saw a U.S. Army staff car pull up, followed by a military police jeep. Out of the car stepped a major and a civilian dressed in the same kind of "simulated" officer's uniform that I was wearing. The civilian was obviously a visiting American VIP, whom the major was going to take on a tour of Berlin's number one Allied junketeer's attraction. I sidled up to the group, while the major was palavering in Russian with the Ivans guarding the site. I asked the MPs about the VIP, and they told me that he was Senator Kenneth Wherry of Nebraska. Wherry was well known as one of the most reactionary and Isolationist Republicans in the Senate, but despite my dim view of his politics, I thought I'd like to join his entourage.

"Would you mind my tagging along, Senator? I'm a graduate student from the University of Illinois. On a short assignment in Berlin."

"Come along. Always happy to oblige a fellow midwesterner. What's your short assignment in Berlin?"

"I work for the Commerce Department and am stationed in Frankfurt. I came to Berlin to screen technical documents for FIAT at a few industrial targets."

"Commerce Department? FIAT? Can't say that I'm too happy with what I've heard about that outfit."

I was surprised by the senator's remark, because I didn't know that the agency had become the subject of a fierce controversy in Washington. So I attributed Wherry's badmouthing of FIAT to his dislike of my employer, Secretary of Commerce Henry Wallace. For a conservative Republican like Wherry, Wallace was one of the worst of the pinko holdovers from the hateful Roosevelt administration. But as I would discover only some forty years later, Wherry's lack of enthusiasm for FIAT probably sprang from his sharing the views of our deputy military governor, General Lucius Clay, that the agency was just a cover for a gang of (mainly Jewish) carpetbaggers who were impeding American efforts to promote the economic recovery of the poor, defeated Germans.

The major was very knowledgeable, not just about the Chancellery but about Nazi history in general. He turned out to be an excellent guide, which was probably his regular assignment in the military government's public relations office.

"In January 1938, Hitler ordered his protégé, Albert Speer, to build him a New Chancellery on a huge scale, Senator. The building was to be more representative of the nerve center of the mighty Third Reich than the Old Chancellery next door. Hitler had said that the Old Chancellery reminded him of a soap manufacturer's home office. Had this awesome palace put up, not for more working office space, but for show, to impress visiting Big Shots and dazzle his henchmen. Gave Speer just one year to put up this colossal marble pile, from designing the building, drawing up the working plans, tearing down houses to clear the site, to raising the structure. With a workforce of 4,500 people, Speer finished the project exactly on schedule, in January 1939. No costoverruns. Speer had an unlimited budget."

"Reminds me of how they built the Pentagon, Major. Colossal waste of money. Voted against the appropriation."

"I wouldn't know, Sir. In July 1945, Winston Churchill was one of the first Allied civilian VIP's to visit the Chancellery, and now, you, Senator, are going to be one of the last. To keep the place from becoming a shrine for any Nazi revival movement, the Chancellery—especially the Führerbunker where Hitler died—is about to be demolished. The site will be razed so completely that no recognizable reminders of its past are going to remain."

The Chancellery, roofless, was a near-total ruin. For symbolic reasons it had been a preferred bombing target of the Allied air forces in the closing year of the war, and during the fall of Berlin it drew heavy Soviet artillery fire. Yet the overall outlines of the oblong building were still preserved.

"Speer designed it in the classicist style of his late teacher and Hitler's favorite architect, Ludwig Troost. Even though admired by the Führer, that style was disliked by many Nazi architects. Paul Bonatz, for instance, thought that Teutonic medievalist buildings, such as Bonatz's own Stuttgart railway station, were more representative of the shared ideals of National Socialism. But being more Nazi than the Führer was not a good way to make a career in the Third Reich. Troost's widow had Hitler's ear. She saw to it that Bonatz's commissions were kept to designing Teutonic medievalist Autobahn bridges."

"Looks to me like the Chancellery was really a beautiful building, Major," said the senator.

We entered, as had Hitler's official visitors, through the monumental flight of marble steps I remembered from wartime newsreels. At the top of the steps I pictured the black-uniformed, white-gloved SS honor guards of the *Leibstandarte Adolf Hitler* presenting arms. I fantasized that the senator and me, his trusted aide, accompanied by our adjutant, the major, were on our way to call on the Führer. What would my Bismarck Gymnasium classmates say if they saw me now, the Jewish schlemiel who couldn't *Sieg Heil!* in the schoolyard, on his way into the holy of holies, a place they could have only dreamt of ever setting foot in?

We crossed a huge doorway into an anteroom, of whose erstwhile mosaic wall decorations traces were still visible. From there we marched through seemingly endless, high-ceilinged, marble-lined hallways to get to what was left of Hitler's ballroom-size office. All the while I was imagining passing heel-clicking SS guards, wondering which of them was Hildegard's lover. Once we got to the office, what came to my mind were not so much old newsreels but some of the scenes of Chaplin's *Great Dictator* playing in that room. Hitler doing a mad, globe-tossing dance, yelling at the hapless

Mussolini in Chaplinesque pseudo-German, *"Einzig strafen nicht die Jüden-Tüten!"* and barking *"Vat took you zo long?"* at an aide who enters seconds after a bell had summoned him.

The major led us through the Chancellery garden to what had been the emergency exit of the *Führerbunker*, buried fifty feet underground. The bunker had served as Hitler's residence and headquarters during the closing months of the war, after bombardment had rendered the main Chancellery building uninhabitable. We went down the emergency staircase, whose concrete walls were covered with graffiti in the languages of the victorious Allied nations, conveying such messages as "Greetings from Ivan Ivanovich," "Kilroy Was Here" and "Fuck Adolf!" The bunker was divided into a couple of dozen tiny rooms, stripped long ago of all their contents. The bunker's walls were blackened by soot, because it had been set on fire on its abandonment a few hours before the Chancellery was captured by the Soviets.

"Looks to me more like an underground rabbit warren," the senator remarked. "What a difference from the huge Chancellery spread upstairs! And to think that it was in these tiny holes-in-the-wall that the curtain came down on the mighty Third Reich in April 1945! What were these rooms used for, Major?"

"About half of them housed service functions. The remainder were Hitler's private apartment, including his bedroom and study, Eva Braun's bedroom, a couple of guest rooms and a conference room in which Hitler and Eva Braun were married a couple of days before the fall of Berlin."

Entering Hitler's study reawakened my feelings of overwhelming relief that he was really dead. Such emotions of deliverance had welled up in me a few months earlier at the University of Illinois library, when I got the heebie-jeebies on running into a display of a copy of Hitler's then recently discovered last will and testament. Standing now in the room in which, according to the major, Hitler had executed this document at 4 A.M. on April 29, 1945, I remembered the gist of his vicious testamentary curse on the Jews.

> It is not true that I, or anyone else in Germany, wanted war in 1939. It was provoked by those international politicians who were either of Jewish origin or working for Jewish interests . . . Centuries will go by, but from the ruins of our towns and monuments the hatred of those ultimately responsible will always grow anew. They are the people whom we have to thank for all

this: International Jewry and its helpers. . . . My reasonable solution of the Polish–German problem was rejected only because the ruling clique in England wanted war, partly for commercial reasons, partly because it was swayed by propaganda put out by international Jewry. I shall not fall in the hands of the enemy, who requires a new spectacle, presented by the Jews, to divert their hysterical masses. I have therefore decided to remain in Berlin and there choose death voluntarily at the moment when I believe that the position of the Führer and Chancellor can no longer be held. Above all else, I demand of the leadership of the Nation and its people scrupulous observance of the racial laws and merciless resistance against the world-poisoner of all nations, international Jewry. (W. L. Shirer, *The Rise and Fall of the Third Reich.* Simon and Schuster, New York, 1960. pp. 1124–1126)

Recalling how Hitler, writing his last will and testament some 30 hours before he and his bride would commit suicide, began and ended it cursing the Jews made me realize how deeply Hitler hated us. He didn't just use anti-Semitism as a populist political gimmick or merely picked on the Jews as a convenient political scapegoat on which to hang all of Germany's troubles, as many German citizens of the Mosaic faith and Hitler's foreign prewar appeasers had wanted to believe. His testament proves that Hitler passionately believed the Jews to be the root of all evil. He regarded their annihilation as his main mission in life. With Hitler alive, the physical survival of all Jews would have been in perpetual jeopardy.

The major continued his lecture:

"The day after Hitler's suicide, Goebbels, who had moved with his family into the bunker's guest rooms, killed his six children, by making them swallow capsules with lethal poison. Once their offspring were dead, he and his wife went up the emergency staircase into the Chancellery garden, where, on Goebbels' request, both were shot by an SS guard. Soon thereafter, the last occupants fled the bunker, after dousing the rooms with gasoline and setting them on fire."

A year and a half after these happenings described by the major, he took Senator Wherry and me up the emergency staircase back into the garden, following in the Goebbels's footsteps.

"This is where the cadavers of the Führer, of his mouthpiece, and of their families were incinerated. The SS guards carried out the open-air cremation of Hitler and Eva Braun so effectively that their remains were never positively identified. But the bodies of the Goebbels family were burnt very perfunctorily. They were found lying unburied in the garden by Soviet troops on the next day."

"Thanks and good-bye, Senator Wherry," I said as the tour concluded. "And thank you, Major, for your expert guidance on this once-in-a-lifetime experience, which I won't ever forget."

Not wanting to reveal to them that I was a Jew (actually, I wondered whether the savvy major wasn't a Jew himself), I didn't confess that having stood where Hitler expired and was cremated helped make the awareness of his death finally penetrate into my subconscious. Hitler had moved victoriously from one success to the next as long as I could remember. The belief in his invincibility was rooted so deeply in my emotions that I kept on fearing the Nazis would still win the war long after they had already lost it. But after this visit to the *Führerbunker*, the demise of Hitler and of the whole Nazi spook began to sink in, at last.

As I left the Chancellery ruin, I heard music coming from the far end of the Wilhelmplatz. A brass band was playing a tune I recognized. It was the Socialist hymn "*Brüder zur Sonne, zur Freiheit*" ("Brothers to the Sun and to Freedom!"), which had been sung by the International Brigade fighting on the Republican side in the Spanish Civil War. I hadn't heard "*Brüder zur Sonne*" since my days in the German–Jewish youth movement. Aroused by the stirring tune, I went over to have a look. A few dozen disheveled Germans were surrounding the U-Bahn entrance, singing along with the band. I joined in with the singing; some of the people look surprised that an Ami knew the words. Two boys were flanking the entrance. They wore a *Kluft* that seemed a dead ringer for that of the only recently defunct Nazi Jungvolk, except for a powder-blue rather than brown shirt and a red rather than black kerchief. One of them was holding the Communist red flag and the other the striped black-red-gold colors of the Weimar Republic.

Throughout my early childhood, the Republican black-red-gold had inspired good feelings, signifying toleration and safety. By contrast, the Imperial black-white-red, configured as stripes on the flag of the reactionary Nationalists and as a black swastika in a white medallion on a red ground on the Nazi flag, inspired terror, signifying Jew-baiting and danger. Flags were ubiquitous in prewar Berlin. People flew them from cars, boats, allotment gardens, beach chairs, sand castles, apartment windows, not just on national holidays, but all the time, as political statements. The Weimar Republic tolerated flaunting the black-white-red colors of its declared internal enemies, and one never seemed to be out of their sight. On coming to power, the Nazis made the swastika banner the national flag and proscribed all further show of the black-red-gold. So I hadn't seen the black-red-gold colors displayed since I was in the fourth grade. Reencountering them here on the Wilhelmplatz was another reassurance that the Nazi spook was really dead.

The uniformed boys were members of the Freie Deutsche Jugend, or FDJ, the Communist youth organization that had been recently founded in the Soviet Zone by Erich Honnecker (who would be overthrown forty-two years later as the last of East Germany's Communist *Bonzen*). The bandsmen were wearing the gray uniforms of Berlin's municipal transport system, the Berliner Verkehrs Gesellschaft, or BVG. I asked one of the bystanders what was going on. He told me that they were dedicating the reconstructed, bombed-out U-Bahn station. It was formerly called "Kaiserhof," but henceforth would bear the name "Ernst-Thälmann-Platz."

After the music stopped, a man stepped forward and gave the dedication speech.

"On behalf of all of us here, I express our thanks to the glorious Red Army for having liberated Germany from fascist oppression. Out of these ruins we are going to build a new Democratic Socialist Germany. Our success is assured because we will be guided by the best friend of the German people, Josef Vissarionovich Stalin. What could be more fitting than our dedicating today one of the first reconstruction projects to be completed to the memory of our dear comrade, Ernst Thälmann? He gave his life for the ideal of building socialism on German soil and will be an ever-lasting inspiration for us in the hard days that still lie ahead."

On concluding his oration, the speaker raised his right arm to give the clenched-fist salute of worker solidarity, to which the audience responded in kind. I recalled my frustrated wish to participate in the Nazi *Sieg Heils* in the Bismarck Gymnasium schoolyard eleven years earlier. But here I felt no desire to join this crowd in the saluting, which my Ami uniform would have made awkward in any case. Everybody then followed the speaker and the flags down the steps to the U-Bahn platform, while the BVG band played another favorite of mine, the march "*Das ist die Berliner Luft!*" by Paul Lincke, Berlin's own Irving Berlin.

I was astonished by the opulence of the rebuilt station, utterly different from the frugal ceramic-tile standard decor of the U-Bahn stops. The walls were faced with polished red marble, taken from the ruin of the New Chancellery lying just above. Glittering chandeliers hung from the ceiling, and "Ernst-Thälmann-Platz" was inscribed in golden letters on the marble walls. I was told that the new design of the station was inspired by the elegant Moscow subway, to provide a preview of the marvelous socialist future that lay ahead for Germany.

No other East Berlin U-Bahn station was ever done over in Stalinesque Moscow style. In the 1970s, the station's name was changed once more, this time to "Otto Grotewohl Strasse," in honor of the German Democratic Republic's first prime minister. And in 1991, after the demise of the GDR, the station was given its fourth name in my lifetime, namely "Mohren Strasse."

Except for the rousing music provided by the BVG band, I was repelled and depressed by the rite onto which I had stumbled. The FDJ flag bearers in their Nazi-style get-ups, the clenched-fist saluting, the demagogic harangue, it all seemed so *déja vécu*, except that this ceremony was more squalid than the political rallies the Nazis used to put on with their totalitarian panache.

The proceedings in which I had just taken part marked a turning point for my political *Weltanschauung*. It suddenly dawned on me that maybe I wouldn't want to live in a socialist society after all, after having devoutly believed in the merits of socialism ever since I was twelve. Here, in the Russian sector, I was on socialist grounds for the first time in my life, and I didn't like it.

Retrieving the Past

I HAD IMPLEMENTED TWO OF THE THREE MAIN FANTASIES featured in my obsessive I'm-finally-back-in-Berlin nightmare. I had seen the familiar places again, and I had made love with a blonde Berlin shiksa. There remained only the fantasy of running into and showing off to an old schoolmate.

As I was riding the U-Bahn back out to Dahlem after my visit to Hitler's lair, a man in a demobilized Wehrmacht greatcoat kept staring at me. He launched some tentative smiles in my direction, which I indignantly ignored. What in hell does this Kraut want from me? Finally, he came up and asked me in English, "Excuse me, Sir, but aren't you Günter Stensch? I'm Heinz Behr. We were classmates for two years in the in the Nachodstrasse *Volksschule* and then for another two in the Bismarck Gymnasium."

"For God's sake, man!" I answered in German, "Of course, Heinz Behr! I didn't recognize you right away because it's been such a long time. How are you, old buddy? How about riding with me all the way out to Thielplatz? We can have drinks and snacks in my room and catch up on what happened to us during the ten years since we lost touch."

Heinz had been my classmate for a longer time than any other boy. Besides, we were connected by another link. His father, who owned an electrical supply shop, Elektro-Behr, in the Nürnberger Strasse, was a client of Ballnath & Stensch's. More than once Mr. Behr had had a terrible row with my father, because he refused to pay for lighting fittings that my father had fobbed off on him without Behr's having ordered them. I continued our conversation in German, even though Heinz wanted to practice his English on me.

Once we settled down at the Gossler to gin-spiked grapefruit juice and peanuts, Heinz told me how he survived the war.

"You probably didn't know that I'm half Jew. I tried to hide it from our classmates and kept quiet about it at school. My mother is an Aryan and my father a Jew. He and I both survived because my parents raised me as a Christian. This made us a 'privileged non-Aryan family,' and saved my father from being deported to Auschwitz. While the war was going well for Germany, they drafted non-Aryan half-breeds of the first degree like me only for civilian war work. But after Stalingrad, I was sent as cannon fodder to the eastern front. I happened to be one of the few lucky stiffs among our classmates who managed to come home alive and intact from that Hell."

"Our Nürnberger Strasse store was bombed out, but not long after the end of the war my father reopened Elektro-Behr in the Joachimsthaler Strasse. We're doing mostly repair work on electrical appliances, which is good business these days since new appliances are virtually unobtainable.

I've started to study electrical engineering at the Technische Hochschule in Charlottenburg."

I filled Heinz in on what happened to me and my family in the meantime and on what I was doing in Berlin wearing an Ami uniform.

"Can you imagine Heinz? I discovered that the apartment building we used to own in Moabit is still standing. All around it, nothing but ruins, but our building seems completely intact."

"You're in luck, Günter. In the Western sectors, formerly Jewish-owned real estate—whether it was directly taken away by the Nazis or sold under duress after 1935—is to be returned to their original owners. Moabit is in the British sector. So you ought to be able to get your house back. Why don't you go and see Dr. Ernst Sachs? He's a lawyer who specializes in restitution claims. His office is in the Konstanzer Strasse, near the Kurfürstendamm."

By running into Heinz, all three of the main fantasies featured in my obsessive I'm-finally-back-in-Berlin dream had materialized. It gave me an eerie feeling of having the power of dream prophecy, which, as a doctoral candidate in science, I had to consider, of course, phonier than a three-dollar bill.

I called on Dr. Sachs the very next day, to find out whether there was anything to what Heinz had told me about the restitution of formerly Jewish property. He turned out to be a well-dressed man of about fifty, smoking a cigarette taken from an American pack lying on his huge, highly polished desk in a nicely furnished, well-heated office, free of the ubiquitous Kraut body odor. Since he was burning up five marks' worth of illegal tender, I supposed that the restitution business must be flourishing. He offered me a delicious, hard-to-get German Bahlsen petit-beurre cookie from a tin on his desk.

"Your classmate informed you correctly about the real estate restitution situation in Berlin, Mr. Stent. I'm willing to take on your case on a contingency fee basis."

"Thank you. I'd like you to handle the case for us. What do I have to do to get started on getting our house back?"

"First thing, we've got to do a title search on the Essener Strasse property, to make sure that the tract book actually shows your and your father's former ownership. Then we've got to find out the name and present whereabouts of the person registered as currently holding title. It's not very likely that the person who bought the house from you is still around for us to reclaim the house from. So many people died, fled the city, or just disappeared. Once I know who the present owner is, I'll contact you at your Army Post Office address."

I left Dr. Sachs's office in high spirits, excited by the prospect of regaining the status of a substantial Berlin landlord, which I had hoped would impress Hildegard.

In the evening, I picked up Hildegard to take her to the Schlosspark Theater in Steglitz. I was lucky to have managed to get two precious seats for *Liliom* from the Gossler receptionist. The reason why it was almost impossible to get tickets was that Hans Albers played the lead role of the fairgrounds roustabout in the play. It was thrilling for me to see Albers in person. By now in his mid-fifties, he had been the most popular movie star of my childhood —a German Clark Gable/Gary Cooper/John Wayne, all rolled into one, whose dramatic range as an actor was as limited as that of his Hollywood analogs. Like them, Albers always acted the same character, always speaking in strongly Berlin-accented German, whether he was cast as a circus *artiste* (*The Blue Angel*), a navy captain (*Bombs on Monte Carlo*), a daring aviator (*No Reply from F.P.1*), a detective (*The Man Who Was Sherlock Holmes*) or a Nordic Faust (*Peer Gynt*).

Liliom, in which Albers had appeared in Berlin in 1931, had been his greatest and almost last theater role before he quit the stage for the movies. He kept on making films throughout the war, most of them in the light, costume-kitsch-comedy genre that Goebbels thought more effective morale boosters than propaganda potboilers. Goebbels reckoned that after working in a factory all day and facing a night in an air raid shelter, no one was interested in spending the evening taking in yet another flick about the nefarious Jew–Bolshevik world conspiracy. Albers had been recently de-Nazified, and in this production of *Liliom* he was making his first postwar comeback to the Berlin stage. Seeing Albers perform in person cemented the cultural bond between Hildegard and me, because he had been the favorite movie actor of her childhood as he had been of mine.

On my last day of my cooked-up FIAT mission to Berlin, I went to the "bartermart" run by the U.S. Army near military government headquarters. Held in a huge, disused, and unheated former Wehrmacht garage, it was a kind of flea market at which Allied personnel could strike black market deals with the Krauts that were strictly forbidden in the street. The Kraut vendors were standing behind card tables, on which they displayed all sorts of second-hand goods—family heirlooms and other personal possessions they had managed to hold on to during the war, as well as items they had latterly "organized" on the black market in the street. (In Berlin slang, the verb "*organisieren*" is used in a wonderfully expressive way to denote acquiring something by irregular, more often than not, illegal means. To people baptized with Spree water, "organizing" does not imply reprobation, as would "stealing," but approbation, as long as it is done shrewdly and to fill some dire need. There seems to be no equivalent metaphor in American slang.)

Most of the stuff offered for "barter" was junk, but it did include some valuable items, such as silver cutlery, porcelain, and oriental rugs. The Army maintained the legal fiction that only person-to-person exchanges rather than "sales" were taking place at the bartermart, since the Amis paid for their purchases in the *de facto* legal tender of cigarettes and coffee, rather than in the *de jure* currency of Reichsmarks (which the Krauts wouldn't have accepted anyhow) or, God forbid, in U.S. dollars. Although the cigarette-based prices were higher at the bartermart than in the street, the real cost to the Amis of even the most expensive items were negligible. A carton of Camels (worth a thousand marks on the Berlin black market) would get you a precious Meissen figurine.

Although I was not, or more likely, *because* I was not a smoker, I had plenty of cigarettes. My sister had ordered them for me in Chicago, from a wholesaler, who sent them directly to my Army Post Office address. Thus I came to the bartermart with virtually unlimited means at my disposal. I wasn't interested in silverware, porcelain, or oriental rugs. What I was looking for were things I had once owned and then had to leave behind when I fled Berlin, or which I had always wanted as a boy but never did have.

To satisfy my desires of the first category, I "bartered" a few packs of Camels for a Zeiss Ikonta camera and a whole carton for a portable Olympia typewriter. They were of the same kind that Friedl had given to me in the summer of 1938. I also got a hold of some old German Electrola records for a single pack. Of the most precious of these records, one had on it the overture to the *The Merry Wives of Windsor*—the first opera I saw with Cousin Walter—and the other the Don Cossacks' *Evening Bell*.

I satisfied the second category of my desires by "bartering" a carton of

Camels for a model steam engine, fired by an alcohol lamp under a big boiler and driving real pistons in real cylinders with real steam. The flywheel, turned by the moving pistons through a camshaft, was connected through a system of belts to a collection of rotating miniature machine tools and a carousel. I had owned a little steam engine as a kid, but it was paltry compared to this magnificent exemplar of the species. It was the kind of extravagant toy that used to be displayed in shop windows to attract the attention of passers-by rather than in the expectation that anyone would shell out the fortune it would cost to buy it. But what was I going to do with my find? I couldn't very well play with it in Champaign in the TEP fraternity house or in the Noyes Chemical Laboratory. But I found a solution. I would send it to my four-year-old nephew, Ronny, in Chicago! Maybe he was still a little too young for such a big steam engine, but surely he would appreciate it when he got a little older. Meanwhile I could amuse myself by firing it up whenever I went up to the City (as Chicago was referred to in downstate Illinois) to visit my sister. Where she was going to put this enormous device in her cramped apartment didn't concern me at the time.

Before long, I had lost the Ikonta camera on a skiing trip; the typewriter didn't work properly; and the steam engine never ran because it was damaged during shipment to Chicago. But I had made one more purchase for a few packs that was driven by neither category of desires but merely by good sense: the unabridged editions of the Muret-Sanders (English–German) and Toussaint-Langenscheidt (French–German) encyclopedic dictionaries published in Berlin by Langenscheidt, each volume of which nowadays retails for hundreds of dollars. These dictionaries turned out to be of invaluable, lifelong service to me. I still keep them next to my desk in Berkeley, by way of a workaday, rather than romantic, souvenir of Berlin.

That evening, I took Hildegard to a concert by the Berlin Philharmonic, held in the Titania Palast cinema in the Steglitz district. The old *Philharmonie* concert hall, located in the center of the city near the Potsdamer Platz, had been totally destroyed in the final days of the war. That night, the orchestra played under the direction of the young Romanian conductor, Sergiu Celibdache, because its permanent conductor, Wilhelm Furtwängler, had not yet been de-Nazified.

Beethoven's Fifth, a.k.a. "Victory" in the Allied countries and "Destiny" in defeated Krautland, was the main item on the program. Hildegard swooned as the music of the divine composer, to whom she simply referred to as "HIM," reverbrated through the hall. I caressed her hand, overcome by a feeling of spiritual communion with her that I had never felt with any other girl. Why, I wondered, should the chaste sharing of a musical experience with a woman arouse so much more erotic excitement than heavy necking at the movies? As we left the Titania Palast, I promised Hildegard that I would come back to Berlin to see her just as soon as I could. That's all I said, though. I didn't tell her anything about my feeling close to her, or about my finding her beautiful, or even about how I was going to miss her.

After the concert we went to our little secret nest at the Gossler. Even though I had not yet managed to fall in love with Hildegard, I found our union so patently natural, so happily relaxed, so wonderfully self-evident, perfectly befitting two Berlin children seemingly made for each other.

A letter from Hildegard reached me soon after I got back to document screening in Leverkusen.

1 December 1946

My dear, big Boy! Günter!

Are you surprised that I am already writing you now? Probably not. What I'm missing most is our witty conversation. Well, it could hardly be otherwise, since you are sooo clever and I come from a "Superman" family. Where's the joke?!!

Today, it's the First Advent. We've just lit the first candle. Everything is already so Christmassy. You've got no idea how sad that can make one. I assume you too had a contemplative hour. Maybe you even thought of me. As I did. The whole day. Forgive me, but thoughts are duty-free!! Yes, Günter, I feel totally different today. I bet you don't even have a little Christmas tree at home. Do you still remember? You laughed when we once talked about it and you said that's just for kids. Believe me, Günter, over Christmas one turns into a kid. One is very happy about every trifle, and—one is also very sad about every trifle. Even though it's the same, year after year. It's all a matter of mood. I'm enclosing a little branch from a fir tree. It's for you; therefore you ought to be a little happy with it. Don't laugh! It all comes from real, genuine feelings. And for you, I think, that ought to be the most beautiful thing about it.

Healthwise, I'm not too well at the moment. I've been in bed for a couple of days. I have a bad cold and a very sore throat. How are you? Got so much work that you don't know what to do with all your spare time? You're allowed to be very busy with your work, but not with women!!!

To everybody's amazement, the flowers you brought me are still in full bloom. But then I'm taking very good care of them. (You see, the flowers, all by themselves, are making me think of you all the time—even when I don't want to!!)

Günter, if you want to do me a big favor, send me some cigarettes sometimes. I can't afford to buy any. And it shouldn't be such a hardship for you since you don't smoke. As you know, it's my only addiction. Are you angry that I'm asking you for this?

Write to me, Günter. Don't forget me that quickly! I am very fond of you!

A kiss from your Hildegard (or Hildelein)

I was happy to get her sweet letter. But I felt uneasy about the cigarettes. This was to be the first of many letters in which Hildegard would ask me to help her, to give her something she needed, something that involved only a trivial effort and expense on my part but would mean a tremendous lot to her. At first, I didn't really know what bothered me about her asking. It had never occurred to me to give a present to a girl in the States, except flowers or corsages, as specified by the etiquette of dating. I wouldn't have known (or been too lazy to rack my brain to think of) what present to give a girl, even if I *had* thought of giving her one. And now, instead of being glad, as I should have been, that Hildegard made it so easy for me to make her happy with a trifle like a carton of cigarettes, I resented it.

It was going to take me a while to work out the abominable reason for my resentment. It was much worse than stinginess: It was my unwillingness to take on any responsibilities for her welfare. Despite the enormous disparity in our means—she being destitute and I living high off the hog—I wanted us to be equals rather than benefactor and beneficiary. Never before having had any dealings with a girl who was materially worse off than me, I didn't know how to handle this situation emotionally. I was afraid that the more I showered Hildegard with gifts, the more I was committing myself to her and tying my hands in case I did meet my ideal dream woman, which she was not. My empathy deficiency prevented me from appreciating the cruelty of my reluctance to provide her and her family with the means for relieving their most pressing needs just because I wanted to remain a free agent in case Dream Woman showed up. Just one carton of Camels would put them on Easy Street.

Yet her letter made me yearn to be with Hildegard. How were we ever going to meet again, with Berlin closed for off-duty travel and asking Freddy to cook up another phony assignment for me there out of the question? Then I noticed that my old travel orders to Berlin weren't going to expire for another ten days! So I got on the Berlin Express on the following Friday, to spend the weekend with Hildegard. The receptionist at the Gossler, who greeted me like an old friend, didn't balk at letting me have a room for one night when I flashed my travel orders. He probably suspected what kind of mission I had turned up for again.

I took the U-Bahn to Wedding, to call on Hildegard. We fell into each other's arms. I lifted her off the ground and twirled her round and round on the staircase landing in front of her apartment's door, too excited to talk, just kissing and kissing. When we calmed down and stopped kissing, I handed her the carton of Camels I had brought along, despite my uneasiness.

"How did you manage to get to Berlin for the weekend, Günterlein? Didn't you tell me over and over again that you can't come back, except on official business?"

"Nothing simpler for your Wunderkind. I just used my original travel orders a couple of days before their expiration date."

"Well, Sir, you must really care for me, if you're willing to devote so much creative energy to enjoying my company!"

There was no performance of *Men in White* that Saturday night, and so Hildegard was free. She would have to work on Sunday night, but by then I would be heading back to Frankfurt on the Berlin Express. We took in the Saturday afternoon concert of the Berlin Philharmonic at the Titania Palast, and then went to the Gossler. Hildegard's number one conversation topic was the imminent start of the rehearsals for *A Day of Rest* at the Schiffbauerdamm.

"There's less than three weeks to go 'til the opening on New Year's Eve, and I haven't even seen the script yet. This farce ought to be a hit, because there's so little else to laugh about in Berlin these days. And isn't New Year's Eve fabulous for an opening? Not to mention that this is going to be my biggest part yet. Can't you come back to Berlin for my opening, Günterlein? Surely, if you put your fabulously brilliant mind to work, you can figure out some way to get here! Wouldn't it be a dream if we spent New Year's Eve together?"

"I'd wish nothing better than coming to your opening night. It might give me another chance at my favorite indoor sport, heaving rowdy drunks out of theaters. But I promised my folks in England that I'm going to visit them over the holidays. Maybe I could cut my visit short and hightail it directly from London to Berlin just after Christmas, in time for your opening. If I work some scam to get on the Berlin Express without valid travel orders and am caught and wind up in the stockade, you've got to tell the Army that I did it all for love."

I was kidding, of course, never imagining that within three months, Hildegard would do exactly that and save me from a disciplinary action I was facing over a violation of Army regulations.

Poor, Lonely, and Bored

TWO DAYS AFTER OUR ARRIVAL IN ANTWERP on New Year's Day, 1939, Friedl had recovered enough from the physical and psychological trauma of our escape across the Green Frontier to drag herself out of bed, and we went to register with the police. Despite the assurances from the people at the Jewish Refugee Committee, I was still afraid that the Antwerp police might send us back to Germany. But they treated us with civility. Our illegal entry into the country notwithstanding, they issued us identity cards that allowed us to remain in Antwerp for the time being, provided we did not engage in business or accept gainful employment. I was proud of my Belgian identity card. It documented that I was not a nonperson. So we settled in for the wait for our British entry visas for which my father was going to apply in London. It turned out to be a long wait.

Most of the next few days I spent writing a long letter to Mausi and Rudi describing our escape, which I had drafted in my mind while traipsing through the snow across the Green Frontier. I enjoyed converting our ordeal into a coherent story with beginning, middle, and end, although I had always hated having to write those tiresome weekly themes in school on subjects like "How I Spent My Summer Vacation." It was my first literary effort, and the first of many travel accounts I would compose for my sister and brother over the years.

I took my meals at the soup kitchen run by the refugee committee, sometimes standing in the back yard for an hour in the outdoor chow line. Friedl came with me the first time, waiting for a free meal in the winter cold in her Persian lamb coat. She decided that lining up for a handout was not her lifestyle and thereafter forewent hot meals rather than suffering the indignity of a nebbish refugee. So she nourished herself in our hotel room with bread and cheese, which she bought with the little spending money we got from the committee.

Thus begun 1939, the year on which I came to look back as the nadir of my life. I no longer had to live in constant fear of being beaten up in the street or being sent to a concentration camp (except in my dreams, which didn't take notice for decades of this change in my situation). But I had traded physical insecurity for economic insecurity, conviviality for loneliness and boredom. Comfortable in our snug Berlin apartment with my books, my typewriter, and my panoramic view over Schöneberg, I had never lacked money for the satisfaction of my material wants. And in the PRIWAKI I had been busy all day in the company of friends. Now I had turned into a destitute refugee, dependent on charity and sharing a bed in a tiny room in a crummy hotel with my stepmother. I was all on my own, with nothing to do.

After a week or so, we moved from our hole-in-the-wall at the Hotel Max into a furnished room in a mid-nineteenth-century, three-story burgher's row house in the center of town. Electric wiring had been installed early in the century, but there was no central heating or running hot water. The landlady occupied the ground floor, on which the kitchen was located. We lived in what must once have been the master bedroom on the first floor, furnished with one French-style, undivided double bed. I slept on a cot in a tiny garret on the third floor, directly under the roof.

I spent most of my time in the movie houses on the Kaiserlei, Antwerp's Kurfürstendamm, which leads from Central Station toward the port. They opened at noon, charging admission prices for double-feature performances that were ridiculously cheap compared to Berlin. Sometimes I went to two cinemas on a single day and saw four films in succession, whiling away more than eight hours at the movies.

This random intake of dozens of American, British, and French movies was not without influence on my linguistic and cultural development. All Anglophone films were shown in their original versions, with Flemish subtitles. So, when we finally joined my father in London, I spoke fluent American movie English, able to come up with phrases like "Put up your dukes, Buster!" "You betcha sweet life, Sister!" and "Hiya, Good-Lookin'! What's cookin'?" I also knew all about the American Way of Life, at least from the Hollywood perspective.

Mornings, before the movie houses opened, I often walked down to the busy Antwerp port, to watch the ships docking. They were unloaded and loaded by big cranes mounted on wide-gauge railroad tracks coursing the single, mile-long quay, and tugged out into the Schelde River toward the North Sea. I yearned to be on one of those ships, preferably on one going to the States. Yet I would have settled for any overseas destination—Canada, Australia, Latin America—as long as it would get me out of Europe, where I obviously had no future.

On the evening of January 30, 1939, we listened to the broadcast of Hitler's speech to the Reichstag on the sixth anniversary of the Nazi *Machtübernahme*. There had been rumors that Hitler would make an important statement. Whereas he had previously promised merely to cleanse Germany of its Jews, now, for the first time, Hitler announced publicly that he was going to kill all of European Jewry. He began by declaring that great progress had been made in recent months in moving toward the satisfaction of Germany's legitimate demands. Only a few issues still awaited their resolution. So the specter of war, which no people on earth wished to avoid more than the peace-loving German nation, had become much less ominous. However, Hitler went on, the final stages of a peaceful resolution of Germany's legitimate demands were being sabotaged by Jewish agitators, who were stirring up England, America, and France against the German people. Hitler made what he called a "prophecy." If international Jewish financiers should succeed in plunging the nations again into a world war, then the result will not be a Bolshevization of the earth, and thus the victory of Jewry, but the annihilation of the Jewish race in Europe. I became horribly frightened. I was sure that there would be war, because I knew that Hitler wouldn't keep his promise to make no further demands once "the few issues that still await their resolution" had been resolved. But since I also knew that he would keep his promise to exterminate the Jews once war had broken out, I realized that I had heard the definitive, irrevocable announcement of the (not-as-yet-named) Holocaust.

The oppressive fear that Hitler's prophecy had aroused was intensified further in the middle of March, when he seized Czechoslovakia. Just as I had supposed all along, Hitler didn't keep his promise to make no further demands once a few remaining issues have been resolved. War seemed inevitable, now that any fool could see that he had shafted Chamberlain and Daladier at Munich and that he would keep on making one demand after another, unless the French and the British were going to let Hitler walk all over them without offering any resistance. In either case, the days of European Jewry were obviously numbered.

Our permit to enter the United Kingdom arrived at the British consulate in Antwerp three days after the fall of Czechoslovakia. It took us another month to get the Belgian identity document that would allow us to travel to England. Finally, on April 20, 1939, we left Antwerp on the Harwich night boat. (That day bore a preprinted command in the diary-calendar I had brought with me from Berlin: "Flags! The Führer's 50th birthday!") What I had yearned for all these months, to be towed out on the Schelde River toward the North Sea, finally came to pass. Admittedly, the night boat was not bringing me to America, but at least it was taking me off the European continent.

One weekend in December 1946, when I was driving back to Frankfurt from my Bayer Plant target in Leverkusen, I made a little 200-mile detour through Belgium. I thought that it might be fun to strut down Antwerp's Kaiserlei again, in Ami uniform this time, wallowing in smug self-satisfaction.

During the late stages of the war, Antwerp's vast, nearly undamaged port on the Schelde River had become the main port of supply for the Allied armies fighting to deal the final blow to Nazi Germany. To make up for not destroying the port before retreating, the Germans showered Antwerp with missiles during the final months of the war. Moreover, the strategic aim of the Wehrmacht's last desperate offensive in December 1944, the Battle of the Bulge, was to drive westward from the Ardennes Forest to recapture Antwerp. Nevertheless, as I was surprised to see, there were very few signs of war damage. Antwerp's monumental Central Station still stood in Edwardian splendor, in stark contrast to the wholesale demolition of almost all major German railway stations. To me, the boy railway expert, the Central Station had been of much greater architectural interest than Antwerp's Cathedral of the Holy Virgin, one of the finest exemplars of Gothic architecture in the Lowlands. Antwerp Central, almost as big as New York's Grand Central, was ludicrously oversized for the sparse train traffic it served. It must have been built as the Cathedral of Our Lady of Progress.

On the Kaiserlei, the shops, cafés and movie houses were all intact. And despite the ravages of the Holocaust, there were a few bearded Jews in their black hats and overcoats walking on the Pelikanstraat, the erstwhile center of the Jewish diamond-cutting trade. I wondered how they had managed to survive. One of the cheap kosher restaurants on that street, where Friedl and I had sometimes eaten, was in operation. I went in to have lunch, feeling as if I were fulfilling an obligation toward the dead.

On the back of a picture postcard I sent to Hildegard, showing the Cathedral of the Holy Virgin, I wrote (in English) "wish you were here," callously intending this trite, famously insincere American greeting as a joke. In truth, I would have enjoyed having my blonde Hildegard hook her arm under mine as we strolled down the Kaiserlei, by way of a badge and witness of my triumphant return.

Driving back to Frankfurt from Antwerp, I passed through Thionville, a town in the Lorraine of which I thought I had never heard. Yet the name "Thionville" struck an uncanny, *déja entendu* chord with powerful erotic undertones. At first, I couldn't figure it out. Then it came back to me. In the fall of 1938, just before Kristallnacht, I had received a picture postcard in Berlin with greetings from my PRIWAKI classmate and dream girl Ursula A. It showed a panoramic view of Thionville, where, on her way to Palestine, she was staying for a while with her elder brother, who had recently emigrated to France. I interpreted Ursula's sending me that card to mean that she must have cared for me more than I had imagined. I was furious with myself for having been such an idiot and missed my chance while she was still in Berlin. I badly wanted to rush to Thionville and see her before she left for the Levant, to make up for all those wasted opportunities.

The excitement I had felt on my return to Berlin eight years later when I looked up to Ursula's former bedroom window from my parked jeep and the subliminal erotic arousal that had just overcome me on her behalf as I was passing through the town with the long-forgotten name "Thionville" obviously meant that I was still carrying the torch for her that she had lit when we were in our early teens. So, while driving the rest of the way back to Frankfurt, I wondered what the meanwhile twenty-two-year-old Ursula would be like and whether, if I happened to run into her now, in 1946, I would prefer her to Hildegard as the object of my love. It would be another forty-three years until I had the answer, when I finally saw Ursula again at the PRIWAKI reunion held in New York City. Even taking into account that by 1989 Ursula was a middle-aged woman who had probably lost some of her girlish charms, I decided that she couldn't have possibly matched Hildegard's outstanding qualities as a desirable woman in 1946. I never would understand why I hadn't fallen in love with Hildegard the moment I met her.

A Happy New Year's Eve

MY PREVIOUS SUCCESS IN SNEAKING BACK TO BERLIN to see Hildegard for a weekend gave me the idea of trying another travel order ruse to attend her opening night on New Year's Eve. My standing FIAT travel orders stated that Mr. Gunther S. Stent, a U.S. civilian attached to the 6800th Field Information Agency, Technical, was authorized to proceed for temporary duty to Leverkusen, Germany, and return from there to his permanent station. Mr. Stent's standing orders did not mention the location of his permanent station. So I reckoned that, using these ambiguous orders, I could book a round trip on the Berlin Express at any time. Going, I'd declare at the Frankfurt RTO (Railway Transportation Office) that I'm returning from temporary duty in Leverkusen *to* my permanent station in Berlin. And coming, I'd declare at the Berlin RTO that I am proceeding on temporary duty to Leverkusen *from* my permanent station in Berlin. So I made up my mind that after spending the 1946 Christmas holidays with my family in London, I would weasel my way to the Schiffbauerdamm Theater in time for Hildegard's New Year's Eve première.

I applied to FIAT headquarters for Christmas leave and travel orders from Frankfurt to the United Kingdom for the period December 19, 1946, to January 2, 1947. I got what I asked for, except that the expiration date of my leave was shown as December 28, 1946. There was no explanation why I hadn't been granted the full two weeks I requested. I supposed that it was just one of those bureaucratic snafus. Never mind! Who was going to check whether I was back in Hoechst during the holidays, when no one was likely to be working, anyhow. And if I was going to make illegal use of my standing Leverkusen travel orders, did it matter all that much that I would also be AWOL?

A few days before I took off for England, a letter came from Hildegard.

Günter! My dear Du!!

Today is the Third Advent. I'm thinking so much about you. Don't be angry, but I have to be a little caressing in my letter to you. Believe me, when I write to you then it always relieves my heart at little bit. Can you understand that? Think a little of me on Christmas Eve. I kiss you in my thoughts (but only because it's Christmas).

Her letter reinforced my resolve to visit her by making fraudulent use of my standing travel orders. I answered by return mail.

Hildelein, mine!

I am thinking of you too all the time and will kiss you in London in my thoughts (and not just because it's Christmas Eve!). I'll definitely try to come to Berlin on New Year's Eve for the opening of *A Day of Rest*. But you better not count on my making it, because it isn't going to be easy. In case I do make it, I'll turn up at the theater before the curtain goes up. Be sure to leave a ticket for me at the box office.

Flashing my legitimate Christmas leave travel orders to London, I got on the officers' car of the Brits' Rhine Army Express at Cologne. In contrast to the U.S. Army's Berlin Express, on which officers and enlisted men shared the same accommodations, they traveled in segregated cars on the Rhine Army Express. Despite my progressive political sentiments, I felt no outrage over this flagrant, undemocratic discrimination. Far from it, I was pleased to be riding in exclusive company since the train was packed with raucous Brits going home for the holidays. Next morning, the train pulled into the British military pier in Calais Harbor, where I was struck by the sight of two pissoirs on the platform, one labeled "Officers" and the other "Other ranks." This scene reminded me of the "White" and "Colored" signs on public facilities I had encountered on a visit south of the Mason–Dixon Line during the war.

I got to Victoria Station in the afternoon and went to meet Ronnie at his office at Empiria Products Ltd. in Holborn. This was the export–import firm he had recently joined, after his return from wartime Army service in India—mustachioed, tweed-suited, pipe-smoking, and every inch the British officer in civvies. Decked out in my "simulated" FIAT officer's uniform, I was proudly introduced by Ronnie to his boss, Mr. Fisher, as his—Allied-hands-across-the-sea—"Yank brother." I stayed with Ronnie, Gabi, and my beautiful, blonde-pigtailed, super-bright, eight-year-old niece Monica, in their apartment just off Ealing Common. Gabi was expecting her second child within a couple of months.

It gave me a wonderful feeling to be able to renew the close ties with my brother and sister-in-law that had been sundered by the war. The high point of my visit was the Christmas dinner at their place, which included also Gabi's mother and my former French tutor, Gabi's elder sister, Steffi. There was so much to tell at the dinner table, to catch up on what happened to the British and American branches of our family since February of 1940, when we last saw each other on my departure for the States.

After the invasion of the Lowlands and France in May 1940, Ronnie and my father were interned as enemy aliens on the Isle of Man in the Irish Sea, along with 20,000 other German Jewish refugees. The British government never explained the reasons for its decision to lock up all male German–Jewish refugees, who, after all, could be expected to be the most die-hard enemies of Nazi Germany and the most loyal supporters of the Allied war cause. By late September 1940, the German–Jewish refugees were beginning to be cleared out of the Manx camp. Some were shipped to POW camps in Australia or Canada. Others, like Ronnie, were allowed to join the Pioneer Corps, the British Army's noncombatant unit for ex-felons and other social outcasts. After two years' service as a private in the Pioneers, Ronnie was transferred to a Royal Engineers Officers Cadet training unit, commissioned as a lieutenant, and posted to New Delhi to serve on the staff of the Royal Engineer-in-Chief for India. Evenutally promoted to captain, Ronnie returned to London for his demobilization in May 1946.

My father was too old to join the Pioneers or to be shipped to an overseas camp. So he stayed locked up on the Isle of Man after Ronnie was gone, being finally released in the spring of 1941. He and Friedl settled in Harrogate, a spa in Yorkshire, where they opened a flourishing little electrical supply and lighting business, the Belmont Trading Company. Toward the end of the war, they moved back to London, where they set up Messrs. Belmont Trading in the basement of a decrepit building off Tottenham Court Road. My father looked after the sales and Friedl after the books (which, as eventually transpired, meant that the vixen had become the guardian of the hen house).

On Boxing Day, I took the train down to Brighton to see my father and Friedl. They were spending the holidays in a boardinghouse, just off the wind-swept, wintry promenade lining Brighton beach. They didn't seem to have aged at all and, being evidently reasonably prosperous, had shed the persona of the destitute refugee. I was glad to be with them. I did have feelings of filial love for my father, and, if not love, at least some sentiments of comradeship for Friedl, in view of what she and I had gone through together before the war.

Yet I was uncomfortable in their company. They treated me like the child I had been when I left them to join Claire in Chicago. I was especially annoyed by my father's telling me again and again what I ought to be doing with my life: Go into business and make lots of money, instead of frittering away your talents as an academic starveling. Coming from him, I rejected this unsolicited advice outright. I probably resented my father and Friedl's tutelage even more than grown children ordinarily resent parental inbutting, because, having been out if touch with me during my critical formative years, I felt they had no idea who I was. When they were not giving me advice, their conversation tended to be boring.

There was some business for me to discuss with my father.

"Just before heading for London, Dad, I received a letter from Dr. Sachs, our Berlin lawyer. He informed me that the title search showed that Erich Schmidt, the bookmaker to whom we sold the Essener Strasse property in the fall of 1938, is still its owner of record and still lives in Berlin. So should I instruct Dr. Sachs to initiate restitution proceedings, to take away the house from Schmidt, who had bought it from us in good faith? Or should we forget all about it and let the poor man keep his house in the wrecked city?"

My father looked at me, totally amazed.

"What are you talking about, Günter? Gone off your rocker? Of course, we are going to ask for our house back! It's just too bad for Schmidt, just as it had been too bad for us when it was our turn to give it up in 1938. If Schmidt has any complaints, let him write to Hitler! Make sure, though, that Sachs isn't going to screw us on the contingency fee for handling the case."

I left London on the day after my Christmas leave expired. Back in Frankfurt's Central Station the following noon, my heart sank when the GI on duty at the RTO frowned as I tried my fits-all-destinations standing travel order scam on him. A few moments later, his face lit up, and he said, "Oh, I see. You're permanently stationed in Berlin, aren't you?" "Affirmative." He cut a ticket for me on that night's Berlin Express. When I got to Berlin on the last day of 1946, a Tuesday, there was a new receptionist at the Gossler. Like the GI at the Frankfurt RTO, he first puzzled over my travel orders and then asked whether I was permanently stationed in Berlin, on my way home from Leverkusen. Only this time, when I brightly replied, "Affirmative," I had fallen into a trap.

"The Gossler is a *transient* officers' billet, Sir. Why are you asking for a room here? Why aren't you staying in your permanent Berlin billet?"

"Because I gave up my permanent billet when I left for temporary duty in Leverkusen, that's why. They're going to assign me a new permanent billet after the New Year's holiday. What am I supposed to do meanwhile? Sleep under a bridge?"

He picked up the phone to call the Berlin FIAT field office. I thought I was done for. When they discovered that I was not only making fraudulent use of my standing travel orders but was also AWOL, I was going to celebrate New Year's Eve in the military police cooler. Fortunately, it seemed that I was not the only Ami with gala New Year's Eve plans. FIAT's field office was already closed for the holidays and would remain closed until Monday, January 5, 1947.

"OK, Sir. You can stay at the Gossler until the FIAT office opens on Monday. Then, your status will have to be cleared up."

Since I was going to scoot back to Frankfurt already on Friday night, I'd be long gone from the Gossler by the time this busybody could check out my bona fides.

In the late afternoon, I entered the Schiffbauerdamm Theater through the stage door. The elderly porter who remembered that I was Fräulein U.'s Ami, saluted me. Hildegard danced a little jig when she saw me.

"Günter, Günter! You came, you came!"

"What's the surprise? I said I would try to come, didn't I? Didn't you mean it when you said I was a genius? Maybe you say this to all the boys?"

"Yes, I do. Tell it to every one of the dozens of my admirers. But until today I wasn't sure that you, Mister Ami Scientist, really *are* one!"

"Actually, there was a little trouble at the Gossler this morning, which could have led to your genius spending New Year's Eve in jail rather than with you, my dear Hildelein. Luckily, I managed to talk my way out of it."

We laughed in our happiness and hugged and kissed. Hildegard looked very attractive, despite being costumed as a slavey with a babushka kerchief hiding her flaxen tresses.

The performance started at half past six, to a packed house. The program identified *A Day of Rest* as a farce in three acts by the Soviet comic playwright, Valentin Katajev. The locale is a small convalescent home, "Butter Mushroom," the physician-in-chief of which is a woman. In Sydney Kingsley's world of *Men in White*, female doctors, let alone female physicians-in-chief, were, of course, unheard of. Hildegard, listed last among a cast of eleven, was taking the part of Shura, a maid at the Butter Mushroom. Shura-Hildegard discovers that Comrade Saizew, who asks the manager for authorization to buy furniture lacquer, is a swindler. Yet it all works out for the best. Saizew turns out to be a swindler with the heart of gold for which Soviet workers are famous. He needs the lacquer for painting some children's beds in a nearby crèche.

It didn't make all that much of a difference to me that the play was a drag. After all, I hadn't come to see *it*; I had come to see *her*. The final curtain fell at about nine, accompanied by tumultuous applause. There were many curtain calls; Hildegard was beaming. I thought she had acted very well, and each time she came out to take the applause, I shouted "bravo!" proud of being her man. At the last curtain call, the two leading ladies, but not Hildegard, were presented with flowers. Damn it! Why hadn't I thought of bringing her flowers?

In his review of the premiere, the theater critic of the Berlin paper *Sozialdemokrat* called *A Day of Rest* a farce that is just as harmless as it is congenial, although he noted sourly that "Hildegard U. had to be shown by the director how a maid behaves, which she then imitated faithfully, without any signs of personal involvement." Of course, Hildegard had to be shown how a maid behaves! Raised during the war in working-class Wedding and gone off to the Eastern Front with the Wehrmacht Truppenbetreuung when she was sixteen, where would she ever have been attended by a maid?

From the theater, we went to a bar with a jazz band on the Kurfürstendamm, to celebrate Hildegard's triumph and ring in 1947. Shortly after midnight, we headed for the Gossler. Hildegard was doubly exhilarated, first, by the biggest acting success of her career, and, second, by my having come all the way from London to witness it. She was sure I must really care for her to have gone to all that trouble! In my near-total bliss I wondered why I didn't fully succeed in loving this wonderful girl. Maybe her face was not beautiful enough to be my Dream Woman's. But it was the kind of magical New Year's Eve I had wished for all my life. Since my childhood I had been imbued with the Berliner's creed that you might as well be dead if you don't have a good time on New Year's Eve. And so I had been painfully aware of my perennial failures, keeping a mental record of these annual disappointments.

My earliest remembrance dates back to New Year's Eve 1933. My mother was already in the sanatorium; my father was out of town on business; and Mausi and Rudi had gone with their future spouses to a party of the KC fraternity. Nine years old, I was left alone in the Fasanenstrasse apartment with our maid and her six-year-old illegitimate daughter, whom she had brought up to Berlin for the holidays from her home town of Sonneberg, the Thuringian toy capital. We were wearing paper hats at dinner, but it was boring—there was no singing or joke telling, and I thought enviously of Mausi and Rudi.

New Year's Eve 1935, I was on a skiing trip with the PRIWAKI in the Bohemian Riesengebirge mountains. Our teachers had slyly moved their watches forward and sent all of us kids from the lower grades to bed at ten P.M., after deceitfully shouting "*Prost Neujahr!*" Next morning, the older kids made fun of us for having been had.

Lacking the proper formal attire for the dinner dance gala at the Cristallo in Cortina d'Ampezzo, my father and I spent New Year's Eve 1936 dining alone, as usual, in our room. It was New Year's Eve 1938 that Friedl and I fled across the Green Frontier to Belgium. New Year's Eve 1944, I was supposed to spend with my fiancée Dorothy S., but a few days earlier, while we were still engaged, she suddenly married another man. So I rang in the New Year alone in my room at the fraternity house, petulantly contemplating suicide to spoil her marriage. And the previous New Year's Eve I had spent getting drunk with a fellow graduate student at Katzinas' Tavern in downtown Champaign, wishing I were with a woman.

But now the spell had been broken, and in Berlin, of all places! I told Hildegard that this was the happiest New Year's Eve of my life, which was probably the most affectionate thing I'd ever said to her. I also performed a symbolic act, of whose meaning among American fraternity men Hildegard was not, of course, aware. Just in case I happened to run into Miss Right in Europe, I took along the golden chain with my jeweled fraternity pin as a pendant that I had given to my ex-fiancée, Dorothy S., two years ago, on the night of our engagement. A few weeks later, she returned my pin with her farewell letter. I now hung it around Hildegard's neck, and she cried from happiness.

When we parted the day after New Year's Day, Hildegard suggested that maybe we could meet somewhere in the Western zones. Although it was very difficult for Germans to travel across the Soviet zone to the West, she thought that she would be able to manage it. I said that I'd let her know where we could best meet.

Before leaving Berlin, I went to see Dr. Sachs about recovering title to the Essener Strasse property. I asked him to initiate the restitution proceedings, despite Erich Schmidt, to whom we had sold the house, still being the registered owner. Dr. Sachs thought there was no question about our getting the house back, but that it would take a while. He knew whereof he spoke. Transfer of the title to my father's and my name did not occur until the summer of 1950.

When I got back to Frankfurt Central Station and caught a ride out to Bad Homburg, I was flabbergasted to find the Park Sanatorium deserted. The sergeant in charge of the staff told me that while I was gone, FIAT headquarters had been moved from Hoechst to Karlsruhe, 100 miles to the southwest. All personnel were assigned new billets in the Karlsruhe area. There had been rumors of a move for a few weeks, but it seemed unbelievable that they would move our headquarters from its central location in the Frankfurt area, within reasonably close reach of most of our industrial targets, to the far southwestern corner of Germany. I now realized why my Christmas leave request had been granted only until December 28, 1946. They wanted to make sure that I would be back and help with the moving, not to mention moving with them.

I managed to get Freddy on the phone.

"Where in hell have you been, man? Your leave expired a week ago. Get your ass down here on the double! Our new headquarters are in an old insurance building on the Kaiser Allee in the center of Karlsruhe. There's a statue of a Kraut soldier on a horse in front of the place."

Freddy, bless his soul, hadn't told anyone that I was AWOL.

Karlsruhe was in ruins, flattened by aerial bombardment. Many of the streets were blocked by debris, and to get from Karlsruhe Central Station to FIAT headquarters my driver had to make one detour after another through a labyrinth of dead-ends. The old insurance office with the mounted dragoon in front seemed to be one of the few buildings still standing.

Our new billets were in Durlach, an undamaged, attractive Karlsruhe suburb in the foothills of the northernmost part of the Black Forest. Several contiguous blocks of modern, two-family houses had been cleared of their Kraut occupants and given over to FIAT officers and civilian investigators. Each of us was assigned, for sole occupancy, a three-room apartment with kitchen and bath, and the furniture the evicted Krauts had to leave behind. I was given the downstairs apartment at Kastell Strasse 19, the most spacious quarters by far that I ever had for my exclusive use since fleeing Berlin eight years ago. The upstairs apartment was assigned to another document screener, Fred Thornhill, a German–Jewish chemist in his mid-thirties, who had brought his Massachusetts-born gentile wife and their baby daughter over from the States. To let Hildegard know that I hadn't forgotten her, I sent her two parcels with food, coffee, and cigarettes as soon as I was settled in.

A letter from Hildegard arrived before I left Karlsruhe for Leverkusen, to continue my document screening at the I.G. Farben-Bayer plant:

Günter, dearest,

Only two days have gone by since we parted, but since I'm suffering from Sunday boredom today, I'd like to chat with you a little. I feel so terribly lonely since you went away. How come one becomes used to another human being so terribly quickly? After all, I've got my profession, my family. Lots of acquaintances! Well, I don't know how this comes about. Is happiness possible only when you are with a person who is dear to you? Why does life tear people apart so often, into the seven winds, why? All that makes me so sad. But I don't want to turn melancholic. That doesn't suit me at all.

Du, it's turned very cold here. Even my hands are numb. Be happy that you have a warm room. It might have been a little warmer even in our apartment, but since we have no window panes, my room is like an ice-box.

Got over the pain of separation? (Ha-ha-ha!).

The critics' opinions of the premiere are a very mixed lot. I don't want to go into details.

Yesterday I sent off the picture of me I promised you. If you really can't come back to Berlin for a long while, maybe you'll forget what I look like? So, picture, please get to my Günter quickly.

I'm waiting for you! Forever!

Your Hilde

Hildegard's veiled, ironic allusions to my emotional coldness, to my never telling her that I really care for her, were lost on me, as was the anguish behind her "Ha-ha-ha," and the delicacy of suggesting that she was writing to me only from Sunday boredom. Ever under the sway of the emotional isolation that I devised for myself, I failed to appreciate, or even take an interest in the feelings she provided for me in her letter. It was all part of my enduring failure to take cues from others. Unless someone was actually trying to kill me, or indicating an interest in sleeping with me, I simply paid no attention to their emotions.

And so in my reply I didn't assure Hildegard, as I should have, that I felt as lonely without her as she feels without me, that I was suffering from the pain of our separation, and that, photo or no photo, I would never forget what she looks like. I just reaffirmed that I was looking forward to seeing her before too long in the West and merely provided her with my factual news.

Holding Pattern

WHEN FRIEDL AND I GOT OFF THE NIGHT BOAT from Antwerp at Harwich on April 21, 1939, I was mighty glad that I had made it to England at last, after hanging around Antwerp for four dreary, lonely months. My father met our boat train at Liverpool Street Station. He had rented a furnished bed-sitter for us in a Victorian row house on Randolph Avenue in London's Maida Vale district, up the Edgeware Road from Marble Arch. Located on the second floor, our room was equipped with a little gas burner for light cooking and a boarded-up fireplace with a gas heater. To make the gas flow, copper pennies had to be fed nonstop into a coin box attached to the main valve. My father and Friedl slept in the double bed. I spent the night on a couch in the landlady's ground-floor parlor, which I could access only after eight P.M. and had to vacate by eight A.M. We had no private toilet or bath. It was only a marginal improvement over our Antwerp digs, but we were out of Hitler's immediate clinch.

Not long after our arrival, my father received a registered letter from a Berlin law firm. They advised him that, in accord with Section 1 of the decree of December 6, 1938, authorizing the forced sale of Jewish interests in manufacturing and trade, Herr Friedrich Ballnath had acquired Herrn Georg Stensch's 50 percent share in the Ballnath & Stensch partnership. Herr Stensch had, therefore, no longer any connection with the firm and was expressly enjoined from representing himself as having such a connection, in Germany as well as abroad.

I was with my father when he opened this letter. He stared at it speechlessly, and then began to cry, which I had never seen him do before. His life's work, the company he had founded more than a quarter of a century before and from which he had derived the main meaning of his life, had been taken away. His trusted partner, Fritz Ballnath, had done him in!

"Those shitty goyim are all alike, Günter. They'll stab a Jew in the back at the first chance!"

My father seemed to derive little comfort from reading this letter in the safe haven of London. He didn't appear to appreciate how lucky he was to have escaped the fate that the Nazis had in store for us, about which the ever-worsening situation of the German Jews since the Kristallnacht had left no doubt. I think that it was only after receiving this letter that my father finally realized that Hitler was not a passing phenomenon and that there would be no return to Berlin for him.

Since the visitor's visa under which we had been admitted to the United Kingdom as refugees did not allow my father to engage in any business activity or accept paid employment, he had no source of income in London. So he supported his family by drawing on the money he had managed to smuggle out of Germany by *Devisenschiebung* over the years since the Nazi takeover or to collect abroad as debts owed by Ballnath & Stensch's foreign customers. I never found out just *how* much that money amounted to, but he never failed to remind me that even on our very penurious style of living it wouldn't last long. Thus I was constantly worried about how we were going to survive once my father's money was spent.

I was as bored in London with my lonely, purposeless existence as I had been in Antwerp. My father didn't want to waste any precious money on sending me, the PRIWAKI's unruly deadbeat-expellee, to school while I was merely on a holding pattern for taking off for the States. So to while away the time, I founded my own Autodidact's Academy of Practical Arts. Its venues were the South Kensington Science Museum and the Speaker's Corner in Hyde Park. Mornings, I would hike from Maida Vale through Paddington, across Hyde Park to the Victoria and Albert Memorial, and then down Exhibition Road to the museum.

Technology, especially railways, had fascinated me ever since my early boyhood, a passion I was to retain all my life. In Berlin, I had hung around Zoo Station on weekends and holidays, watching the trains pull in and noting the coded inscriptions painted on the rolling stock. They provided experts like myself with such vital statistics as the home base, model type, and service record of the locomotives and wagons. To deepen my taxonomic knowledge of the iron horse, I liked to visit the Berlin railway museum at Lehrter Station.

Now I discovered that the Lehrter Station's collection couldn't hold a candle to the South Kensington Museum's treasure of ancient steam engines, locomotives, and rolling stock. They had some engines built by James Watt in the late eighteenth century, including one of the first with Watt's centrifugal, whirling-steel-balls-on-a-stick governor of piston cycle frequency. Despite my total ignorance of physics, I managed to work out an intuitive understanding of how this device works, and was astounded by the brilliance of Watt's invention. When, in later years, I had to teach college sophomores the principle of homeostasis as one of the defining aspects of living creatures, I always presented Watt's ingenious steam engine governor as an exemplar of an automatic regulator. This was a peculiar pedagogic quirk of mine, since most biology professors trot out the baking oven thermostat for this purpose. The museum also had a big library, in which I browsed through books on the development of the iron horse and the automobile. Thus I invented for myself the life of the research scholar, who spends his days in museums and libraries.

At closing time, I would recross Hyde Park to Marble Arch, where a dozen or so speakers would be setting up podia, ladders, or soap boxes at the Speakers' Corner for their evening harangues. Dynamic orators and effective demagogues, they held forth on a wide range of topics—Mormonism, Jews for Christ, Eastern mysticism, scientific atheism, as well as Irish unionism and antiunionism, anti-imperialism, and antiroyalism—each of them giving more or less the same spiel, night after night. Many speakers had their loyal opposition—faithful hecklers who would interrupt them with the same rude remarks night after night, evoking the same repartee, laced, more often than not, with obscenities.

Most evenings, the speakers included also British Fascists and Communists. Although there was plenty of verbal abuse of and by them, it never led to any physical violence between these rabid political adversaries. Such physical restraint would have been unthinkable in Weimar Republic Berlin, where any similar confrontation of Nazis and Communists on the Alexanderplatz would inevitably have led to mayhem.

Hyde Park's best-known speaker and my personal favorite was the African Prince Monolulu, who made his living as a freelance bookie at the London racetracks. The prince wore a farrago of African and Asian togs, topped by a Native American, multicolored, feathered head dress.

"Each item and each color I wear symbolizes some feature of the brotherhood of man: Red, white, and blue stands for the British Empire, which comprises Jews, Muslims, and blacks. Red also stands for revolution, as white does for peace, green for Ireland, and yellow for the press. My face stands for Germany and its imperial colors: black (my skin), white (my teeth) and red (my tongue)."

I found the prince's talks highly entertaining, despite my failure to grasp most times what point he was trying to make. To my surprise, he occasionally lapsed into fluently idiomatic German. It was only on the subject of Jews that I thought I understood him. His feelings were ambivalent:

"On the one hand, I admire the Jews because, when God commanded "Let there be light!" and there was light, it was the Israel-lite. On the other hand, I dislike the Jews because they are too much like the Germans. In fact, the Germans hate the Jews so much for just that reason. Contrary to the shallow idea that racism is rooted in 'otherness,' one can truly hate only someone who is like oneself and whom one therefore understands."

It was probably the Prince who gave me the first inkling that we German Jews were stuck with a double dose of an especially onerous heritage.

A political bombshell exploded on August 24, 1939. Incredible as it seemed, the archenemies Hitler and Stalin had made a nonaggression pact. The Soviet Union would not join Britain and France if they honored their treaty obligation to come to the aid of Poland in case she were attacked. A week later, Hitler annexed Danzig, and German troops crossed the Polish frontier. Were the British going to make good their guarantee and fight Germany? We had good reason to hope that they would, because the prime minister, Neville Chamberlain, announced that Britain would declare war unless the Germans stopped their attack on Poland. The Germans ignored the British ultimatum and kept on savaging Poland. So Chamberlain went on the radio on Sunday, September 3, 1939, to announce that a state of war now existed between Great Britain and Germany. At last, the years of knuckling under to Hitler were over!

Ronnie explained to me that the present strategic situation of the British and French was unfortunately much worse than it had been at the outset of World War I. Nevertheless, he was sure the Allies were going to give Hitler a run for his money. Besides, there was always America to reckon with. Roosevelt wasn't going to stand by idly and let Hitler win! I badly wanted to share my brother's optimism, but just couldn't. I was still, and would yet long remain, obsessed by Hitler's prophecy that the Jewish race would be annihilated in Europe if international Jewish financiers succeed in plunging the nations once more into a world war.

By the time war broke out, a year had passed since I filed my application for a U.S. immigration visa at the American consulate in Berlin and was told that it would take a few months until my waiting number is called. I still had heard nothing about my visa, and the London consulate, to which my waiting number had been transferred, would not provide any information about the progress of my case. I was getting more and more disheartened, fretting all the time about how life was passing me by while endlessly waiting for joining my sister in glorious Chicago. Maybe—God forbid!—I was going to be stuck in dreary London forever.

Finally, finally, at the end of January 1940, I was notified by the U.S. consulate that my visa application had been approved. Before issuing the document, the vice-consul made me swear that I would do certain stipulated things for the United States, such as defending it against its enemies, and not do others, such as pimping or operating a house of prostitution. The Jewish Refugee Committee helped me to get a British certificate of identity, in lieu of a national passport. Within a week, it was stamped with the world's most precious endorsement, the U.S. immigration visa. Feeling on top of the world, I sent a telegram to my PRIWAKI classmate Günter Steinberg in New York, to announce proudly that I would be landing there before long on my way to join my sister in Chicago and that I was looking forward to seeing him. He wired back, suggesting that I stay with him and his mother in New York for a few days before heading West.

Because of the threat of German submarines, no transatlantic sailings out of British ports were publicly announced. To cross the Atlantic from Britain, one had to make a booking for an undisclosed departure date on an unidentified ship. My father booked a third-class steamer berth to New York for me at a little Jewish travel agency, the Palestine and Orient Lloyd (P&OL). Its owners had probably so named it in the hope that people would confuse it with the mighty Peninsular and Orient (P&O) Line.

Three more weeks went by until Messrs. P&OL called to let me know that I was to take a boat train to Liverpool from Euston Station on the morning of February 22, 1940. For security reasons, no one was allowed to accompany me on the train and see me off at the ship. My father, Friedl, and Ronnie just came with me to Euston Station, where we made our farewells on the platform. It was clear that we weren't going to see each other again for a long time, certainly not until the war was over and unless Hitler had been beaten. My father's and Friedl's eyes were moist, but I felt little pain over parting from them, totally absorbed by the thrill of being on my way to a new life in God's Own Country.

PART TWO

Changing Cultures

The Soda Jerk

THE BOAT TRAIN I BOARDED AT EUSTON STATION one February morning in 1940 went directly to Liverpool Harbor, stopping on the quay alongside an old-time Cunarder, *R.M.S. Scythia*. As I stood on the railing while the *Scythia* pulled away from the dock and moved down the River Mersey toward the Irish Sea, feelings of utter happiness welled up in me. I was finally making my getaway from that rotten old Europe, which held no future for me. At last, I was shipping out to the New World on an ocean-going boat, which I had so ardently longed for on my daily visits to Antwerp harbor.

I shared a cabin with three of the twenty or so young German–Jewish refugees (almost all male) who were emigrating to the States, among whom I—at fifteen—was the youngest. We speculated about the careers that awaited us in the American Land of Unlimited Opportunities. I had no doubt about mine. I was going to finish high school in Chicago and then study engineering at the American equivalent of a *technische Hochschule.*

The crossing of the wintry Atlantic was stormy and slow, because of the ship's zig-zag course to evade Jerry U-boats. There were no movies or other forms of organized entertainment for the third-class passengers; so we passed the time playing parlor games, telling jokes, and singing German folk songs. After ten days at sea, the news spread that the American coast was coming up on the horizon. I rushed on deck. An icy wind blew into my face. I could see that we were heading toward a city, which I took to be New York. But where were the tall skyscrapers of lower Manhattan? The buildings I saw in the distance weren't even as tall as the houses in downtown Berlin. Could it be that the photographs and movies I had seen of the Manhattan skyline had been doctored to make the scene look more impressive?

The city on the shore turned out to be Halifax, Nova Scotia. The quay where the *Scythia* docked was guarded by dozens of Canadian soldiers and policemen, and no passengers were allowed ashore. Freight was being unloaded from the *Scythia* day and night. No one explained why we had been taken to Halifax rather than New York. Not to worry, though. Eventually the *Scythia* would go on to New York. After two days in port, the ship steamed back into the Atlantic. It now transpired—I don't remember how—that we had carried a substantial fraction of the gold bars of the Bank of England. It was taken to Canada for safekeeping in anticipation of a possible German invasion of the British Isles.

When the *Scythia* entered the real New York harbor on March 8, 1940, the Manhattan skyscrapers looked just like the forest of giant stone trees of which their pictures had always reminded me. I cried a little when we passed the Statue of Liberty, for joy that I had made it to the New World at last, without having been done in at the last moment by one of Hitler's U-boats.

The Immigration and Public Health inspectors came on board to process the passengers' travel documents in the dining room. After I passed muster as a person without loathsome diseases, I was admitted as a permanent resident and, God willing, future citizen of His Own Country.

No sooner had I become a proud American resident, a forlorn refugee no more, than I was accosted by a woman from the American Joint Distribution Committee. She had come on board to help the Jewish refugee greenhorns get a footing in the New World. I had evidently been fingered to her as an unaccompanied refugee minor requiring her special attention. She asked me to go with her to the AJDC office in Manhattan, where she would arrange for my stopover in New York and my onward travel to Chicago.

Resentful of her re-relegating me to the category from which I thought I had just escaped, I wanted to tell her that she was making a big mistake: I wasn't a Jewish refugee at all. But she no doubt knew from the inspectors that I was carrying a British identity document issued only to passport-less refugees. So I tried to get rid of her by protesting that I was staying with friends in Manhattan and that I was a seasoned traveler who could find his way to Chicago perfectly well on his own.

"How are you going to get to Chicago?"

"I'm going to take the Greyhound bus. That's the cheapest and most scenic way to get around in the States, isn't it?"

My being such a know-it-all wise-guy obviously increased, rather than lessened, her concerns, and she insisted that I did need her help. So I promised that I would come to see her the next day, not really meaning to. While the baggage was being unloaded from the steamer, I called Günter Steinberg to let him know that I had just arrived from Liverpool on the *Scythia*. He offered to pick me up at the Cunard Line pier.

As I stood next to my gear in the vast dockside customs shed, wondering how to get hold of one of the few perambulating inspectors, the woman from the AJDC showed up with an inspector firmly in tow. After he cleared my belongings, she had a porter wheel them into the street, where Steinberg was waiting. He had to give her his address and telephone number before she finally let me go.

When Günter emigrated to New York with his mother in the summer of 1938, I envied him. He sent back enthusiastic reports about American high school life, in which his accounts of the free and easy way of dating girls were of paramount interest to me. He expected to graduate in little more than a year and then go on to college. How long would it take me, who had been out of school for more than a year and a half by now, to catch up with him? As soon as he got to New York, Günter had changed his first name to Gunther, and I got the idea from him to Americanize mine as well, or, rather, to revert it to its Wagnerian, premodern German spelling.

I stayed with the Steinbergs in their Upper West Side apartment, outfitted with the tasteful modern black-and-red lacquered furniture that I remembered from visits to their place off the Breitenbach Platz in Dahlem. Unlike our crummy rooms in Antwerp and London, there was nothing refugeeish about it. Gunther showed me the New York panorama from King Kong's perch atop the Empire State Building. I was intoxicated by the reality of the America that I had imbibed in so many photographs and movies. But I also remembered the vivid description of Harlem street scenes that the PRIWAKI's director, Dr. Selver, had given us in the spring of 1938, after his return to Berlin from a quick visit to New York: hordes of unemployed, down-and-out blacks lolling in various states of undress on the steps of delapidated row houses lining garbage-strewn streets. Since this picture hardly jibed with my vision of the American paradise, I asked Gunther to take me there. But Mrs. Steinberg vetoed our traipsing through Harlem.

"It's too dangerous! Better you spend your short stay in New York visiting the AJDC lady at her office and stop acting like a jerk. You *are* a Jewish refugee, aren't you? Why pretend otherwise? I bet you didn't even thank her properly for the help she gave you in clearing customs! You go and make up for it and have her help you with the arrangements for the trip to Chicago!"

My unwanted AJDC benefactress told me that traveling to Chicago by Greyhound was out of the question, for reasons she didn't detail. At the time, I indignantly suspected her of not trusting me to find my way back on the bus after one of the several meal stops en route. But in later years, after I got to know more about the Greyhound system, I thought that she probably feared I would get picked up by one of the pederasts or pimps who were thought to be lurking in big-city bus depots.

She had made a reservation for me on the New York Central's all-coach Pacemaker streamliner to Chicago and telephoned my sister to let her know my arrival time. She refused my offer to pay for the ticket from the fifty or so dollars that were left of the pocket money my father had given me when I left London. I *was* grateful to her for what she had done for me, but I probably didn't express my gratitude properly.

I was excited to ride in one of the Pacemaker's luxurious coaches with their reclining, upholstered easy chairs mounted in big salon cars. They were so much more classy than the austere European third-class railway wagons, which crammed their passengers into small compartments with wooden benches. It was a big disappointment for me, though, that a puny electric locomotive pulled this fabulous train out of Grand Central Station. Whatever happened to the New York Central's world-famous, futuristically streamlined, Raymond-Loewy-designed "Hudson Class" steam powerhouses, whose pictures I had admired at the South Kensington Museum? Fortunately, when the train made its first stop at Harmon, we were hitched to one of those fabulous giants. The little switch engine had merely hauled the Pacemaker out of Manhattan through the tunnel under Park Avenue.

Mausi, the person dearest to me in all the world, was waiting for me with Heinz at Chicago's La Salle Street Station when the Pacemaker pulled in next morning. They looked very American: Mausi like a Hollywood glamour girl, in makeup and a chic hairdo, and Heinz like a dapper G-man, got up in a snazzy, double-breasted, broad-lapeled suit and highly polished shoes. I arrived wearing a European-style, narrow-brimmed fedora hat, which was part of the grownup wardrobe Friedl had selected for me in Berlin. Heinz suggested that I get rid of that monstrosity, lest I wished to be instantly recognized as a German refugee, fresh off the boat from the Old Country. Wanting nothing less, I threw my fedora into the first garbage can I saw at the station, and have never worn *any* hat since.

They had changed their surname to "Hines" because no one at Neisner Bros.—the Woolworthesque chain of five-and-ten-cent stores where Heinz worked—could pronounce "Münsterberger." So they simply called him "Mister Hines." Needing a new first name, he chose "Robert" (or "Bob"), because it seemed a euphonious conjunct for "Hines." Mausi had anglicized her given name "Clara" (which she used to hate because in her Berlin girlhood it was totally out of fashion, evocative of wizened maiden aunts) to "Claire" (which was very á la mode in Chicago). They welcomed me to Chicago in Yankee-accented English, and none of us ever spoke German to one another. And they started calling me "Gun," to help conceal my ignominious German origins. Nor did I ever call them "Mausi" and "Heinz" again.

Bob pointed to a parked, brand-new, tan 1939 Chevrolet sedan that they had bought just a few days earlier. In Berlin, I saw photos of them in front of their first car, a secondhand 1936 Chevy, which I had found plenty impressive. But this new, sleek, hypermodern job—it was one of the first models with the revolutionary horizontal rather than the traditional vertical radiator grille—was something else! Obviously Claire and Bob must be prospering in America.

When they had first arrived in Chicago, they discovered that their sponsors, Alice and Jules Weil, did not own the big department store with which family lore had always endowed them. Instead of being a downtown emporium on the order of Marshall Field's, the Weils' store turned out to be a tiny ladies' accessory shop under the 63rd Street El. Their affidavit of support notwithstanding, the Weils didn't offer Claire and Bob any help, financial or otherwise. They held it as a matter of principle that since nobody had helped *them* make their way when they first came to the States as impoverished hoofers, there was no reason why they should put themselves out for the next generation of immigrants.

Claire and Bob managed their own start. They rented a modestly furnished apartment in the Hyde Park district on Chicago's South Side, which had turned into one of the "Fourth Reich" settlements of German and Austrian Jewish refugees, similar to Washington Heights in Manhattan or Swiss Cottage in London. I slept on a Murphy bed, which was stored upright behind a door during the day and swung down into the living room at night.

Bob did not resume his medical studies in Chicago, because he hadn't completed the *Physikum* (that is, preclinical semesters) in Berlin and therefore couldn't transfer any credits to an American medical school. He would have had to start his medical education from scratch, and it would have taken him about seven years before he could have begun to earn any money. So he had to find a paying job—not easy in the Chicago of 1937, where the Great Depression was still lingering—and he had no marketable skills. Finally, he found employment as a stock boy at Neisner Bros. He worked long hours—the store was open until late in the evening several days a week and sometimes on Sundays—and he had to drive a long way to work. By the time I arrived, he had been promoted to assistant manager.

Because she liked fixing hair, Claire had taken a nine months' course in the Loop's Burnham School of Beauty Culture. After her certification as a licensed beautician, she opened Claire's Beauty Salon on 53rd Street, the Kurfürstendamm of Hyde Park's Fourth Reich. It was a large store with a neon tube twisted to spell Claire's name in cursive letters in the window, a cheerful place, except that it was pervaded by the nasty odor of permanent wave chemicals. I took over the after-hours janitorial chores of mopping the linoleum floor and washing the big shop window.

I achieved my dream of going to an American high school when I enrolled at Hyde Park High. Built in Greek Revival style with a flight of monumental steps leading up to a portico held up by three pairs of majestic Doric columns—reminiscent of the Brandenburg Gate—the school stands on Stony Island Avenue, facing Jackson Park. Hyde Park's vice principal thought I should enter as an upper-level junior, or maybe even as a senior, since I spoke fluent English and had enjoyed the benefits of the world-renowned German secondary education. Just to confirm the undoubtedly high level of my academic skills, he gave me a few tests and quickly discovered that I had none: I couldn't do fractions, let alone geometry or science. Although I was almost sixteen, he made me start high school as a freshman, Hyde Park Class of 1944, keeping company with fourteen-year-old little shrimps.

I came to the States as "Günter Stensch," but I enrolled at Hyde Park High as "Gunther Stent." From then on, I called myself by the sanitized surname that my brother, Ronnie, had adopted, first informally soon after he settled in London in 1935 and later officially when he joined the British Army. I never told anyone in Chicago, or later, at the University of Illinois, that "Stent" was not my original surname.

Hyde Park High didn't disappoint the high hopes I held for American high school life. Many of my coed classmates seemed pretty. They wore makeup and reminded me of Deanna Durbin and the girls in the Coca-Cola "The Pause That Refreshes" ads. After Claire taught me the basic fox-trot and waltz steps, I started going to the weekly dances held at the school gym, where I learned to jitterbug.

It took me some months to work up enough nerve to ask one of my dancing partners for a date. Her name was Perle M. She wasn't exactly a raving beauty, but still a not unattractive, well-dressed blonde from Hyde Park. Having dreamt about it for such a long time, I was thrilled by the prospect of finally going out with a girl—and ridden by anxiety. What were we going to talk about? Maybe we could talk about opera. Should I hold her hand while we were walking side by side or sitting next to each other? Should I try to kiss her when we parted at her front door?

Perle's parents seemed surprised when I bowed slightly to them. We went downtown by electric train, to take in a movie at one of the vast, garishly decorated Balaban & Katz first-run theaters in the Loop. Most of the evening passed in silence. I racked my brain for conversational openings. But the few I found never led to sustained dialogue. Perle didn't know anything about, and didn't seem interested in, opera, or any other topic that I could bring up. Once or twice, after we stopped talking, Perle tried to restart the conversation by asking me what I was thinking about. Annoyed by her trying to pry into my private thoughts, I answered, "Nothing special." I didn't have enough nerve to take her hand, either while walking or while watching the movie in the dark theater, let alone to try to kiss her when we said goodnight.

Emboldened by having crossed the dating threshold, however, I asked out other Hyde Park High girls, but I fared no better with them. To my surprise, Perle's successors also wanted to know what I was thinking about when I fell silent. I never did work out a satisfactory answer to this, to me very obnoxious question, which I would never have dreamt of putting to anyone. Unfortunately, while I found it hard to have any rapport with authentically American girls who attracted me erotically, I wasn't interested in the German–Jewish girls. There were quite a few at Hyde Park and with whom I would have shared more to talk about.

Hyde Park was reputed to be one of the two best public high schools in Chicago, Senn High on the North Side being the other. Miss Rubovits, who taught English, demanded clarity as well as grammatical perfection in the weekly themes we had to write for her. The most heinous transgressions in her book were comma splices, split infinitives, and prepositions placed after the words they govern. Assigned themes had to be rewritten again and again, until her exacting standards were met. I owe her an obsession that would plague me throughout my later professional life: I became a compulsive editor who rewrites every text that is submitted to me for criticism. Miss Rubovits's legacy not only made me waste much time but also, as I took on more and more editorial responsibilities for scientific books and journals, brought me much ill will from rewritten colleagues.

In her American history class, Miss Kurrie left the teaching to the students. She posted a schedule that specified the days on which given students would be responsible for making fifteen-minute presentations of given chapters in our textbook. Depending on its quality, the presentation would earn the presenter from one to three credit points. In case of default of the designated presenter, any student prepared to talk for fifteen minutes on the chapter *du jour* would be guaranteed the maximum credit of three points, regardless of the quality of presentation. Miss Kurrie assigned course grades on the basis of the total number of points accrued in exams, book reports, and presentations.

My first scheduled presentation mortified me. I had practiced it in front of Claire on the previous day, and yet I could hardly mouth my carefully prepared spiel when I stepped in front of the class. But on hearing my own voice, I gained confidence. Next time, I spoke more easily, and a few days later I earned my first automatic three points as a pinch hitter. Eventually, I made most of the presentations, as my indolent classmates increased their rate of default. Thus I gained greater and greater facility as a windbag capable of holding forth on any subject without knowing what it's all about. This rhetorical skill proved mighty useful in my later academic career. I managed to accrue more points than any previous student in Miss Kurrie's classes without troubling to take any exams or turning in any book reports.

I noticed a curious feature of the textbooks provided to us at Hyde Park High. The name of the Chicago superintendent of schools was listed as a coauthor of almost every one of them, whatever its subject—mathematics, English, history, science.

"Say, Bob," I asked my brother-in-law, "isn't the Chicago school system lucky to have such a universally knowledgeable superintendent who can write a textbook on almost any subject?"

"This is one of the funniest questions I've ever been asked, Gun. Lucky, my foot! Everybody knows that this guy was put in his superintendent's job by Mayor Kelley's corrupt political machine. He's too dumb to write a book on *any* subject. The royalties that the real authors of the book have to share with him are part of the graft that runs through the whole city government."

My background as a juvenile shoplifter in the larcenous Berlin street and son of an entrepreneur with a penchant for sharp business practices ought not to have led me to expect very high levels of civic virtue in Chicago. Yet I was amazed by Bob's revelations about the openness of rampant corruption in Chicago. He told me that slipping a five-dollar bill to the examiner would guarantee passing the driver's license test. Moreover, once you'd gotten your license, you could avoid getting stuck with a citation by wrapping the license in a couple of dollar bills as you handed it to the cop who stopped you for a traffic violation.

Another example of rampant graft was a dingy little store not far from Claire's Beauty Salon that, as suggested by the objects haphazardly deployed in its show window, seemed to be selling cigarettes and cigars. One couldn't look directly into the store from the street, but I peeked a few times through the opened door as someone was leaving. I was surprised to see a partition running across the store, just a few feet behind the front door, through which another door led to a back room. The path to the inner door was blocked by a counter with a glass display case underneath, containing a few more diffidently scattered packs of cigarettes.

I wondered how that place managed to stay in business, when, as far as I could make out, everybody bought their smoking supplies at the corner drug store at 53rd and Woodlawn. Bob told me not to worry about the poor tobacconist, who probably wouldn't sell you a pack of cigarettes even if you asked him. He was making a fortune, running an illegal, off-track bookmaking operation in the back of the store.

"The bookie isn't fooling the cops with that phony cigarette display, is he, Bob? They must have noticed what I noticed!"

"You bet, they have! Which is why they're being paid off to leave him alone."

Within a few months after my arrival in Chicago, Bob became involved in one of those cops-and-robbers shootouts for which the City was—not for nothing—world renowned. The police had learned that, on a designated evening, a gang was planning to hold up the Neisner Bros. store where Bob was working. Even though it was likely that the gangsters planned to force him or his boss at gunpoint to open the safe, the cops didn't share their information with the store's management. Instead, they quietly staked out the place with armed plainclothes detectives. When the gangsters ran into the store waving pistols, the insouciant cops took cover and started firing, evidently ignoring the danger of the shootout for the customers and staff in the store. Bob managed to dive under one of the sales counters at the start of the fusillade, and no one, except one of the robbers, was hit. Neisner's didn't lodge an official protest with the city government over the way the police had handled this incident. It was just part of doing business in Chicago.

I found myself a job: washing dishes on the evening shift at the dining room of the Goldblatt Bros. department store in the Loop. The pay was $12 a week, plus meals. The work was much easier than I had expected. I had always pictured the Horatio-Alger-esque protagonist of the American success story starting out on his career bending over a sink and handwashing mountains of plates for hours on end. As it turned out, I just had to load the dirty dishes on a conveyor belt, which slowly moved them through an enormous dishwashing machine. I worked as part of a four-man team with three blacks: Two of us handled the input and two the output end of the belt.

My coworkers' dialect was well-nigh incomprehensible to me. Since their eyes twinkled most of the time they spoke to me, I guessed that they were asking me obscene questions, to which I replied yes or no, at random. Whatever my answer, it would provoke peals of laughter. These first American blacks with whom I came into personal contact seemed to me to correspond fairly closely to the happy-go-lucky, jolly Negro servant stereotype whom I had seen in many a Hollywood movie. It didn't occur to me at the time that my black coworkers had little to laugh about. Grown men in their thirties or forties, probably with families to support, they were earning the same measly wages that I was paid as a teenage high school kid. Moreover, for me this job was at the bottom rung of a tall ladder that I, an immigrant greenhorn, could expect to climb to higher and higher levels. However, for them, native-born Americans all, there was little hope of ever stepping up even onto the next rung.

I quit Goldblatt Bros. at the start of the 1940 fall semester, because the bus trip from Hyde Park to the Loop and back took too much time. So Claire arranged with the druggist at the corner of 53rd and Woodlawn, a few doors down from her salon, to take me on evenings and weekends as a delivery boy. In-between deliveries, I sat at the drugstore's soda fountain and watched the fountain men do their job. When the night man suddenly quit, the druggist asked the afternoon delivery boy and me whether we thought we knew enough about fountain work to split the night shift. My partner was a fellow German–Jewish kid at Hyde Park High. (In later life, he became a multimillionaire furniture manufacturer.) Since neither of us evinced any doubt about his competence as a soda jerk, we took over the fountain in the slack evening hours.

By the summer of 1941, I was ready for the big time. I was hired as a short-order man at one of the busiest fountains in town, Liggett's Drug Store at the Chicago & Northwestern Railroad Terminal, just north of the Loop. I was paid 70 cents an hour, the highest rate of pay I had drawn up to then, and worked harder than I have ever since. Liggett's was located in an entrance way to the station, through which the hordes of commuters and intercity travelers passed. Its fountain was huge: an elongated oval counter with dozens of stools, enclosing a central area where the food was prepared. The customers were served by waitresses, who called out their orders in the argot of the profession: "burn one" (hamburger), "two over easy" (fried eggs), "draw one" (Coke), "shake a choc" (chocolate milkshake), "split one" (banana split). From the opening of the store in the early morning until its closing at midnight, every stool was taken. During the rush hours, there was a candidate customer waiting behind each occupied stool. Three or four of us short-order men worked at a fiendish pace to keep up with the deluge of orders, which we had to memorize.

The job at Liggett's was physically as well as psychologically exhausting.

There was unrelenting pressure to fill orders at top speed, so that the impatient customers on the stools would not miss their trains. And just as, in his masterpiece, *Modern Times*, Charlie Chaplin kept on turning an imaginary wrench during his time off from the assembly line, so did I keep on frying imaginary hamburgers after work and in my dreams.

I was terribly disappointed that in all that year and a half I had been a student in a coeducational Chicago high school, I still hadn't found a lovable American girl friend. And even if I hadn't had the luck to come across Miss Right in school, why hadn't I at least found a less-than-perfect woman with whom to have sex? How come was I still a virgin despite my daily contact with so many nubile girls at Hyde Park High? I figured it was because I didn't have a car. According to the studs at Hyde Park, the way you get laid is to borrow your dad's car to take a girl to a Saturday night movie. You do some heavy necking on the theater balcony, and, after the movie, park with her on the dark lakeshore. Eenie-meenie-mo, you lay your date on the back seat.

I couldn't very well ask my surrogate dad, Bob, to lend me his car, especially since I didn't know how to drive. So I simply had to get my own car, even if it took most of the $150 I had saved from my drug store jobs. I had finally reached the age of eligibility for a driver's license—one reason why I had been waiting impatiently for my sixteenth birthday ever since my childhood days in Treptow. Bob came with me to a used car place on Stony Island Avenue. The cheapest car on the lot was a decrepit 1934 Ford sedan, on sale for $60. Bob drove it around the block and ascertained that it was capable not only of moving under its own power but also of being stopped by depressing the brake pedal. So I decided it was just the right automobile for me. Its back seat seemed adequate for the purpose for which I was making this substantial investment. Once Bob had taught me how to drive the derelict chariot, I discovered that its front wheels shimmied, its brakes were very weak, its tires were worn smooth, and its tailpipe belched clouds of blue smoke. Worst of all, it would hardly ever start on its own, and I usually had to get the motor running by having the car pushed.

When I arrived in Chicago in March 1940, the Phony War, or *Sitzkrieg*, was being sat out on the western front. But a month later, the Nazis occupied Denmark and Norway, and having become masters of the Scandinavian coast, began their *Blitzkrieg* in the west in May. Within two weeks, they had conquered the Lowlands and invaded France, which surrendered in June 1940. A year later, the Nazi armies invaded Russia, of which they had conquered vast areas by the end of 1941. I was distraught by this nearly uninterrupted series of Hitler's victories, but not surprised, since I believed in his invincibility. Being out of his reach, under the protection of the all-powerful anti-Hitler, FDR, I was no longer terrorized. And I was less afraid for my family in England after the Luftwaffe had lost the Battle of Britain and Hitler had called off the invasion of England. Yet Hitler and I still believed that he would be back to cross the Channel once he had finished off Russia. The invasion was *aufgeschoben, nicht aufgehoben* (postponed, not canceled).

What did catch me with surprise was the incredible news—Claire, Bob, and I heard it on the radio while we were having our Sunday dinner on December 7, 1941—that the Japanese had bombed Pearl Harbor. Didn't those sneaky little creeps realize that America was going to beat the living Jesus out of them? I never thought that Hitler's invincibility applied also to his accomplices, Hirohito and Tojo, no more than it applied to Mussolini. In fact, I didn't believe that there was any invincible Nazi other than Hitler. All my fears of perdition had always focused on his person as the owner of magical powers, and I imagined that only his death would bring salvation.

We had no doubt that the United States would declare war on Japan. But would FDR succeed in persuading Congress to declare war also on Germany, which was merely Japan's ally and had not attacked America directly? We were colossally relieved when Hitler did FDR the yeoman service of declaring war on the United States first, because all the might of America would now be put to work to defeat Nazi Germany. Maybe Hitler wasn't invincible after all. It might take a few years to bring the Japanese to their knees, but we weren't all that concerned with what was happening in the Far East. What counted for us was America's immediate material and military support of Britain and Russia in their war against Hitler.

The entry of the United States into the war in December 1941 brought few immediate changes to our everyday life. For me, the first—paradoxical—consequence came on my eighteenth birthday in March 1942, when I had to register both at the U.S. district attorney's office as an enemy alien and at my local draft board as a prospective member of the U.S. Armed Forces. The draft board issued me a number for the lottery that would determine the order in which registrants were called up to report for induction for military service. Since men were being drafted very slowly, I thought it probable that I would still be able to enter college in the fall.

The Smoothie

ALTHOUGH I ENTERED AS CLASS OF 1944, I graduated from Hyde Park in February 1942, after working hard to get through high school as fast as possible, to make up for all that time I wasted during the eighteen months I had been out of school waiting for my U.S. visa in Berlin, Antwerp, and London. I signed up for extra courses each semester, went to summer school, and asked for and received credit for knowing French (they wouldn't give me any credit for knowing German, because it was my mother tongue). I also finagled credit for geometry by passing a proficiency examination after I alleged (falsely) that I had taken geometry in Berlin. In fact, I had taught it to myself in Chicago by working my way through the textbook's proofs of Euclid's theorems. Thus by accumulating credits at a furious rate, I made it through Hyde Park High from beginning freshman to graduating senior in twenty-one months, a shining example of the motto of my graduating class of 1942, *Ad astra per aspera*. Yet, I was still appallingly ignorant of most academic subjects, except for English composition and geometry.

A few weeks before graduation, a cute little blonde, Mary Jane W., had passed me a note in English class:

My dear Mr. Stent,

Staring at the back of your head inspires me no end to write you. And besides, I never get a chance to talk to you. What are you going to do Christmas vacation? Do you have a telephone? Not that I would call you up (don't flatter yourself). I was just wondering.

Lots of Luck, Little Love, Not many regards, and no kisses. (Aren't you sorry?)

Mary Jane

This first letter I ever received from a girl presaged the artful advance–retreat style that almost all my future lover-correspondents, including Hildegard, would adopt to preserve their dignity in dealing with a cad who would not reveal his feelings.

Aroused by Mary Jane's unexpected come-on, it didn't take much nerve to start schmoozing her right after class. On our first date, I gave her the maxi-treatment within my resources. I took her dancing on the Allerton Hotel Roof Bar on North Michigan Avenue, which I had once visited with Claire and Bob. I ordered the only cocktail I knew by name—whisky-sour—because it was their favorite.

Mary Jane wasn't exactly a great conversationalist, but her sex appeal more than made up for that deficiency. I asked her to be my date for my graduation prom, held on a ship anchored on Lake Michigan, off Grant Park in downtown Chicago. She looked luscious in a long white formal dress, while I, unable to afford a dinner jacket, wore the only one of my Friedl-selected Berlin suits—with a carnation in the jacket buttonhole—that Claire could still alter to fit me. As usual, my Ford wouldn't start on its own. So I pushed it, while Mary Jane, in her formal, sat at the wheel and let out the clutch, which showed that she was a good sport and not one of those stuck-up dames.

The outlay for the prom night—the ticket for the dance, the orchid corsage for Mary Jane, and the cover charge at the Chez Paree nightclub to which we went after the prom—ruined me financially. But it was another step up *ad astra*. With Mary Jane I had found at last the good-looking, sexy, 100 percent American girlfriend I had been searching for ever since coming to Chicago. We would sit in my Ford, parked on the lakeshore, doing some heavy necking. Alas, not knowing how the Hyde Park studs managed to lay their girls on the back seat, I didn't accomplish the project for which I had invested my capital in the jalopy.

Not long after the prom, Mary Jane invited me to a dinner dance in her neighborhood, at which she presented me to her parents. I thought she enjoyed the evening as much as I did. Our jitterbugging had become mutually attuned, and our conversation—mainly about movies, food, and school gossip—reasonably fluent. It was my best date so far, but, as it turned out, my last with Mary Jane. After that night, she always fobbed me off whenever I called to ask her out. Finally, she confessed that her parents hadn't realized that I was a Jew. They thought I was just one of your nice German boys, until they met me at the dance. They were furious and threatened her with dire consequences if she continued to have anything to do with me.

This was an awful blow, not so much because of losing Mary Jane. I wasn't really in love with her. It was because it made me painfully aware that even though I had managed to evade Hitler, I hadn't evaded my shameful origins. The psychic lesion I suffered that evening was a long time in healing. After all these years, resentment of Mary Jane's anti-Semitic parents still wells up from the deeper recesses of my soul whenever I hear the Golden Oldie, "The Johnson Rag," which the band was playing while she introduced me to them.

In February 1942, shortly after my high school graduation, we moved from Hyde Park to the North Side. Located on a quiet little side street near the Lake, the Hines's new apartment had two bedrooms, instead of one. From my point of view, it was a move up on the social scale, because fewer German Jews were living in this ethnically nondescript middle-class neighborhood than in Hyde Park. Claire's doctor had told her that her general exhaustion due to working too hard for all those long hours was probably responsible for her failure to implement her wish of starting a family. He advised her to become a housewife, and so she closed her beauty salon on 53rd Street. In due course, my nephew, Ronald Hines, was born the following year.

I was looking forward to starting my career as an engineering student in the fall. Since there was no way for me to pay for tuition, room, and board at college, I had to put all my hopes on getting a full scholarship. I didn't think that this was going to be too much of a problem. The ease and speed with which I, the academic ignoramus, had breezed through Hyde Park gave me the conceit that I was a genius who didn't need any concrete knowledge for academic success. My fondest wish was to go to the University of California at Berkeley, on San Francisco Bay in the Golden State of my dreams. I had come to admire its scenic glories and beautiful coeds in many a magazine article. Unfortunately, I couldn't find any listing in the California catalog of a scholarship for which I would have been eligible. So I lowered my sights and applied for admission and scholarship support to Harvard, Yale, and the University of Chicago.

To build up some financial reserve to cover my incidental expenses at college, I took a job as an office boy at the Chicago Title and Trust Company in the Loop. The tract books of Cook County had been destroyed when the courthouse burnt down in the Great Chicago Fire of 1871, and the only records of real estate transactions that had survived the fire were copies of the books stored in CT&T's vaults. So the company had a monopoly on the real estate title search and insurance business in Chicago. Every would-be buyer of real estate in Cook County had to use CT&T's services, to ensure that the seller was really the legal owner of the property.

I was one of a dozen of the company's teenage "runners," whom the title examiners sent to fetch the precious tract book copies from their storage in a closely guarded vault. The examiners checked and rechecked all the diverse encumbrances on the title, such as mortgages and contested probate. They were also looking for racial, ethnic, or religious restrictions on ownership and occupancy. The latter were called "race docs" and gave me the willies, because many of them mandated the exclusion of Jewish occupants. It was easy work, compared to the nerve-wracking short-order cook drudgery at Liggett's Drug Store fountain.

Smarting from the ruination of my romance with Mary Jane, I decided to try Jewish girls. Not Jewish girls from my old Fourth Reich Hyde Park haunts, mind you, but from our new, swankier North Side neighborhood. So I asked a German–Jewish boy I had befriended at Hyde Park High to trek up to the North Side and go with me to a B'nai B'rith dance at Temple Sholom on North Lake Shore Drive.

There I met Rita L., a good-looking brunette, clever and incomparably easier to talk to than the girls I had dated at Hyde Park. She was a junior at Harriet Smalley, an exclusive private school on Chicago's Gold Coast. To impress this classy girl, who was Jewish but didn't look it, I fibbed that I was a freshman at the University of Chicago. By way of making small talk while we were dancing, I posed her two geometry brain teasers I had picked up at meetings of the Pythagorean mathematics honor society at Hyde Park. Before the evening was over, Rita found out that I had lied to her. My good friend, who evidently wanted to cut me out of the deal and date her himself, ratted on me. He let on to Rita that I had only just graduated from Hyde Park High and was working as an office boy downtown, merely *hoping* to go the university in the fall. Nevertheless, a few days later I received a letter from her,

Dear Gunther,

Through some lucky accident I somehow got your address. You'd never guess what our geometry class has been wasting time on. I brought those two problems you gave me to school, and we spend approximately 15 minutes of each period proving that people who think they have a solution are wrong. Mr. Barnes, our teacher, has finally come to the conclusion that the angle and circle problem is impossible. For all I know, you just made it up on the spur of the moment while we were dancing, and don't even remember any more what it was all about! Maybe it's like your "going to Chicago"?

The other problem, about the sides of the triangle, Mr. Barnes pretends to know and says that he only knows one student who ever did it by himself. He promised that anyone in the class who can solve it doesn't need to take the final exam.

I am wasting paper, so good-bye.

Rita L.

P.S. If you happen to remember the proof of the problem about the triangle, you might drop it by. (You don't want me to have to take the final exam, do you?)
P.P.S. Please excuse my penmanship, as I have writer's cramp from too much homework.
P.P.P.S. I met some people who know you (Fancy that!?!)

Rita

Surprised and aroused, I responded with the first letter I ever wrote to a girl. I enclosed the solutions to the two geometry problems, indicating my expectation of her proper show of gratitude for my having got her out of taking the final exam. Whatever gave her the idea that I might be the kind of phony who makes up things on the spur of the moment? Rita answered within a few days.

Dear Gunther,

    The day after I wrote to you, Mr. Barnes came to school with a smile from ear to ear; he had found a solution to your "impossible" angle and circle problem. So he really isn't as dumb as I thought.

    So long,

Rita

P.S. I just remembered: See you on New Year's Eve in the year 2000—on the moon.

Rita was born in Czechoslovakia, but her first language was German, as it was for most Czech Jews (including also my future FIAT supervisor, Freddy K.). She had emigrated with her parents and her younger sister to Chicago in 1938, after Daladier and Chamberlain's sellout to Hitler at Munich. It so happened that she lived just around the corner from our new apartment on Brompton Avenue. Her folks were civilized, intellectual people, without any commitment to or interest in Judaism, which suited me fine. They reminded me of my mother's, Breslau, side of my family. Rita was going to Harriet Smalley on a scholarship.

    On one of our first dates, she took me to a party at the home of one of her Smalley School classmates. They were a bunch of swell kids, who made clever conversation about movies and books; danced to jazz records; and sang folk songs to guitar accompaniment. What impressed me most, though, was that they were an ethnically desegregated lot, gentiles (McCormicks and Johnsons) as well as Jews (Kahns and Goldbergs). *This* was the kind of crowd I had been looking for.

    Rita and I obviously seemed right for each other, an opinion shared also by most people who knew us, including her parents and Claire. I liked her better than any girl I had met during my two years at Hyde Park. As far as all criteria of desirability other than sexy American shiksahood were concerned, I had obviously traded way up from Mary Jane. But I always remained on the lookout for some other, fabulously ideal rather than real, woman. For the next eight years, Rita would be my loyal girlfriend-in-reserve, who would always take me back whenever one of my other romances went sour.

    My confident expectation of getting a scholarship took a nosedive after I took the College Entrance Exam Board. As it turned out, doing well on the exam *did* require concrete academic knowledge, of which this genius from Hyde Park High was in short supply. I knew that I had done very poorly on most sections, except, maybe, for English composition. Toward the end of May, a letter came from Yale, informing me that my application had been rejected. A few days later, there was another rejection letter from Harvard. Finally, I heard from the University of Chicago that it was willing to admit me, but without any scholarship support. Since paying for its tuition was beyond my means, Chicago was out of the picture as well.

Mindful of how I had bombed the College Entrance Exam Board, I was not taken by total surprise at this dire turn of events. Yet I was crushed. After sulking for a few days, I remembered that there still was a safety school, which in my fatuous snobbishness I had been too haughty to consider: the cornball University of Illinois in Champaign-Urbana. Being the state university, it *had* to accept me, because of my excellent high school grade-point average. Moreover, for residents of the state of Illinois, the tuition was very reasonable. There was even a slight chance of my not having to pay any tuition at all at Illinois, provided I won one of the Cook County scholarships. These scholarships were awarded to the highest twenty scorers in a competitive examination given to graduates of Chicago high schools.

Because I wasn't competing against the top teenage brains of the country, I did better on the Cook County than on the College Entrance Exam Board. So I was awarded a four-year tuition and fees scholarship at the university, worth all of $90 per year. As for meeting my living expenses and incidental outlays, such as books and health plan fees, I would be on my own. I had about $200 to my name, and as an experienced fountain and short-order man from the City, I ought to be able to find a part-time job that would keep me afloat in Champaign-Urbana.

Anyhow, it was high time that I stopped being a burden on Claire and Bob. For the past two-and-a-half years, they had never asked me (and I had never offered) to contribute any of the money I made in my after-school jobs or at CT&T to their household. Instead, I frittered away my earnings on dates and on the luxury of running my car. Moreover, sharing their cramped quarters with me, they had virtually no privacy.

Rita played in a different league. As one of the star students at Harriet Smalley, she had been spotted and interviewed by talent scouts from the Seven Sisters schools on the Eastern Seaboard. At about the time that I was notified of my modest Cook County scholarship award at the poor man's state university, she heard from Bryn Mawr and Vassar that either of these elite colleges would admit her on a scholarship in the fall of 1943, without her even having to take the College Entrance Board. Far from being jealous, I was proud of having a girlfriend who succeeded where I had failed. It may even have taught me a little, much-needed humility.

On Labor Day weekend 1942, I moved down to Champaign, with all my belongings stashed in a couple of suitcases. Although I knew I was going to miss Claire, my surrogate mother, I felt good about to starting life on my own. Henceforth I would be returning to her household as a guest rather than a dependent. My childhood was over.

Playing House

HAVING FAILED TO FIND THE AMERICAN GIRL OF MY DREAMS in my two high school years in Chicago, I was looking forward to discovering her among the many attractive coeds that I expected to encounter at the university. I did meet lots of girls there in the four years between my arrival in Champaign in 1942 and my return to Berlin in 1946, but I never developed the kind of spiritual closeness to any of them that had quickly sprung up between Hildegard and me, not even to the one girl with whom I fell in love and who jilted me after our brief engagement. Hildegard didn't need to ask, "What are you thinking about?" because she could read my mind—even if I couldn't read hers. In fact, she got to know me better than I knew myself.

In mid-January 1947, not long after we had parted in Berlin, two letters Hildegard wrote on successive days arrived together in Durlach.

Günterlein,

I'm sitting in the theater. It's freezing at home but nice and warm here. I have some time, and what else is there for me to do, other than writing to the man of my dreams? (Ha-ha-ha!) First of all, a thousand thanks for the package you sent me. It really gave me great joy. You can't imagine how bad things are with me. Terrible. At the moment I'm on a starvation diet. I haven't had any bread in my stomach since Sunday. Lived only on soups. No potatoes; there aren't any because it's so cold. Can you make yourself a rough picture of how things are in Berlin? The only solace is that I'm not the only one who is so badly off. These days I'm really desperate. What haven't I tried to get something to eat! Unimaginable. It certainly doesn't sound very pretty what I am bending your ear with, but Günter, I am really at my wits' end. Forgive me that I'm not writing you about more pretty things, but I don't have any one but you to whom I can really talk. You probably have no idea what it means to me to have someone who totally belongs to me, don't you understand, Günter? Even if it's just for a short time. Listen, Günter, I won't forget it!

Your Hildegard

Enclosure
My Figure Dissected into Numbers

Dress	Size	44
Shoes	Size	38
Stockings	Size	9
Slip	Size	42
Nail polish	Size	red
Purse	Size	??

That's all! But terrificly chic.

My dear Günter,

I am longing for you. I feel so terribly alone. We're putting on *A Day of Rest* today, and I'm sitting in my dressing room, waiting to go on stage. There's a reason, of course, why I'm writing to you again so soon. I'm ashamed that I wrote the letter I sent you yesterday. I don't want to make you sad and bother you with my troubles. Forgive me. I didn't know how selfish one can be when one is in a desperate mood. I'm feeling much better already today. I got a gift, a half-loaf of bread and 4 pounds of potatoes, and you can't imagine, Günter, how the world looked totally different right away. If you were with me,—oh, Günter—.

Our play is slowly getting a stupendous reputation. House sold out every evening. Beautiful, isn't it? It's bound to run for quite a while yet. It certainly gives me a lot of joy.

I've been dreaming about that trip we are going to make one of these days, you and me, shut off from that big, motley world. A fairytale? I don't know. Anyhow, it would be much too beautiful to be true.

My dearest Günter: Be happy and write me something dear, soon!!!

Your Hildelein

Hildegard's letters further increased my uneasiness over having taken on, willy-nilly, a responsibility for her body and soul far transcending the obligation to show my girlfriend a good time when we could manage to get together. It couldn't have been that meeting her not at all unreasonable expectation that I would supply her with the necessities of life, of which she was in dire need, was a financial burden on me, especially since the food and clothing that meant so much to her would cost me so little. But the feelings of resentment that had arisen in me when she asked me in her very first letter to send her some cigarettes were only intensified when I saw her jocose, yet heartrending new wish-list with her measurements and her happiness over a gift of half a loaf of bread and 4 pounds of potatoes. It wasn't that I was indifferent to her suffering—I would like to believe that I did want to help her—but I felt paralyzed, torn by the antinomies of compassion and fear of commitment to a woman I didn't love. I did buy a few of the items on Hildegard's "My Figure Dissected into Numbers" wish list at the Karlsruhe PX and sent them to her, by way of a compromise between my conflicting emotions.

In any case, my response to her pathetic letters was not the kind she deserved. I didn't write the "something dear" she'd asked for, such as "My poor, brave darling. I can't bear to think of you cold and hungry. If only I could be with you and take care of you!" Nor did I promise that I would never let her down. Instead of declaring my tender feelings for her, or at least expressing compassion for her plight, I merely provided her with an account of what I had been up to lately.

As for the trip we are going to make one of these days, I wrote that I had been thinking about it, and maybe it's not too beautiful to be true, after all. Why don't we take a week's holiday in about a month's time, in the part of the Harz Mountains that lies in the British Zone? How about her getting on the interzonal train from Zoo Station to the British zone, and getting off at Braunschweig, just on this side of the interzonal border? I would meet her at the Braunschweig RTO with my jeep, and off we go to the Harz's Brocken

Mountain, whither the witches fly on their broomsticks on Walpurgis Night for their annual convention and engage in unnatural sexual acts with the Devil.

Three days after I sent off my letter to her, I received a telegram from Hildegard, asking me whether I am ill. She is worried and needs me badly. To reassure her that everything is OK, I tried to get her on the phone at the Schiffbauerdamm Theater, but she wasn't there. So I left a message for her that Gunther S. Stent was in the best of health (and failing to show any oncern for her welfare by expressing my hope that all is well with her too). I also telegraphed her my invitation to play in the snow of the Harz Mountains and proposed February 21 for the date of our meeting at the Braunschweig station RTO.

I didn't have to wait long for Hildegard's response.

> Hurrah, Günterlein,
>
> We are going to go on a trip, all alone, you and me! Can there be anything more beautiful?! Are you also looking forward to it as much ? I just can't tell you what my feelings until the 21st will be. I had been ill, just when your telegram came. I was cured immediately.
>
> Your Hildelein

Skiing had been one of my childhood passions. As an eight-year old I had been dreaming that I was schussing downhill in the snow-covered Treptow Park, even though that park was flatter than a pancake. All those years since fleeing Nazi Berlin, I had had no opportunity to ski—not since my last trip to the Bohemian Riesengebirge in 1937, when I was thirteen. I had always fantasized about skiing in the Alps, but never got beyond the Riesengebirge's modest slopes and its puny peak of peaks, the 5,000-foot Schneekoppe. In America, skiing had not yet become the popular sport it would turn into in the 1950s, with the first major Stateside skiing resort, the Union Pacific's development at Sun Valley, then still being an exclusive playground of the superrich, totally unaffordable for a hard-up student like me. So I was eager to to visit the Alps during my European tour of duty with FIAT and make up for all that lost skiing time.

A couple of weeks before my proposed meeting with Hildegard in

Braunschweig, the US Army Rest and Recreation Office at Karlsruhe told me that the most desirable skiing resort in the whole European theater of operations was a French officers' lodge on the Ehrenbachhöhe above Kitzbühel in the Tyrol. Fabulous food and terrific skiing. It was very hard to get accommodations there, but if I was willing to wait six weeks, they could book a week's stay for me, from March 15 to March 22. I asked them to make the booking, looking forward to realizing another childhood ambition at last.

I started out from Karlsruhe in the early morning of February 21, 1947, for what I thought would be an eight-hour drive to Braunschweig. I reckoned that I'd get there just about at the time that Hildegard would be arriving on the interzonal train from Berlin. But because of a series of breakdowns of my jeep and because the Autobahn, of which many war-damaged sections were still impassable and had to be detoured via slow side roads, was covered by deep snow, it took me about thirty hours to get to Braunschweig. The Brits at the Braunschweig RTO knew all about me. Hildegard had waited for me at their office for nearly 24 hours. Finally she gave up and, in desperation, decided to look for me in Karlsruhe, for which she left on a train a couple of hours before I arrived.

By now, it was early evening, and I was too tired to drive right back to Karslruhe on the dark, snow-covered Autobahn. So I checked in at the British transient officers' billet for the night. I managed to get one of the FIAT investigators who lived in a house neighboring mine in Durlach on the phone, and told him that a Fräulein Hildegard U. from Berlin might show up at Kastell Strasse 19 tomorrow, looking for me. Would he please paste a note on my front door and let her stay with him until I return tomorrow afternoon?

I finally got back to Durlach the following evening. After my jeep was hit by a snowplow on the Autobahn, I had to hitch a ride to Karslruhe from Kassel, where I left my jeep at the U.S. Army Ordnance Garage for repairs. Someone seemed to be in my apartment, because the lights were on. It was Hildegard, asleep on the sofa in my living room, wearing the golden chain with my recycled TEP fraternity pin around her neck.

I was tremendously relieved to find her safely installed in my well-heated digs. Dreadful visions of her wandering shelterless all over the freezing, snow-bound British and American zones looking for me had haunted me all the way on my harrowing trip back from Braunschweig. Riven by guilt feelings I asked myself how I could I have been so careless, trying to cut my driving time so close and leaving no slack for contingencies?

I woke Hildegard with kisses, and without wasting any time on small talk, we celebrated our reunion on the sofa. When the time did come to talk, I apologized:

"Hildelein, you can't imagine how happy I was to find you here! I'm terribly sorry for messing up our meeting in Braunschweig. I got there too late because of car troubles and the deep snow on the goddamn Autobahn, which I couldn't help. But it's my fault, all the same, because I should've done better planning."

"Never mind, Günterlein! Didn't Shakespeare say 'Everything is good that finishes well?'"

"Something like that. Actually, it's the name of one of his plays. But never mind Shakespeare. Your trip down here from Braunschweig must have been horrible. How'd you manage it?"

"My trip *was* a nightmare alright, but I don't want to bore you with the details, Günterlein. After all, what does it matter, now that we are together again?"

I left it at that. Hildegard had brought me a pair of cufflinks—my first—each made from a silver Swiss 50-Rappen coin.

"I hope you'll feel I'm at your side whenever you wear them, as I feel I'm with you whenever (which happens to be always) I wear your fraternity pin."

I doubt that anything I said by way of thanking her for her precious gift adequately expressed my awareness and appreciation of the sacrifices she must have made in order to rustle up this token of her love. What sacrifices was I prepared to make for her? None, except maybe playing games with military government regulations to enjoy her company and recycling the fraternity pin my ex-fiancée Dorothy had once worn.

Yet for the rest of my life, I did keep my promise to think of Hildegard whenever I saw her cufflinks. Years ago, I lost one of them, but I kept the other in a little jewel box, together with my shirt studs, rosettes, and the several successor generations of cufflinks. Whenever I opened the box on dressing for a formal event and beheld this bittersweet keepsake, I remembered her. Half a century later, on finally reading Hildegard's letters attentively in preparation for writing these pages, I had a silversmith replicate the lone surviving cufflink, to restore the original pair.

On the day I had set out to meet Hildegard in Braunschweig, Gen. Lucius Clay, newly promoted to the post of U.S. military governor of Germany, announced that FIAT would be closed down, effective June 30, 1947. Clay, deputy governor since VE Day, had finally won his battle to liquidate the agency's operations. None of us document screeners in the industrial branch had any idea that the War and Commerce Departments were struggling against General Clay to keep FIAT going and that this ruckus had required the intervention of President Truman. I only learned of it more than forty years later, after the publication of John Gimbel's history of FIAT, *Science, Technology and Reparations: Exploitation and Plunder in Postwar Germany*. (Stanford University Press, Stanford. 1990)

According to Gimbel, General Clay shared the opinion of the German industrial managers that the fiendishly efficient and rapacious gang of FIAT carpetbaggers (who, as it happened, included a goodly number of German and Austrian refugee Jews) were robbing German industry of its precious trade secrets and thus obstructing the economic recovery of the devastated American zone. Hence the puzzling Christmas 1946 move of FIAT's headquarters from centrally located Hoechst to out-of-the-way Karlsruhe was probably contrived by Clay to sabotage our operations. Clay's arguments played well enough in Washington (as indicated by Senator Wherry's hostile reaction to my introducing myself as a FIAT investigator on our meeting at the ruin of Hitler's New Chancellery) to force the War Department to set a date of FIAT's termination, which the victorious Clay wasted no time in announcing.

Within a couple of weeks of Clay's announcement, I was notified by the TIIB London Office that my appointment would be terminated at the end of June 1947. Moreover, in case I wished to resign at any prior time, I would be free to do so without having to reimburse the government for the cost of my overseas travel. In fact, if I resigned well in advance of the final termination date, commercial (cushy) rather than military (spartan) transportation would be provided for my return trip to the States.

I was delighted, because I had begun to wonder whether I couldn't cut short my year-long commitment to serve FIAT in Germany. After all, I *had* managed to turn my I'm-back-in-Berlin nightmare into reality, blonde shiksa and all. Hadn't the time come to worry about my career and get on with completing my doctoral dissertation research, by getting back to Champaign in time to sign up for the next summer semester? A few days later, TIIB had booked me on the Holland–America Line's *M.S. Noordam,* sailing from Rotterdam for New York on April 26.

Unimaginable as this would have been for me when I headed for Europe six months earlier, I had grown homesick for Illinois. America had evidently become my real home, whereas despite my steadfast sentimental attachment to Berlin, Germany now seemed alien. This estrangement could not be attributable to my regarding every Kraut I saw as a putative Nazi Jew-killer. I had finally come to believe that most of the Nazis were dead, or in prison, or hiding in Argentina. The Hitler spook was gone forever, never to return, not even after the Germany that now lay in ruins had risen from its ashes— and I reckoned that might not happen within my lifetime. It was simply that I had become used to living among the free-and-easy Americans and could no longer brook the mean-and-uptight Krauts who were holding up a mirror to me that reflected my own personality deficits.

Most of FIAT's field operations were being shut down, including the document-screening project at the Leverkusen plant. In the few months remaining before the agency became history, as much as possible of the huge backlog piled up at headquarters of miles of unprocessed 35mm film with documents indiscriminately photographed under the direction of feckless screeners like myself was to be indexed and translated. So I was reassigned to work during my last two months with FIAT at its documents branch in Karslruhe. This meant that I could stay in Durlach and play house with Hildegard in my commodious billets.

The English translation of liberated German documents was done by the cheap labor of German personnel paid in worthless Reichsmarks. No doubt, the translators regarded the three square meals they were provided each day at their separate and unequal FIAT canteen as their main wages. Another work incentive for them was the occasional access via their Ami contacts to precious staples from the PX, such as cigarettes, coffee, and chocolate. Allied and German personnel worked in separate rooms at the headquarters building. It was easy to tell, even after working hours, who worked in which rooms, since the space used by the Krauts was permanently pervaded by the unwashed body stench.

The incompetence of the German translators was well-matched with that of the Allied document screeners. Most of the translators had neither the scientific background nor the linguistic skills necessary for rendering the German originals into technically literate English. But even if all of them had been lingustically competent, it would have been impossible for those few dozen people to translate within their lifetime the miles of microfilmed documents that had rolled in from the field.

To establish a minimum of quality control for the translations—just in case someone in Washington *would* ever have a look at any of them—some of the document screeners from the Industrial Branch were detailed to review and copy edit the translators' output. This turned out to be the task that I was given. It was the only serious work I ever did for the Department of Commerce during my whole stay in occupied Germany, and I did it reasonably well. But it was a painstaking chore, since I had to rewrite most of the screwed-up English texts. Thus by delaying even further the processing of the documents, I increased, rather than decreased, the backlog of unprocessed material.

At first, it was thrilling for me to live night and day with a woman. I brought food and kitchen supplies from the PX, and, as an experienced hash-slinger, made home-cooked meals. Many evenings we went into Karlsruhe, to have dinner at the FIAT Club, and take in a play, an opera, an operetta, or a ballet, which Hildegard usually declared to be backwater-provincial by Berlin's metropolitan standards. Sometimes we went to reruns of UFA films I remembered from my childhood.

After I retrieved my repaired jeep from the Kassel ordnance garage, we made excursions to the Black Forest and other sights in the Karlsruhe region. To avoid running into the highway patrols of the army constabulary, who were making spot checks to catch unauthorized users of military vehicles—especially those supersuspicious drivers with Fräuleins aboard—we went on deserted backroads rather than the busy Autobahn. Our trips included a visit to Heidelberg, famous for being the setting of Sigmund Romberg's *Student Prince* and headquarters of the U.S. Third Army. I smuggled Hildegard into the Heidelberg PX, said to be one of the best in the European theater of operations. Ever the actress, she impersonated a jaded Allied personnel customer, hiding her excitement over having entered this shoppers' paradise and taking in with feigned ennui its fabulous display of merchandise, unobtainable for Germans. We bought the items on her "My Figure Dissected into Numbers" wish list, such as a dress, shoes, and stockings, that I hadn't picked up at the Karlsruhe PX when she had sent it to me from Berlin in January. But compared to the sacrifices she must have made to get the silver cufflinks for me in Berlin, the few shekels I shelled out for these goods amounted to next to nothing. Our radiantly happy faces on a snapshot of us standing arm in arm on Heidelberg's Neckar River embankment show our bliss.

Hildegard made friends with Mrs. Thornhill in the upstairs apartment and sometimes baby-sat her infant daughter. She also managed to get to know a German family, the Breyers, living in one of the few nonrequisitioned houses in the neighborhood. We had settled down to the ideal life of the newlywed, middle-class couple in the suburbs: cozy, modern apartment; young husband working at the office downtown; young wife doing light housekeeping and whiling away her time with the neighbor ladies; evenings on the town.

What should I do about Hildegard when I return to the States in April? Should I ask her to join me in Illinois next year and marry me when I get my doctorate? Unlike me, Hildegard had no trace of the undesirable Kraut characterological traits; she was kind, considerate, and easy-going, a wonderful companion. They don't make 'em any better in the States, although I wondered what Claire would think of my presenting her with a Nazi-tainted sister-in-law.

Yet, far from proposing marriage, I didn't even tell Hildegard about the TIIB letter. I simply made the arrangements for my return to the States without discussing them with her. After it was all done, I just mentioned casually that I would be sailing for New York from Rotterdam on April 26. She said she supposed that I was glad to get back to finishing my doctoral work, to which I responded affirmatively.

In any case, how long could our idyllic setup in Durlach last? When Hildegard had first proposed that we meet somewhere in the West, she hadn't mentioned, and I hadn't troubled to ask, how she could take time off from appearing in *The Day of Rest*. I just supposed that the theater would give her a week's leave. But when her second week in Durlach had gone by, I became uneasy.

"What arrangements did you actually make at the Schiffbauerdamm to get time off to meet me in the West? Don't you have to get back to Berlin pretty soon to play Shura?"

"Tell the truth, on the night before I left for Braunschweig, I simply called the theater to report sick. You're dead right, Günterlein; I really ought to be heading for home before they find out at the theater that I've gone to meet my Ami in the West."

My questioning Hildegard was driven also by a gradual waning of the initial thrill of domesticity. I still enjoyed Hildegard's company, but I had begun to resent being fenced-in. I kept thinking of my hard-to-get reservation for a week's stay at the fabulous French officers' skiing lodge in the Tyrol, which I would have to kiss good-bye if Hildegard were still with me in the middle of March.

The awful fact was that the pull of the chance of enacting at last my childhood fantasy of skiing in the Alps was stronger than keeping company with a good woman who loved me and whom, even though I didn't truly return her love, I was considering marrying. I mentioned the skiing trip to Hildegard, but I emphasized that I was mainly worried about her losing her job at the theater. This worry was not feigned, since I was afraid that I was about to become wholly responsible for her. So we agreed that Hildegard would go home just before I took off for Kitzbühel.

"I'll take you to Frankfurt and put you on the train to Berlin."

"Sure. Your plan is all for the best."

On the day before her projected departure, Hildegard told me that she had gone to see a doctor with whom the Breyers had put her in touch.

"He says I need to have a minor operation, nothing terribly serious. Since he can do the operation here in Durlach, I'd prefer to have it done here and stay just a while longer."

Hildegard didn't explain what the operation was for, and believing (or maybe wanting to believe) that she was reluctant to discuss it, I didn't press her for details.

"I insist that you go ahead with your skiing trip, Günterlein. Please don't worry! I can take care of myself perfectly well in the apartment. And if I need any help, I can get it from the Breyers. I'll be on my way to Berlin as soon as you get back from Kitzbühel."

With a callousness that appalls me as I now reflect on my conduct as a twenty-two-year-old, I abandoned my obviously ill Hildegard in Durlach, merely because I didn't want to lose the chance of enacting my childhood fantasy of skiing in the Alps. I simply wished her *Hals und Beinbruch* for her operation, kissed her good-bye, and got on the train to Salzburg. In bizarre contrast to the worries that had been driving me out of my mind after I showed up late for our rendezvous at Braunschweig station, I was insouciant about leaving Hildegard to her own devices, all alone in my billet while facing unspecified surgery. I must have felt that while I had been responsible for standing her up in Braunschweig, her having to stay on for an operation was none of my doing. As would eventually transpire, the need for her operation was attributable even more to my negligence than her Braunschweig ordeal.

Jewish Greeks

I FOUND THE UNIVERSITY OF ILLINOIS CAMPUS VERY APPEALING when I first saw it as an entering engineering freshman on the 1942 Labor Day weekend. Straddling the border between the Twin Cities of Champaign and Urbana, the campus was laid out as a central broadwalk shaded by a canopy of huge elms, lined on either side by a single row of architecturally hodgepodge classroom and laboratory buildings. The elegantly furnished, colonial style Student Union cut across the northern end of the broadwalk and the circular, Pantheonesque auditorium, fronted by a statue of Alma Mater stretching out her arms to embrace her returning children, across its southern end.

In view of my passion for the iron horse, civil or mechanical engineering were obviously the most fitting majors for me. However, I was intrigued by the option of chemical engineering listed on the registration form, a calling of which I had never heard. Whatever chemical engineering was all about, it couldn't have much to do with railroading. But to me it suggested something brand new, something futuristic. I hadn't been keen on chemistry at Hyde Park High. In so far as I cared for science at all, I preferred astronomy, until one of my teachers warned us against it as a profession. He said that astronomers, as they are star-gazing through their telescopes night after night, tend to be driven crazy by thinking of the vastness of the universe compared to the tiny, insignificant dimensions of their bodies. So I signed up for Chem. E., because, with a name like that, it just *had* to be more interesting than plain, boring chemistry, besides surely offering more opportunities for travel. "Gunther Stent, chemical engineer," had a nice ring to it, something that would set me apart from your run-of-the-mill engineers.

The university had no student residence halls then, except a recently completed one for men, which was occupied by the Navy for one of its training programs. So Illinois students, a.k.a. "Illini," lived either as "Greeks" in fraternity or sorority houses or as "independents," in furnished rooms rented by townspeople or in privately operated communal roominghouses. (At Illinois, the appellation "independent" implied pitiful social inferiority rather than freedom of spirit.) The clerk at the university housing office, where I had gone for help to find a room, asked,

"Are you gentile or Jewish?"

"I'm Jewish. Why are you asking?"

"Because gentiles and Jews live in separate accommodations, that's why. Here's a list of landladies who rent rooms to Jews."

Had I known about the university-sanctioned residential Jewish apartheid beforehand, I might have lied and answered "gentile."

I would have dearly loved to join a fraternity, preferably one that wasn't Jewish. But joining a gentile fraternity was not an option for this self-hater, no more than it had been for my elder brother at the University of Berlin, of which the hermetic social isolation of Jews at the University of Illinois evoked sinister remembrances. In any case, joining a fraternity, gentile *or* Jewish, was beyond my financial means. Nonetheless, I signed up for Greek rush week, just to see what I would be missing as an independent. I was also going to freeload some of the excellent meals to which the fraternities were said to be treating their rushees. After the rushee dinner at the Jewish Tau Epsilon Phi house, one of the brothers asked me how I felt about pledging TEP. I fessed up that I was too poor even to think of joining *any* fraternity. All the income I had was $40 a month, plus tips and meals, from my part-time job as a fountain man at a drug store near the campus.

A few days later, the brother turned up at the drug store and proposed that I move into the TEP house at a cut rate. The brothers were willing to offer me this deal because they reckoned that, as a Cook County Scholar, I would be bound to make top grades. This would save the house, which was teetering on the edge of academic probation, a heap of trouble. So I moved into Psi Chapter of Tau Epsilon Phi. I felt that as a Greek, even if only a *Jewish* Greek, I had gone up another step in my climb out of the pit of refugee-hood. I was poised to look down on the underclass of independents.

The Jewish sorority sisters at Illinois were from well-to-do families, mostly from the Chicago area, though some had come out to Illinois from the East Coast. Probably they hadn't managed to make it into one of the Seven Sisters schools, who severely restricted their intake of Jewish women. Many of the sisters were attractive and well dressed. I thought that, in view of my lucky rise into the social upper crust of Greeks, it should be easy to find the girl of my dreams. She might even be the rich kind of girl with whom, according to my father, it was so easy to fall in love.

It turned out that looking for her was not a voluntary but an obligatory aspect of my status as a fraternity pledge. Three weekends out of four, I *had* to have a date with a sorority girl. If I couldn't make it on my own, the TEP house social chairman would fix me up with one with the help of his equivalent in one of the three Jewish sororities. Dating independents was forbidden for pledges—and frowned on for brothers—as slumming.

I didn't do very well under this system. I was a broke, gauche, engineering nerd who talked funny with a German accent, hardly "fraternity material." I hated to call a sorority girl I barely knew and ask her for a date. With each turn-down it became harder for me to work up enough nerve for the next call. I never did get a date with an Alpha Epsilon Phi or Sigma Delta Tau, and was usually fixed up at the last moment with a leftover wallflower at the bottom-rated Phi Sigma Sigma house. Before long, I began to loathe the Greek dating game. Although these were lofty social circles—American-born, albeit of the Hebrew Faith—romance was still eluding me.

On a dateless Saturday night, I thought of my Chicago girlfriend Rita, from whom I had just received a postcard. It was an ad for a production of *Porgy and Bess* at the Studebaker Theater in Chicago, on which she had scribbled,

"They pass these out for nothing. So I thought I'd send you one."

Wasn't Rita better than any of those stuck-up Jewish Greek dames, just as good-looking and a hell of a lot smarter? And weren't two of the Seven Sisters, Bryn Mawr and Vassar, vying to admit her next fall? So I wrote to Rita, to tell her how I was making out at Illinois, boasting that I had become a fraternity man. She replied immediately,

> I'm definitely convinced now that you are really at college, not because your envelope was postmarked "Champaign, Ill.," but only because I heard so from other sources. I'm sure that you're doing a minimum of homework, so you can enjoy the social life you always said you cared so little about. It's too bad that, according to the term formerly used by G. Stent to describe fraternity men, you turned out to be one of those pitiful social-life-mad monomaniacs with moronic aspects.

I was surprised that my TEP brothers didn't seem to be bothered by the social segregation of gentiles and Jews on the Illinois campus, which I found so oppressing. It wasn't that they were unaware of the all-pervading anti-Semitism, but they seemed to accept it as merely one of many regrettable features of an imperfect world. They never mentioned any disappointment over not being able to live in integrated housing, nor did they show any overpowering yearning for dating the out-of-bounds gentile Greek sisters. They didn't seem to be ashamed, as I was, of belonging to a reviled minority.

Long puzzled by their apparent freedom from Jewish self-hatred, I came to believe that it was attributable to my TEP brothers having been raised in an ethnically diverse society. The members of this society identified themselves as "American" only vis-à-vis foreigners. In a domestic context, by contrast, they defined themselves in terms of their ethnicity of origin. In any particular case, this was only one of many possibilities, such as English, Italian, Irish, Polish, or German. So being a Jew in America didn't stigmatize one as an outsider—a foreign body contaminating an ethnically homogeneous nation—as did being a Jew in Germany.

At the start of my second semester, my Chicago draft board notified me that I had been included in a "national quota" of technical university students. Such students received a 2-A deferment from military service until graduation, so that they would become available for war service as expert civilians. This unexpected news presented me with an emotional, as well as ethical, dilemma. I was aware, of course, that none of my American fellow-students carried as great an obligation to fight against Nazi Germany as did I as a German Jew. So wouldn't it be the morally correct thing for me to do to refuse the 2-A deferment, for which I hadn't asked? Shouldn't I request my draft board instead to process my induction papers as fast as possible? Besides, hadn't I always *wanted* to be a military man, ever since, as little Krümel, I played soldier in the Schwarze Fähnlein?

I would have liked to serve as an officer, of course, but with less than a year of college credit under my belt, I didn't qualify for officers' candidate school. Perhaps my chances of becoming an officer would be greatly improved if I hung on for another couple of semesters. And maybe my contribution to the defeat of Nazi Germany *would* be more effective if I really did do my war service as an expert civilian scientist after graduation, rather than as a run-of-the-mill GI, far away from the action. Not to speak, of course, of the eventual career advantages I would derive from having managed to complete my undergraduate education. So, not without being ashamed for being yellow-bellied, I rationalized away what, in my heart of hearts, I knew to be my moral obligation to serve in the Army and accepted my 2-A deferment.

I got my first scientific job at the start of the 1943 spring semester, working for a professor of plant pathology in the agronomy department. He was studying a disease of maize that caused individual kernels on the cob to shrivel, with the fraction of shriveled kernels increasing with the severity of the disease. In the previous fall semester, my boss had collected untold thousands of kernels from diseased plants that he had subjected to various putatively prophylactic or therapeutic treatments. To assess the efficacy of the treatments quantitatively, he had hired me to separate and count, one by one, shriveled from nonshriveled kernels. Sitting all alone in the professor's small laboratory in the little agronomy annex in the experimental field dedicated to growing his plants, I felt as if I were one of the pigeons helping (the Grimm Brothers' German) Cinderella. While her stepsisters were living it up at the Prince's ball, Cinderella's feathered assistants separated good from bad wheat kernels according to the algorithm

> *Die Guten ins Töpfchen, die Schlechten ins Kröpfchen.*
> Good ones in the pot, bad ones in the crop.

Although I wasn't totally thrilled by sorting corn kernels, I was proud to have advanced from fountain man to scientific assistant, indeed to employee of the state of Illinois. Yet my pride in my rising career notwithstanding, the drudgery of my assignment made me ever more careless in my differential diagnosis of normal and abnormal kernel morphotype. I don't think the professor ever controlled the numerical results I delivered to him, for instance, by giving me a set of kernels for analysis that he had previously diagnosed himself. He just incorporated my data into his reports, and, I presume, eventual publications. This, my very first contact with research, gave me an early warning of the perils of basing scientific conclusions on student-collected data.

At the end of the spring semester, Rita invited me to be her date at the Harriet Smalley School graduation dance. The girls were going to wear white formals and the boys white dinner jackets. Rita said that it wasn't absolutely necessary for me to dress up in a tux, but it would be sort of nice. Thrilled by this opportunity to mix with high society, I decided to fritter money I desperately needed to keep me going in Champaign on the cheapest white tux I could find. I reckoned it would be a wise investment, since I was looking forward to going to many more formals. As it turned out, I had no occasion to wear the white tux ever again. It would be forty years—by which time the old relic no longer fit me—until I bought another dinner jacket, to wear at the annual black-tie ball of the National Academy of Sciences in Washington.

The Smalley School graduation dance was—and, for many years would remain—the most grandiose social I had ever attended, by comparison with which the fraternity and sorority parties at Illinois seemed cornball shindigs. The white-gowned girls gave the impression of a bevy of angels, and Rita looked strikingly attractive in her long formal. The evening started out with an opulent buffet supper, after which there was dancing to a band until dawn, with breakfast being served when the music finally stopped. It was like a dream: I was playing the role of Willy Fritsch in the UFA movie, *Congress Dances,* waltzing the night away with Lillian Harvey at the ball in Schönbrunn Castle.

Rita left for Vassar at the end of the summer. For the next four years, we wrote to each other about once a month. Our correspondence, whose style aped that of *The New Yorker*'s "Talk of the Town," had little overt emotional content. The closest she came to any terms of endearment was to call me a louse or a conceited hypocrite. My letters were matter-of-fact news bulletins about myself. Occasionally she hinted that she was going out on dates, but never provided any specifics. Except for the time of my stay in occupied Germany, we saw each other at least once every three months, whenever Rita was back in Chicago from Vassar during her vacations. Our relation was far from the purely platonic one that a reader of our letters might have surmised. We would engage in the full range of heterosexual boy–girl activities, including necking and petting, except for actually "going to the limit."

As I began my sophomore year, I had to sign up for horribly tedious quantitative chemical analysis. It turned out to be one of the most useful of my undergraduate courses, because it was my first encounter with meticulous laboratory technique. Our instructor was the corpulent Professor G. F. Smith (a.k.a. "G. Fat" to his students). I admired him, not because he was a great teacher or renowned scientist, but because he was a millionaire, who taught just for fun. He was the owner of the G. F. Smith Chemical Company, which sold fine chemicals, including—as their labels revealed—most of the reagents that we used in class, including the highly explosive perchloric acid. Moreover, G. Fat was the inventor and patent holder of "Reddy Whip," the ersatz whipped cream extruded from the nozzle of a spray can containing skim milk under nitrous oxide gas pressure. It occurred to me that the life of a millionaire chemistry professor might even beat that of a peripatetic, globe-trotting engineer.

Nelson Leonard, a young assistant professor who took a personal interest in me, taught introductory organic chemistry. One day, he dropped a hint that I was too intelligent to become a chemical engineer and suggested that I change my major to chemistry. He knew I didn't care for organic, but, in view of my professed interest in mathematics, he recommended that I concentrate on physical chemistry. So I declared chemistry as my major. As Leonard had predicted, I liked physical chemistry. In contrast to inchoate inorganic or organic chemistry, it's a logically coherent discipline, whose theories are expressible as mathematical relations.

Introductory physical chemistry was taught by Frederick T. Wall. This was my first contact with my future research supervisor and *Doktorvater*. His lectures were dynamic, lucid, and well prepared. His captivating presentation of chemical thermodynamics introduced me to the subject that raised my juvenile passion for steam engines to a higher conceptual plane. I decided to make thermodynamics my special field of concentration.

In the summer of 1943, I got my second scientific job. I was hired as a laboratory helper by Herbert Laitinen, Professor of Analytical Chemistry, whose group was part of the War Production Board's synthetic rubber research program at Illinois. His four or five research assistants were housed in one big laboratory in Noyes Laboratory, the main chemistry building on the Illinois campus. Thus began my connection with the synthetic rubber program, which would last for five years, until I got my Ph.D. in the summer of 1948. At first, my duties consisted of washing the research assistants' glassware. Before long, I was trusted with more advanced tasks, such as preparing reagent solutions. This required weighing out exact amounts of chemicals and dissolving them in exactly measured volumes of liquid. I was issued keys to Noyes Laboratory, which I proudly looked on as a symbol of my having risen to membership in the chemistry department staff.

I had found my true vocation in pure science. It wasn't that I was fascinated by the problems on which Laitinen's group was working. I had no understanding of them, and even if I had, I probably wouldn't have found them interesting. What enchanted me was the ambience of the lab. I felt more at home there than in the TEP house. My lab mates and I worked at night and on weekends, when the building was mostly deserted. We listened to classical music broadcast by the university's radio station, WILL, and argued about politics and religion.

One night, there was a tremendous explosion in the lab's fume hood, in which one of the assistants was boiling some of G. Fat's perchloric acid. Luckily, the assistant was on the far side of the room, and I happened to have gone to the toilet, but the lab was in a shambles. The exhaust fan of the hood had been hurled half-way across the lab and smashed everything along its trajectory. Although I was scared, I did feel a sense of pride in being such a daredevil engaged in obviously dangerous war work.

Bob, my brother-in-law, was drafted in the fall of 1943. After finishing infantry boot camp at Fort Jackson, South Carolina, he was transferred to Grenada, Mississippi, where Claire and Baby Ronny went to join him. A few months later, Claire asked me to drive down their Chevy, which she had left behind in Chicago. I was excited by this opportunity to explore the States, especially to drive along the Mississippi, of which I carried many romantic visions formed by my reading of *Tom Sawyer* and *Huckleberry Finn*.

To my disappointment, I found Memphis no more enchanting than Chicago. The Deep South *was* very different from the North, but not in the way I had expected from the descriptions of gracious plantation life in *Gone with the Wind*. I felt as if I was visiting a different country from the United States, a spooky, depressing nation like Nazi Germany. It came as no surprise to see how badly the blacks were treated south of the Mason-Dixon Line. Nevertheless, I hadn't been able to imagine the oppressive atmosphere that was associated with their legally mandated segregation. The "WHITE ONLY" and "COLORED" signs posted on restaurants, water fountains, and other public facilities reminded me of Nazi's "JEWS NOT WANTED HERE" signs.

A lone, white clerk served two separate ticket windows, one labeled WHITE and the other COLORED, at the Jackson, Mississippi, railroad depot. When I strode up to the WHITE window to buy my return ticket for the trip north, the clerk stopped dealing with the black people waiting at the COLORED window and came over to serve me. I found this very distressing, not because I was morally outraged by the demeaning treatment accorded to the blacks, or because my traumatic childhood experiences under Nazi oppression had taught me sympathy for the victims of racist hate. On the contrary, as a one-time outcast I would have welcomed the chance to join the master race. But I couldn't help feeling that as a Jew I couldn't *be* a member of the master race, even in America. I was afraid that in a place where the law requires putting up those Jim Crow signs, there would be some other signs pretty soon reading "JEWS" or "JEWS AND COLORED." Then the ticket clerk would take a closer look at me and say,

"We don't wait on kikes at this window. You go over and line up with the niggers, Buster!"

I had been happy to be reunited with Claire, Bob, and my little nephew in Mississippi. Yet I felt as if I were coming up for air when, on my return trip to Champaign, the northbound train from Jackson recrossed the Mason-Dixon Line, as we entered my home state of Illinois.

By the end of 1943, the Nazis were beginning to lose the war. The great turning point had come in February 1943 when the German Sixth Army encircled at Stalingrad surrendered to the Soviets. Nightly bombing raids by the British and American air forces had brought the war directly to German cities, with Hamburg destroyed in a gigantic firestorm on July 29, 1943. In the fall of 1943, Italy surrendered after Allied troops had landed on the Italian mainland, and FDR, Churchill, and Stalin met in Teheran to plan the invasion of Western Europe. The Allied leaders declared the unconditional surrender of Germany as the war's goal, the attainment of which now seemed only a matter of time—to everybody except Hitler and me.

Discovering Science

ABOUT A THOUSAND GIS assigned to the Army's Advanced Specialized Training Program (ASTP) arrived on the Illinois campus in the fall of 1943, replacing some of the male students who had left school for military service. The ASTPs were draftees who had obtained very high scores on the Army General Classification Test. They were sent back to school to study science, engineering, foreign languages, and other subjects the Army had declared useful for the war. ASTPs did not attend regular classes, but took specially designed, intensive courses, to which they marched in military formation along the broadwalk, calling out "*hup*-two-three-four, *hup*" to keep in step.

Because of the shortage of graduate students available for service as teaching assistants in ASTP courses, the chemistry department recruited undergraduates to help train the brainy GIs. So I made my debut as a college instructor: I was appointed as an undergraduate teaching assistant in the ASTP quantitative chemical analysis course taught by G. Fat. I felt like a big shot, lording it over GIs most of whom were older than me. I handed them coded samples of unknown chemical solutions for analysis, showed them how to work a burette and to pipette fluids, and graded their laboratory reports. A few of my soldier-tutees even addressed me, the nineteen-year-old enemy alien, as "Sir." It was almost like being an officer.

The ASTP lasted less than a year. By the end of the 1944 spring semester it was disbanded, and the GIs left the campus without having completed any course of study that could have been useful for the conduct of the war. Concurrently with the dissolution of the ASTP, the Selective Service System also canceled all "national quota" 2-A student deferments. The War Department never did provide an official explanation why the ASTP was suddenly disbanded or why the national quota of student deferments was suddenly abolished.

According to a then prevalent rumor, however, the training of experts in skills that would eventually be of service in the war effort had not been the real reason for establishing the ASTP and the national quota in the first place. The secret purpose of both programs was supposed to have been averting the wholesale killing of the nation's future intelligentsia at the very outset of the hostilities. But when, in the late spring of 1944, the war was entering its final stages, the smart boys' turn to fight had come. Although we didn't know it at the time, D-Day was just a few weeks in the future. General Eisenhower would need the hitherto husbanded young brains as replacements for the heavy casualties that the imminent invasion of the Continent was likely to wreak on the Allied forces under his command.

Before long, I received greetings from the president and commander-in-chief, directing me to report for a preinduction physical examination at the Armed Forces induction station in Chicago. He advised me that if I were found qualified for service, I would be immediately inducted into the Army. So I went up to the City to become a soldier, like everybody else. I was hoping that the three years of college credit I had meanwhile accumulated might get me into officers' candidate school, or qualify me at least for the Army's chemical warfare service, in which one of my professors was serving.

At the time of my preinduction physical, I happened to be suffering from one of my recurrent attacks of bronchial asthma. The asthmatic episodes that had distressed me during my Berlin childhood had subsided in Antwerp, London, and Chicago. They came back with a vengeance in Champaign, probably set off by the pollen released by the corn plants in the fields surrounding the campus. In view of my bronchial wheezing, the Army doctor at the induction station categorized my fitness for military service as "undetermined," and sent me home to await further instructions. These instructions were a very long time in coming, and so while awaiting them I simply continued my studies at Illinois.

Just as I had deftly manipulated the course credits at Hyde Park High to make it in twenty-one months from entering freshman to graduating senior, so did I contrive at Illinois to complete the normal four years of college in twenty-nine months. To this end, I switched my major for a second time, from chemistry to "General Studies in Letters and Science," in my last-but-one semester. The chemistry credits I had accumulated by that time already satisfied the graduation requirements for a primary field of concentration in science. All I still needed to meet the breadth requirement for the general studies baccalaureate were credits for a secondary major in the liberal arts. I chose philosophy, because a TEP fraternity brother had advised me that philosophy was a cinch.

To get started on my secondary major, I signed up for "Philosophy of Science," taught by Max Black, the Russian-born, British-trained analytic philosopher. He was only thirty-five years old at the time and had not yet gained his later fame. The title of his course intrigued me. None of my chemistry or physics teachers had ever mentioned that you need philosophy to do science, and the pairing of philosophy and science in the course's title seemed weird. It turned out that my science teachers were justified in ignoring philosophy. You *don't* need to know anything about philosophy to do good science. But if you want to expound on the history, sociology, or ethics of science then you better *do* have some philosophical knowledge!

In my last semester, I took another philosophy course, "Great Books in Modern Thought," taught by Professor Frederick Will, the father of the (then still subteen) political commentator and newspaper columnist, George Will. Under Will's tutelage we read David Hume's *Enquiry Concerning Human Understanding,* Charles Darwin's *Origin of Species,* Karl Marx's *Capital,* and Sigmund Freud's *Interpretation of Dreams.* All I knew when I signed up for Will's "Great Books" course was that Darwin's book was about animals, Marx's about Communism, and Freud's about sex. I did know a little more about Hume's book. Professor Black had pointed out to us that Hume's critique of the logical validity of inductive reason was a mere misuse of ordinary language. Hume's critique happened to be Professor Will's own topic of special interest, and he tried to make us appreciate (successfully, in my case, at least) the shallowness of the kind of analytic philosophy then represented by Max Black.

Incredible as it may seem, it was in Professor Will's discussion of Darwin's *Origin* that I, a science senior at a major university, heard the word "gene" for the first time. I still spelled as "jean" in my lecture notes the name of the hereditary element whose study would occupy me for twenty-five years. As elucidated by Will, Marx's *Capital*, which I found too hard to read, reinforced my belief in socialism, and Freud's *Dreams* provided me with my first contact with psychology. Professor Will did not become as renowned a philosopher as Professor Black, but he was by far the better teacher. He expressed complex ideas more clearly, notwithstanding Black's professed concern of analytic philosophy with clarifying the meaning of language.

It was also in that summer of 1944 that I began my career as an independent scientific investigator, when I signed up for undergraduate honors dissertation research in physical chemistry with Professor Wall, a.k.a. "The Doc" to his students. He gave me a very modest problem. He just asked me to make a series of simple measurements, whose results seemed unlikely to lead to a tremendous advance of our understanding of nature. I was to shake a bottle containing equal volumes of pure water and of a solution of one of the three isomers of toluic acid (ortho, meta, and para) in benzene and then determine how many molecules of that isomer had gone into the water and how many had stayed behind in the benzene. Big deal! The Doc wanted to have these data because he would be able to calculate from them the strengths of the bonds formed between pairs of molecules (or "dimers") of each toluic acid isomer.

A graduate student of the Doc's, Fred Banes, had just completed a doctoral dissertation on the toluic acid dimer bond strengths, which he had ascertained by means other than the water–benzene distribution method. So the main, albeit unstated, purpose of my modest research project had been to provide an independent confirmation of Banes's data. To the Doc's disappointment, the bond strengths implied by my data did not agree with those obtained by Banes.

Did we infer that I, a mere undergraduate, had disproven the findings of the august Doctor Banes? No, Sirree! The bond strengths inferred from Banes's data were closer than mine to the values that the Doc's had predicted on theoretical grounds. So he put forward the entirely plausible, yet wholly ad hoc, hypothesis that an error was probably introduced into *my* measurements by the entrance of a small amount of water into the benzene phase. I could have explored the Doc's hypothesis experimentally, but there was no time for me to do so before graduation.

This, my very first experience with the interpretation of original research data, should have alerted me to a joker in the deck of widely held notions about the nature of scientific truths. As an innocent scientist-in-the-bud, I had been imbued with the commonsensical belief that there are facts of nature that exist (and are accessible to observation) independently of any preconceived notions of the observer. I had heard, moreover, that there exists a so-called scientific method for using these facts to confirm (or, at least, to falsify) scientific theories objectively.

Suppose I had gone on to test the Doc's exculpatory ad hoc hypothesis and found that, contrary to it, not enough water had gone into the benzene to account for the discrepancy between Banes's results and mine. Would the Doc now have concluded that my bond strength determinations were right after all and Banes's wrong? No, Sirree! The Doc, clever man that he was, would no doubt have thought up another entirely plausible ad hoc hypothesis to explain away any of my data that were not in accord with his theoretical expectations. The Doc would have been acting (just as most experienced scientists would have, but few will admit openly, *surtout pas devant les enfants*), in conformity with the maxim put forward by the astrophysicist Arthur Eddington in the 1920s:

"It is unwise to put over-much confidence in observational data until they have been confirmed by theory."

Halfway through my last pregraduation semester in the fall of 1944, my draft board finally let me know that I had been rejected as physically unfit for military service and was permanently classified as 4-F. This was good news, as far as my professional future was concerned. I was now free to go on to graduate school right after graduation and work for a Ph.D., without which, as I had learned meanwhile, one could not make a career in science. Yet, at the same time I felt dreadfully ashamed for having been declared physically unfit for military service, as well as being troubled morally by not fighting against the Nazis. So I tried to conceal the humiliating fact of my 4-F classification. I pretended (and continued to pretend long after the war was over) that I was deferred from the draft because of the important defense work I was doing as an expert civilian on the War Production Board's synthetic rubber research program.

This charade almost came true in the fall of 1944, when I received a letter from the personnel director of the Metallurgical Laboratory at the University of Chicago. He wrote that my name had been given to him by one of my professors at Illinois as someone well qualified to join their research staff engaged in interesting and vital defense work. I had a fair idea about the nature of this work, because it was suspected, if not known for certain, at the Illinois physics and chemistry departments that "Metallurgical Laboratory" was a cover name for an outfit that worked on the development of an atomic bomb. So I was thrilled by the invitation to join this fabulous undertaking, which, if it succeeded, would be sure to bring Nazi Germany to its knees.

Alas! In response to the application form I had rushed up to the City by return mail, the personnel director regretted that, inasmuch as I was not a U.S. citizen, the Metallurgical Laboratory couldn't hire me. (After the war it transpired that many of the superstar scientists who worked on the atomic bomb project, such as Enrico Fermi and Edward Teller, were no more U.S. citizens than I was. Evidently, Dr. Frisch, my Latin teacher at the Bismarck Gymnasium, had it right: Q*uod licet Jovi, non licet bovi*—What is allowed for Jove is not allowed for oxen.) Deeply disappointed, I firmed up my decision to go to graduate school after getting my AB from Illinois in January 1945 and work for a Ph.D. in physical chemistry.

By now it seemed clear that Germany was done for and that the war in Europe would be over in a few months. On the eastern front, the Soviet Army had taken Warsaw and entered German soil in East Prussia and Silesia, poised for the capture of Berlin only a hundred or so miles to the west. On the western front, the British and American armies had reached the Rhine, and the Wehrmacht was unlikely to be able to prevent the Allies, with their vastly superior air power, from crossing it. But my belief in Hitler's invincibility was too deeply ingrained. His string of military defeats notwithstanding, my fear of his evil omnipotence had been boosted further by his miraculous survival in the bomb plot against him in July 1944. I was still afraid that he had some tricks up his sleeve by means of which he might yet win the war, or at least fight the Allies to a standstill. Maybe he already had the kind of atomic bomb on which they were working at the "Metallurgical Laboratory" up in the City.

Jilted

WHEN I RAN INTO HILDEGARD at the doors of the Admiralspalast in Berlin in November 1946, I was still pining for my ex-fiancée Dorothy S., with whom I had fallen in love two years earlier. Before falling in love with her, I had met Dorothy a few times in Champaign at dances held at Keeler House, the Jewish Independent girls' house where she lived. She was a tall, long-haired brunette, with a beautiful face and a sultry expression. Her home was in Winnetka, an upscale North Shore suburb of Chicago, where her father, Dr. Herman S., was a cardiologist. After graduating from Illinois in June 1944, Dorothy went back to Winnetka to teach elementary school. I don't recall why I had never asked her for a date. Maybe I was afraid that she would turn me down; or maybe I had been put off by my TEP fraternity brothers denigrating her as a slut. It may seem paradoxical that I wouldn't have wanted to take out a girl with the reputation of an easy lay. Yet, in my grotesque self-absorption I must have wanted a woman who considered me someone special, even if I didn't reciprocate her feelings.

I ran into Dorothy again in Champaign in November 1944, at the Homecoming football game. That game was long remembered on campus because Illinois lost to Notre Dame 7 to 13, with the Illini's only touchdown having been scored on a 74-yard run by our black ("Negro," we said then) superstar halfback, Buddy Young. Many people thought that Buddy was a second Red Grange, the Illini's immortal "Galloping Ghost" of the twenties, proclaimed as the greatest runner in the history of football. (After graduation, Buddy did go on to play professional ball. However, his main historical renown would rest, not on his prodigious running, but on his becoming a director of the National Football League, the first black executive in a major American sport.)

Dorothy, meanwhile an alumna and looking gorgeous in her coed-style sweater, woolen skirt, socks, and saddle shoes, had come down to Champaign for the Homecoming weekend. She gave me a big hello, and I was struck by *a coup de foudre*, a lightning bolt of love. What a damn fool I had been not to try to date her! And now it was too late.

On the way home from the game I thought that maybe it *wasn't* too late. Perhaps I could manage to see her before she went back to the City. I summoned my nerve and phoned Keeler House.

"No, Dorothy S. isn't in. She's out on a date."

"Could you please ask her to call Gunther at the TEP house when she comes in?"

Dorothy returned my call late at night, after I had already given up hope that I would hear from her.

"Thanks for calling back. Do you remember who I am?"

"Of course. I wouldn't have called back otherwise!"

After Sunday breakfast, we walked along the harvested corn fields, whose endless, bare expanse began at the campus's southern edge. It was a sunny, crisp autumn day. To keep up the conversation, I reviewed my academic career for Dorothy, and she described her teaching job. Then we drifted toward more personal topics.

"Never found a girl in Champaign I really liked."

"Well, I was in love once with a boy called Tony. He graduated a couple of years ago and is now serving overseas. But that's all over now. He's awfully immature."

Eventually, we ran out of talk. We stopped to gaze at the horizon and to admire the non-existent scenery. Standing very close to her, I was excited by her perfume, which, she said, was Chanel No. 5. I wanted badly to kiss her, but didn't have the nerve to do it. As we walked back into town to Keeler House, I was furious with myself for being such a wimp.

A couple of days later, I wrote to Dorothy, telling her how much I had enjoyed our walk in the fields and how much I would like it if she came down as my date for the fall semester dance the TEP house was going to throw on Saturday, November 17th. I would get her a room at the Inman Hotel for Friday and Saturday nights.

Since I couldn't really expect that for a first date she would make the long trip all the way down from Chicago, I provided for a backup for the very likely case that she would turn down my invitation. I mentioned that I would be in the City on the weekend before the dance to meet my sister, who was coming up from Mississippi to visit friends in Glencoe, not far from Winnetka. Would it be possible for me to see her and maybe take her to the Saturday Evening Pops concert of the Chicago Symphony?

Dorothy's response arrived by return of post, addressed to "Mr. Gunther."

Gunther,

It shall undoubtedly occur to you in due time that your name is somewhat of a puzzle to me yet. Mom thinks perhaps it's even your last. I don't know whether to regret whatever emotion it was that prompted me to call you Saturday nite, but I'm inclined to think that it would be silly to regret it simply because it might have been a poorly constructed tactical maneuver. I sense that you have a rather clever system all your own. You dilute all your most flattering phrases by contradicting yourself immediately afterward. Perhaps that's not quite the word to express it, but I believe I have gathered the technique involved.

I wasn't certain whether I would go down for the TEP dance the 17th, for various reasons. I didn't really know why I should. I mean by that, of course, that traipsing all the way down to Champaign—well one should have a better than average reason. I'm not particularly frantic 'bout going to college dances, and a weekend at school isn't especially a devastating thrill. In other words, I'm not schoolgirlish enough to be intrigued simply by a fraternity affair. But Gunther, I really do want to see you again, and I feel that a weekend down at school would afford us more time together than under other circumstances. Therefore my answer will be a very positive yes, but I'll be staying at Keeler House. It's sort of a concession to the folks for my dashing about.

I'm quite happy 'bout your coming into Chi in just another week, and it should be quite simple to dispose of numerous people for the sake of the evening. The Symphony would be fine, just name the time and place.

I've wondered whether the realization that you're better versed in many things than I am would ultimately disturb me. I don't think it will unless a tiny fear continues to grow that you really think me a scatterbrain. I don't believe I am, actually. Suddenly, I seem to be thoroughly steeped in your theories on modesty.

You know, Gunther, this is just about the time o'day when a feller sort of needs someone. Ever feel that way? Honestly, it would be sort of nice just to sit comfortably with someone you felt close to, and not talk about anything. Any more of this and you'll be convinced I'm wacky. So 'til another evening

Your Dorothy

Her guardedly positive, cautiously crafted letter, artfully laced with disclaimers, set me aglow by the wild thought that she might requite my passion. I didn't read the letter very carefully—didn't notice that she was letting me know in a delicately self-protective way that she was considering our entering into a long-term relation, not to speak of picking up that she even gave warning of possible factors that might eventually trouble it. Just like all her subsequent letters, I quickly scanned this one for the gist of its nuts-and-bolts information about the technical arrangements of our next date and didn't trouble to scrutinize it for its emotional content.

I called Dorothy in Winnetka, to tell her how happy I was to receive her acceptance. "Gunther, by the way, is my first name, and my family and friends call me 'Gun.'" Five days later I got my second letter from Dorothy,

Tuesday, Election Day!

Evening Gunther,

Well, Gun, I'm not known to have an undying love for the Symphony, and it's too soon to denote my true feelings 'bout you; so since allegedly I'm crazy 'bout neither of these I must be just plain crazy. I left school yesterday at a quarter to four. . . . Went down to the Loop, got tickets for the concert, back home to change, and arrived at the school at seven-thirty in time for the PTA meeting. Or don't you know how long the trip from Winnetka to the Loop takes?

Perhaps if you can come out to the house 'round twelve-thirty we can just sit a while and listen to music on the Vic combination. Would that be all right?

Usually the weeks fly by for me, but this one is definitely dragging. Very peculiar of it, don't you agree? I'm getting into a very sentimental and drippy sort of mood, and before it penetrates the letter, I'd best close. See if you can't move the time along a bit 'til you're with

Your Dorothy

The S.'s place in Winnetka was a substantial, two-story, white clapboard house on a tree-lined, quiet street. Dorothy opened the front door, just as beautiful as the portrait of her that I had projected in the theater of my mind a hundred times in the past week. I simply grabbed her hand and shook it. She introduced me to her parents and to her kid brother. Dr. S. and Junior soon left for a high school football game, while we stayed behind, with her mother keeping us company in the living room.

Mrs. S. asked me about myself, since Dorothy, who hardly knew anything about me, couldn't have provided her with much information about her new beau. I told them about my youth in Berlin and my family. Somehow I managed to work in also Hume, Darwin, Marx, and Freud. Mrs. S. seemed impressed, too impressed unfortunately, as would eventually transpire.

Presently, Mrs. S. went into the kitchen. Dorothy turned off the phonograph, and we talked about ourselves, while making slight but ever-so-exciting lateral body contact. The pre-concert kissing and caressing to which I had been looking forward didn't materialize, because—ever the wimp—I felt inhibited by Mrs. S. being down the hall. Dorothy told me about her defunct romance with the immature Tony. He had graduated from Illinois in aeronautical engineering in 1942, just at the time I went down to Champaign as a freshman. He had been enrolled in the Army ROTC program, commissioned a second lieutenant on graduation, and was now serving in Italy as a captain in the Army Air Corps.

Tony was awfully handsome, but not Jewish: a Catholic from a poor, working-class Polish family in Chicago, unapproved by Mrs. S. He had always treated Dorothy very badly, and now everything was definitely and irrevocably finished between them. Well, I thought, that's just fine; no reason for me to worry about Tony. Had I been a more experienced lover, I would have worried about her bringing up Tony for the second time.

When Dr. S. and Junior came back from the football game, Mrs. S. served us a light supper. Dorothy changed into an outfit more dressy than the coed getup in which I had seen her hitherto. The sight of her in black gown, high heels, pearl necklace, and wrapped in a fur coat further inflamed my passion. We went downtown on the Chicago North Shore Electric.

Once seated in Orchestra Hall, I took her hand and squeezed it tenderly, which took less nerve than kissing her. She responded with a tender countersqueeze and turned toward me for deep eye contact. She would be mine! I had goose pimples. The program included the overture to *The Magic Flute*, Schubert's *Seventh Symphony*, and a piece that I heard for the first time that night and would provide for me an everlasting souvenir of that first amorously fulfilled evening of my life, Bach's *Double Violin Concerto in D minor*. I had never been happier.

As we walked up Michigan Avenue after the concert on our way to the North Shore Electric, I stopped and slid my right arm under Dorothy's fur coat. Pressing her body against mine, I kissed her for the first time, a kiss she passionately returned. I was beside myself: She *is* mine! On the train, we held hands, sneaked in an occasional kiss, but didn't talk much.

Mrs. S. had left the living room discreetly for our exclusive use, and we kissed and petted rapturously on the sofa. But by the time I had to leave to catch the last train for Glencoe, I hadn't been able to bring myself to declare that I loved her. I just said that I could hardly wait to take her in my arms next weekend at the TEP dance.

When Dorothy stepped off the train Friday evening at Champaign's Illinois Central depot, we hugged and kissed like long-separated lovers. As I was taking her to Keeler House, Dorothy announced that she would be staying there only that night, to satisfy her mother; in the morning, she would be moving to the Inman. My torrid arousal knew no bounds.

The theme of the TEP dance was "The Roaring Twenties." The house had been done up as a speakeasy; the locked front door would be opened by a dinner-jacketed brother only on three knocks followed by the password "Joe sent me." Except for an unusually good band—Joe Hayes, His Trombone, and His Orchestra doing Glenn Miller-style swing, from the forties rather than the twenties—the whole shebang was just another one of the many house dances I had endured, at which everybody always pretended to have such tremendous fun, but which were insipid, unless you managed to smuggle your date past the dean of students' vice squad chaperones for some illicit action upstairs. But this time, overflowing with bliss, I was dancing cheek-to-cheek with a girl I loved, for our first full-length frontal body contact. She had written in my dance bid "Gunther—Just a thought—Sincere tonight—It's you—I'm sure of it—Your Dorothy." So much for Tony!

We sneaked out early, trying to avoid making the reason for our precocious departure too transparent. The brothers probably had a fair idea of what we were up to, since word had been put out by the girls at Keeler House that Dorothy had moved to the Inman. As we walked arm-in-arm to downtown Champaign, we stopped on every other block of the dark street to embrace.

When we got to the Inman, Dorothy went on the elevator straight up to her room. Meanwhile, I ambled down the stairs to the men's room and took the elevator directly from the basement up to Dorothy's floor. She let me in; we flew into each other's arms and just fell down on the bed. She was an experienced lover, while I, who, despite an overpowering erotic craving, still had only a general theoretical idea of what to do to satisfy it. She showed me.

I had brought with me a thin golden chain, from which my TEP fraternity pin hung as a pendant. In the Greek world of college fraternities and sororities, the bejeweled fraternity pin was of great symbolic moment. The brothers wore it on their shirts or sweaters, just over the heart, never—God forbid!—on a jacket lapel or necktie, as the cachet of their superior social status above the underclass of independents. A fraternity man gave his pin to the woman with whom he was going steady. She would proudly wear it on her blouse or sweater, to advertise her (more often than not, temporary) romantic commitment.

In view of Dorothy's now being a schoolmarm, I thought it would be more appropriate for her to wear it as a necklace. As we were sitting up in bed, I hung the chain around Dorothy's neck, without saying anything. She responded to my silent, symbolic act with wild kisses.

I was an eager apprentice, and so we stayed in bed Sunday morning, beyond the posted checkout time. Suddenly, the maid turned the door key, intending to make up the room. Dorothy jumped out of bed and put her foot against the door, while I was hiding under the blanket. She told the maid to come back a little later, as she'd be checking out momentarily. I was lost in admiration for Dorothy's sangfroid. We left the Inman by reversing the stratagem we had used to enter it last night and walked to the depot across the street. It was snowing.

On the platform, we ran into a red-haired dude, whom Dorothy introduced to me as "Howie" and who (she whispered to me) had been, or still thought he was, one of her beaux. He was a student at Northwestern University in Evanston, just north of Chicago. Howie had come down to Champaign for the weekend and was taking the same train back up to the City. In view of what Dorothy had written in my dance bid last night, I saw no cause for jealousy. We clung to each other, kissing until she boarded her train.

Ever since I moved down to Champaign as a freshman, the little depot had played a role in my emotional life as the place where I kept up my childhood romance with the Iron Horse. But this weekend, the Illinois Central Station had transmuted into the hallowed portal of love. My perception of myself had transmuted as well. I felt I had become a man, at last. Dr. Rosenthal, of the Reformgemeinde temple, was supposed to have conferred manhood on me. But all *he* had done for me was to provide me with Goethe's instructions about how to climb Mt. Olympus.

At the TEP house, we were all boasting about laying broads all the time. Some of this braggadocio may not even have been wholly counterfactual, but no one ever brought up the word "love" in this connection. Although I thought about love all the time, I would have been too embarrassed to mention it in front of the brothers. I wondered what I was going to tell them when they asked me how I made out with Dorothy after we sneaked out of the dance last night. Probably it was going to be

"Well guys, mark down another broad laid for Old Gun."

Yet this horrid debasement of my feelings could not help boost my self-confidence. For the first time ever, I would be understating my erotic success, rather than wildly exaggerating it.

Dorothy wrote the following night:

Monday evening

Golly darling,

What a perfectly miserable departure. There you were in that awful sleet storm, and there I was with Howie. I'm not even certain which of us were worse off, but I'm very much inclined to think it was me. We were off to an exceedingly bad start. He started by telling me to smile, and, doggone, I just didn't feel elated at that moment. And then he wanted to talk, and my thoughts were all with you. So what else was there to talk about? Things not being bad enuff, he started in with the very old line: "Dorothy, you surprise me. He's not even in the League." That's very good, Gun, and you see that you don't even get into the running. I don't want to pit you against the others. I want you to be exactly just the way you are, the lad who comes first without rhyme or reason. But Howie's remark infuriated me nonetheless, because you're the guy whose qualities I'd never bothered to enumerate. So I was completely at a loss for a convincing retort, leaving but one alternative; and so now Howie knows I'm in love.

I'm kind of glad too, 'cause I'm tired of playing games. I don't know; perhaps it's insane to burn one's bridges behind one, but I don't ever want to go back over them. I feel as tho' I'm stepping forward all the while to meet you in some lonely spot called "happiness."

And Gun, I wanted to tell you about this thing, happiness. I feel it right down to my very bones today. I'm not restless and nervous any more. I feel just right about us, and it makes me awfully light-hearted. I even began singing right out loud in one of my classes today. It's a wonderful life . . . 'Tis you.

Yes, Gun, this is the way I want to feel. It's belonging, but not to the point where it hurts, confuses, and stifles one's soul, but rather when one emerges feeling honestly good. I want so badly to try to explain this to you, but that's really sort of foolish, for I'm certain you understand without my having to tell you. That's part of it too.

Oh Gun, 'tis a wonderful thing . . . this love of ours.

For the next two weeks, we wrote to each other almost every day. Dorothy was telling me about her daily life, her school and her family. But the main topics of her letters were her thoughts regarding us and our love, and her need to know more about my feelings. I did not respond to Dorothy's emotional concerns, maybe because I didn't understand them, or didn't see any need for responding to them, or didn't know how to respond. I merely provided her with factual details about my daily life, my courses, my research, and the goings on at the TEP House. The central focus of my letters was not on emotions but on when and how I would see her next. We arranged that I would come to Winnetka on Saturday, December 1, and stay at the S.'s house over the weekend. And I was also going to spend the Christmas holidays, including New Year's Eve, up there together with Dorothy.

Thanksgiving vacation. Yippee!

Congrats on your making Phi Beta Kappa, but I'm not the least surprised. I'm terribly proud of you. . . . You can't even begin to imagine how very proud. But I intend to be quite used to the fact that my future husband will accomplish no end of achievements. You're quite a scholar, my sweet, and, oh, so terribly intelligent. 'Course that's not why I think you are so wonderful. It's the hair on your arm. Okay, Gun, time out while you get that one. It's very, very subtle.

Marj was talking about her forthcoming marriage on the way home today from school. She's all excited about it and informs me that February is only a little more than two months away. It only then occurred to me how very soon your graduation was. Wondered how you were beginning to feel about it. I was terribly excited, but I think it was 'cause neither I nor anyone else fully ever anticipated witnessing the event.

I'm still getting backhanded compliments from you, but if telling me the lads thought you were quite a guy is a form of flattery, I accept it very graciously. Think I preferred your rather astounding statement to me when we were quite alone, "Dorothy, you're quite pretty." A scientific discovery with just the element of surprise!

Your Dorothy

Surprised by Dorothy's reference to me as her "future husband," it dawned on me that she considered us engaged, probably because I had hung my TEP pin around her neck. And she was linking her colleague's imminent wedding to my imminent graduation in little more than two months as well!

Barely able to support myself as a student and facing at least four more years of graduate school, I had never considered marriage. But then I thought that since we are so much in love, why *not* get married? Yes, let's get married! I supposed that I would have to look for a job after graduation to support our household, instead of going to graduate school. Or maybe Dorothy could continue working, while I would try to get some kind of teaching assistantship or fellowship for my graduate studies.

I thought I really ought to give Dorothy an engagement ring, as the fraternity pin only symbolizes going steady. Perhaps I could find something affordable in a pawnshop, and tell her that though the ring might not seem like much, it had been worn by my late mother. It so happens, I would say, that this ring is one of the few keepsakes we managed to bring with us from Berlin. But the developments presaged in her next letter would, before long, obviate perpetrating this swindle on my beloved.

Friday morning.

Gun dearest!

It's absolutely imperative that I write to you this very minute. Tony is back from overseas! It all happened so very suddenly that I scarcely believe it yet. Yesterday he came over at nine in the morning, which is evidence of his changed nature. Punctuality was never one of his virtues, but he warns me he's a different lad. Well, we went downtown, window shopping; had lunch at *Le Petit Gourmet,* and went out to Evanston to see the football game at Northwestern stadium. We've always had fun and been happy together, but he's always stepped in and out of my life, just as he pleased. That he can't do anymore.

It's about time I make up my mind once and for all 'bout what I want in life and where I'm going. I suppose it would have all been pretty simple if I hadn't made a phone call late that Saturday night, and hadn't got interested in one G. S. Stent. Tony couldn't possibly destroy my love for you, I'm sure of it. But Gunther, dearest, are you very sure you want me to feel that way about you? Do you really want us to be in love, for I've no time anymore for mere infatuation. I warned you about all this and I guess I'm rather expecting you to help me 'bout it. Oh, Gunther, write and tell me 'bout things. It's strange but after last week, I feel so terribly close to you. It's a sort of belonging feeling, Gun. I just can't get everything down that I want to say.

I don't know what to do about your coming to the City. I wanted you to come in very much next weekend, because I miss you so, but it would be such a rotten deal for him. You see, I'm not really such a bad kid sometimes, and I do try to think of the other fellow, if I can.

And yes, we still have a date for New Year's Eve. You know, I didn't tell you, but altho' I've had some supposedly very nice times on that occasion, I've never been really happy that night. How about breaking that spell for me? Darling Gunther, I love you

Your Dorothy

I was dumbfounded. Are we engaged or are we not? If we *were*, then why doesn't she tell Tony to buzz off, instead of telling me that my coming up to the City to see her next weekend would put the kybosh on her tossing a bone to the other fellow? Did she think of the other fellow when she declared us engaged? Was she wondering which of us, Tony or me, was The Beau Who Can Show a Girl the Best Time on a Date? Had she warned me about all this when she told me on our first date at the symphony concert that everything was definitely and irrevocably finished between her and Tony? What did she mean by asking whether I really wanted us to be in love? What was she agonizing about? Was she asking me for help as an umpire in the contest with my rival?

I was too proud to whine in my response to Dorothy's news about this wholly unexpected threat of suddenly losing my beloved. So I made no mention at all of Tony's sudden return, nor let on how much her letter had upset me. All that she could have implied from what I wrote was that Tony was not a subject of any interest to me, including any feelings she might have about him. The closest I came to mentioning our relationship was to indicate that I would really like to come up to Winnetka next weekend as planned, because I was dying to see her.

>   Tuesday, very late
>
> Gun dearest,
>
> Brrrrrr, that letter of yours gave me quite a chill. I don't honestly think it was what you said, but rather merely the fact that it was typed. 'Tis indeed sad, Gun darling, that the world has made us such a wary and cautious twosome.
>
> Please come up to Winnetka next weekend. Mom promises that there'll be a decent dinner on Saturday, after which my parents are going to go out and will leave the house to us for the evening. My emotions aren't quite as mixed as I anticipated. I was terribly rattled 'bout doing the right thing by every one, but as fate would have it, everything worked out beautifully for you and
>
>   Your Dorothy

Dorothy met me at my train at the Illinois Central Station in downtown Chicago. Our kisses were just as passionate as they had been when we parted two weeks earlier. There was no mention of Tony, not even of how fate had seen to it that it all worked out beautifully for us. Maybe she discovered that he also had someone else on the string. I didn't ask her to explain or reassure her that I really wanted us to be in love. All of this seemed so obvious that I thought it didn't need saying. I was simply happy to be with my Dorothy.

She thought that June would be a good time for our wedding. By then, I would be settled in graduate school—in California, if all goes well—and she would be finished with her teaching in Winnetka. In view of the wartime teacher shortage, there couldn't be any trouble with her finding a job wherever I was going to do my doctoral studies. Before taking the North Shore Electric out to Winnetka, we surveyed Marshall Field's department store in the Loop for furnishings for our future ménage—clothes, linens, flatware, china, furniture.

The guests at the decent dinner for Dorothy's nice (Jewish, thank God) fiancé included the Glencoe rabbi and his wife. At the table the rabbi asked me,

"Do you know Dr. Silverman at the Hillel Foundation Student Center in Champaign? He's an excellent rabbi. Do you attend his services?"

"No, I don't know Dr. Silverman. Never been inside the Hillel Center. Haven't been at a Jewish service since I was confirmed in Berlin six years ago. I'm an atheist. I think that all religions are baloney, including Judaism."

"It's our religion that holds us Jews together, that provides us with our identity. Never forget that!"

"I'm not interested in my Jewish identity, for which I certainly didn't ask. The biggest favor the Jews could do themselves is to assimilate as totally and as fast as possible."

The rabbi looked disgusted and discontinued the conversation. No doubt he was wondering how this self-hating smartass could be the S.'s idea of a nice Jewish fiancé. They all left after dinner, and having the house to ourselves, we made love in Dorothy's Chanel-No.5-scented bedroom.

Now I noticed that there *was* something wrong: She didn't seem to have her heart in it, at least not in the way she obviously had at the Inman Hotel two weeks earlier. I wondered whether she hadn't liked the way I argued with the rabbi. We firmed up our previous plan to spend the Christmas holidays together; her parents had already agreed that I could stay at their house. I was to show up in Winnetka on December 20 and remain until New Year's Day.

Once back in Champaign, I sent a box of candies to Mrs. S., with a note thanking her for her hospitality. In the following days, Dorothy wrote me several letters, in which she tried to make it clear that although over the past weekend our love may have *seemed* unproblematic and our marriage a sure thing, she wasn't at all certain of it any longer. Faced with so many of life's complexities at once, she wondered whether she could grow up fast enough to handle them all.

> My mother thinks that you're quite the finest young man and that you ought to be warned about what an awful lass I am, because she won't let you be hurt by her. When your chocolates arrived, my mother proudly said, "Look, Dorothy, that's what I got from my son-in-law today!" She wants me to start buying a trousseau, but I still haven't committed myself.
>
> Perhaps I shouldn't say it, but knowing definitely where you are going to teach would facilitate much in the way of an answer to one rather pertinent question before us. You see, Gun, I don't want just to be engaged. I want to know that we could and would be married very soon. Is that so wrong of me?
>
> Darling, I'm completely happy with you, but I'm always troubled by so many things about you of which I have no knowledge. I want to know you so much better. I'd like to know you more as a person. One reason why I don't really know you is that we've had so few unamorous moments together.

In her next letters Dorothy revealed to me her most troubling emotional travail: domination of her life by her mother and her struggle to become an autonomous person. Not only did I—motherless since I was ten—have little appreciation of this problem, but as Mrs. S.'s favorite prospective son-in-law, I had become part of the problem rather than part of the solution.

There had been a row about Dorothy and her brother not wanting to go to Chanukah services with their parents, and then her mother had a conniption fit about some unspecified matter relating to Tony (and, I supposed, myself). At the height of their brouhaha, Mrs. S. became very ill, and at times, Dorothy was dreadfully worried.

> It seems terribly unfair to me that I'm being considered the unique cause of my mother's illness, although no doubt her aggravating herself over me did have a lot to do with it. What a way to hold one's grown children in check! In case you're worried, your dear mother-in-law is feeling much better meanwhile, thank you. I've always had to give in to my parents because they manipulated circumstances to twinge my conscience. If I really were as inconsiderate and thoughtless a person as my mother says I am, then you'd be well rid of me. Actually, you probably would be anyhow.

I didn't give Dorothy the backing for which she was asking. I thought it wasn't my place to come between mother and daughter, all the more because I liked the person of whom I had come to think as my mother-in-law. Motherless myself, I thought Dorothy was lucky to have such a devoted parent who seemed to care so much about her welfare. Tony, by contrast, who had good reason to dislike Mrs. S., undoubtedly supported Dorothy in her efforts to throw off the yoke of maternal domination.

> My mother wants you to know that she is very much looking forward to your staying with us in Winnetka over the Christmas vacations. But I don't want you to show up here until December 21st, because the 20th, the day on which you were supposed to come up from Champaign, will be Tony's last night in Chicago. And, by the way, should I make what might seem like a rash decision, it would actually have been based on a lot of premeditated thought. People change, and so, when I told you about having once been in love long ago with a very immature man, I forgot to take into account that he too might have become more mature. And if things work out differently from what we had anticipated, then it'll all have been due to some of the wisdom that you rather innocently and yet purposely gave me.

This ominous letter didn't reach me until December 18, two days before I was going to leave for Winnetka. It finally dawned on me that I might lose Dorothy to Tony. But then he was about to leave Chicago, and I could work hard over the Christmas holidays at trying to keep her.

Just to confirm that everything was still all right, I called Winnetka the day before I was going to head for the City. Dr. S. answered.

"No, Dorothy isn't at home, Gunther. I don't know just where she is, but she ought to be back sometime in the evening."

"Am I still expected in Winnetka, as arranged?"

"Of course! What are you talking about? Mrs. S. and I are very much looking forward to seeing you. I'll ask Dorothy to call you as soon as she gets home."

Dorothy didn't call, but in the morning I left for Chicago anyhow. When I got to her house, suitcase in hand, Mrs. S. opened the door, looking about twenty years older than when I had last seen her two weeks ago.

"Gunther! Don't you know?"

"Don't I know what?"

"Dorothy was married yesterday!"

I was totally flabbergasted, dumbfounded, speechless. It wasn't that I hadn't begun to fear during the last few days that I might lose Dorothy to Tony. But I never imagined that this could happen so suddenly, just overnight, without any prior announcement. When I recovered my speech, I asked,

"Why didn't Dr. S. tell me when I spoke to him on the phone yesterday?"

"Because *he* didn't know! Dorothy called last night from Union Station. She and Tony were married earlier in the day and got on the Super Chief for Los Angeles. Probably spent their wedding night in a Pullman lower berth. Day before yesterday, she smuggled out a suitcase. Left home for her wedding in street clothes, toting just a big handbag."

In between sobs, Mrs. S. was asking why God was punishing her with such an unruly, ungrateful daughter.

"Her whole aim in life is to spite me. Only spite could have made her marry this Polack jerk, when she could have had a prince like you, Gunther!"

"You're too hard on Dorothy! She is a wonderful girl, which is why I love her. I haven't met Tony, but he can't be as bad as all that, if she's been in love with him for all those years."

"But he's a goy! We haven't got anything in common with him. How's she going to raise her kids? As Jesus-lovers, I suppose."

"You wouldn't have been much better off with me. As I said to the rabbi at your dinner the other night, I don't think much of Judaism myself and haven't been inside a temple since my bar mitzvah."

I went back to Champaign on the afternoon train. A letter from Dorothy was waiting for me at the TEP house. As I tore open the envelope, the gold chain with my TEP fraternity pin pendant fell out.

Gun my dear,

I rather think you are one of the few people that I feel quite keenly about in regard to what I am doing. I can't quite understand yet how we came to be so very close in the short interval of time in which we really knew each other. I recall the evening that I mentioned that possibly the way we felt could be attributed to the fact that we wanted, both of us, so very much to fall in love. I don't know just why we felt that way, but we truly did, and almost did. Or perhaps we really did. It seemed very beautiful, the way we felt together, and altho' it wasn't a feeling of blissful happiness, it was terribly exciting and new. I don't think I really understood it. Sometimes I imagine I was even a little afraid of feeling that way.

So it seems now that what I'm about to do will seem quite inconceivable, but yes, I am going back with him. We'll be married out at the chapel at the University of Chicago, because like the couple of romantics that we are, it'll have to be as sacred and beautiful as two people with confused faiths can somehow find for each other. Gun, I've loved him for such a long time in such a very sweet and happy way, and it's a wonderful way to be. It's charming in its simplicity, but somehow stronger than all the most thrilling escapades or romances that I'd ever encountered.

I'd have married you, Gun, 'cause my folks favored you terribly. It seems so very ironical when their chief objection to him was his religion. What I feel for you is so completely different, but you outclassed everyone, it seemed. Funny, but all the things I told you that Sunday came tumbling through my mind. I even told you that very day how it had been and how I had changed, but it somehow never occurred to me that he would change too. And so you see, it completed the picture that I was searching for. I was so very confused, remember?

Oh Gunther, it makes me feel so very badly that I even called you that night. For if I hadn't called, then none of this would have happened, and I shouldn't have disappointed you this way. And yet, are you really sorry it happened? I wish I could better tell you just what you lent me of you, what type of person you've helped me to become, just a little. Oddly enough I rather think it was you who gave me the courage of my convictions and the strength to carry them out.

I've never in my life made a decision that my parents didn't mull over, thoroughly deliberated, and digested for me, and so in a sense it's been like coming into one's sense of being. Gun, this just must be the one thing I decide by myself, and I wish so terribly that they won't be hurt by my decision. I wished that you might have talked to them for me, but it's too much to even think of asking. I just know that they respected your ideas so very much. And yet, you really couldn't tell them anything. If they truly love me, and I have made the right choice, in time everything will be right once again. Time is a great healer, they tell me. And it will itself tell many things for both you and

Your Dorothy

Devastated by Dorothy's farewell letter, I fell into a long stupor. Perhaps Mrs. S. had been right in suspecting that her daughter had used me merely as a stooge in a fiendish plot to spite her. To punish Dorothy for her crime, I considered killing myself. But this seemed impractical. And so I drafted a pathetico-petulant letter. I declared that although I had truly loved her as I had never loved any woman before, she shouldn't get the idea that she had broken my heart or that she would be the only one I would ever love. This more moderate stratagem also turned out to be impractical, since I didn't know where to send my letter. Moreover, I realized only then that I didn't even know Dorothy's new last name. She had never told me, and I had never asked for Tony's.

The truth was that she *had* broken my heart, since I was still in the thrall of the moral code laid down in the 1930s by the Hays Office for the Hollywood movie industry. According to that code, romance (or life as we know it) ends with marriage, and there is no such thing as divorce. So, I was stunned when, a few weeks after her elopement, Dorothy called me at the TEP house.

"Guess what, Gun! Tony and I are in Champaign! They gave him a generous leave before he's got to report to his new duty station in California. To reward him for the many dangerous combat missions he flew in the European Theater."

"Holy Smokes! Sure didn't expect to hear from you. What are you doing out here? By the way, congratulations on your marriage. I guess the best man won!"

"No need to be so sarcastic, Gun! We went back to Chicago to make peace with Mom and sashayed down to Champaign for a couple of days to revisit the scene of our college romance. Howzabout having lunch with us at the Illini Union?"

Captain Tony turned out to be a personable guy with a gift for gab, tall and handsome just as Dorothy had said he was. He and I got on famously at lunch, talking shop about aircraft tires and synthetic rubber, leaving Dorothy mostly out of the conversation. I couldn't figure out why she had wanted to arrange this meeting. All I could think of was that she wanted to prove to both the man she married and the man she nearly married that the other fellow was not a jerk. A few months later, I got a picture postcard from her from Carmel, California.

> Dear Gun. You've got to see the West! We're discharged and considering living out here. Will probably select LA.
>
> Love, Dorothy

It would be many years until my love for Dorothy totally subsided. Nor, as I eventually learned when I met her a few times after I did see the West and moved to California myself, had the torch she carried for me gone out totally after her marriage to Tony. But it was not until writing these pages and rereading her letters that I fathomed that it was my inability to share souls and attend to the inner life of others that had caused the shipwreck of my first great romance.

Upgrading the Dream

I STEPPED OFF THE ARLBERG EXPRESS AT KITZBÜHEL, in the French occupation zone of Austria, on the morning of March 15, 1947. Situated in the eastern Tyrol, about half way between Salzburg and Innsbruck, Kitzbühel had been one of Europe's most fashionable skiing resorts in the prewar years, on a par with St. Moritz and Cortina d'Ampezzo. Its Grand Hotel was famous in the 1930s as the venue of the illicit romance of Edward, Prince of Wales, and Mrs. Simpson. The Hahnenkamm Ridge, from which most of the ski runs course down into the Kitzbühel Valley, rises to the southwest. There was plenty of snow and brilliant sunshine: ideal skiing conditions.

It gave me an exhilarating feeling to arrive in this gorgeous place where I was about to recapture another of the prized childhood amenities of which I had been deprived when I fled from Hitler. In fact, it was going to be better than anything I had enjoyed as a child: I was about to enact my longstanding fantasy of skiing in the Alps.

I took the cable car up to Hahnenkamm and hiked for about half an hour on a snow-covered trail from the upper cable car terminal to the Ehrenbachhöhe, where the French officers' skiing lodge was located. The view from the lodge's sun deck provided a spectacular panorama of snow-covered high peaks, which included, far to the southeast, the Grossglockner, at 12,500 feet Austria's highest. The lodge supplied me with ill-fitting boots and oversized, white U.S. Army hickory skis—surplus gear from the disbanded wartime ski troop battalions. I was tired from the overnight train ride and not yet adapted to the altitude of the Ehrenbachhöhe.

Although anxious to get on with skiing, I figured—wisely, as it turned out—that I was in no condition to try any of the runs down into the valley from the Hahnenkamm. So I just worked the practice slopes near the lodge and discovered that I had lost all my skills in the ten years that I hadn't been on skis. I couldn't control my downhill speed by any means other than snowplowing, not least because I found the oversized GI skis hard to handle. Wobbly and lacking confidence, I took one spill after another. My ankles hurt, and within a couple of hours I had broken off the tip of one of my skis. But I persevered; I got a replacement for the broken ski, returned to the slope, and by the end of the day, I had regained my capacity to perform the most elementary and inelegant of skiing maneuvers, the stem turn.

Next morning, I thought I was ready for some serious skiing. The conditions were perfect—powder snow on a hard base—and as I succeeded in getting more and more speed control through stem turns, I gained confidence. Yet I still took a lot of spills. The expert run headed more or less straight down from the Hahnenkamm along the cable car route, whose quasi-vertical descent was out of the question for me. I was barely able to negotiate the easiest run, which wound up in Kirchberg, a village about three miles from the lower cable car terminal. But I was glowing with pride that I managed to get down to the valley at all.

On my last cable car trip back up the Hahnenkamm before lunch, I stood next to a sun-tanned, svelte blonde. Holy Smokes! She was the most stunning woman I had seen in Europe. The perfectly molded, regular features of her beautiful face matched those of the dream woman of my fantasy, as the face of the undoubtedly very attractive (and no less blonde) Hildegard did not. I overheard her saying something in German to another skier in the gondola, and her Berlin accent revved up my heartbeat even beyond the high pitch to which her beauty had already boosted it. Maybe she was a movie star, which would explain how she managed to get to Kitzbühel from Berlin for a skiing holiday. She smiled at me, as Hildegard had done when I bumped into her at the door of the Admiralspalast. I used the by-now-proven gambit for picking up a Berolina by asking an inanity in Berlinesque German:

"Aren't the snow conditions terrific?"

Being unexpectedly addressed by an Ami in her own idiom did not fail to startle also this gorgeous specimen.

"You bet! They certainly are. The powder is fabulous. How come you talk like a Berliner? You're an Ami, aren't you?"

"Yeah. I'm an Ami. Actually I'm an Ami from Berlin."

Just then, the cable car got to the upper terminal.

"Auf Wiedersehen, Fräulein. I'm heading for lunch at the Ehrenbachhöhe lodge. But I'll be back on the slopes in the afternoon."

She didn't seem interested in my plans for the afternoon and simply took off on the expert run, schussing down like an Olympic champion, wiggling her hips for an occasional turn. I was angry with myself: You're a real schnook! Why didn't you follow her down, even if you risked breaking your neck? So what if you missed lunch at the lodge?

During the hike from the cable car terminal to the lodge, I rationalized away my self-directed pique. So, OK, I'm a hopeless schnook, letting this fabulous creature slip through my fingers. But then, I haven't done all that badly with Hildegard, have I? She's something really special too! And in any case, I'll be leaving for the States in a few weeks, and I haven't sorted out yet what to do with her. The last thing that I need is another romantic entanglement. And more likely than not, Gorgeous Berolina already has a man and wouldn't give me the time of day. Why don't I just concentrate on the fabulous skiing and focus my erotic fantasies on Hildegard?

After lunch I rushed back on the slopes, pretending to myself that my haste was merely due to my wanting to get in as many afternoon runs as possible. To my deep disappointment, there was no sign of Gorgeous Berolina on the slopes or the cable car. Well, that's that, I thought, she's gone. So what? I'll make just one more run down and then head for the lodge for an *après-ski apéritif.*

It's a maxim of prudent skiing that one should quit *before* one makes the last run of the day, before one gets too tired and overconfident. My punishment for disregarding this maxim was not long in coming: I lost control when I hit deep powder, and fell, head first. Unable to extract my head from the snow, I thought I was done for. But then I felt someone yank my shoulder and pull me up before I suffocated. My savior was none other than Gorgeous Berolina!

"Are you OK? I saw you take off from the cable car terminal. You were so wobbly on your oversize boards that I wondered how you're ever going to make it down to the valley in one piece. Figured I'd better follow you down, just in case."

By the time I got back up on my feet, my soul was overflowing with a never-before-felt wild meld of gratitude, dependence, romantic arousal, and, well, love. Here she was: the beautiful dream woman of my fantasy! The passion that I hadn't been able to conjure up for Hildegard in months of trying had overwhelmed me in seconds. But all I could say was

"I'm OK. Thank you for saving me, Fräulein."

We skied down together, and when we got to the valley, introduced ourselves. Again I used the formula I had tried out on Hildegard.

"Despite my U.S. Army ski troop outfit, I'm really a civilian. An American scientist. I work for the military government."

Her name was Lore T.

"Why did you say in the cable car that you're an Ami from Berlin, Mr. Stent?"

"Because I *am* one, that's why. Born in Berlin, on the green banks of the Spree, but grown up in Chicago, on the shore of blue Lake Michigan. You're a Berliner too, aren't you, Fräulein T.?"

"Yes, Mr. Stent. Happy to meet a fellow-Spree-Athenian. Don't run into too many up here any longer, especially not any who are Amis."

Unlike Hildegard, however, she didn't ask whether I left Berlin because of Hitler.

"How'd you manage to get out of Berlin to come to Kitzbühel for the skiing season?" I asked her.

She laughed.

"I didn't come here for the skiing season, exactly. I've been hanging out in Kitzbühel for three years now. I came here in 1944, just like Leni Riefenstahl, to get out from under all those bombs you Amis were dropping on our hometown. When the Amis came to Kitzbühel, they arrested Leni for making propaganda movies for Hitler. Took her back to Germany. Me, they left alone. I wasn't keen on going back to Berlin, to those awful living conditions there. Nothing to eat. Ivans all over the place. So I just stayed on in beautiful Kitzbühel, a little leftover from the war."

"May I invite you for an *apéritif* before I take the last cable car back up to the Ehrenbachhöhe?"

"Thanks. But I can't. It's already way past when I'm expected at home. But why don't we meet for a bit of afternoon skiing tomorrow afternoon at two, at the upper cable car terminal?"

Trying another line that had worked with Hildegard:

"Fantastic! Wild horses couldn't keep me from being there tomorrow at two!"

Unlike Hildegard, Lore didn't laugh, possibly because (as I would later discover) she had lived in New York before *I* ever got there and probably knew I was merely translating an American cliché.

"Could you give me your address, Fräulein T., so that I don't have to depend once again on Divine Providence to meet you in case we fail to connect tomorrow?"

Lore did laugh at my mention of Divine Providence. "I live at the Postkutsche Hotel."

I headed back to the lodge in a superagitated state. Is it possible that this fabulous beauty might return my love and that I could make her mine? Most likely, she won't even show up for our date tomorrow afternoon. She *had* followed me down from the Hahnenkamm, but probably she was, just as she had said, moved merely by Good Samaritan compassion. She *did* tell me that she was living at the Postkutsche; that ought to count for something! Who, by the way, was expecting her at the Postkutsche? Let's hope that it wasn't her boyfriend! What with my having only a few days left in Kitzbühel, I better work fast, if I do manage to meet her tomorrow afternoon. But how the hell am I going to work fast?

Early next morning, I went on a half-day ski-mountaineering climb up Steinbergkegel Mountain, led by the lodge's ski instructor, Fritz. I was too tense to enjoy this excursion, because I was terrified that we might not get back to the lodge in time for my date with Lore. But we did return soon enough for me to wolf down lunch and ski to the cable car terminal half an hour before the appointed hour. While waiting, I developed theories to account for why she wasn't showing up, not giving much credence to the most plausible, namely that it was still too early. She stepped out of the gondola that pulled into the terminal at exactly 2 P.M., just as gorgeous as I had remembered her. The angel-savior with whom I had fallen in love had come back to me.

We took off our gloves for an enthusiastic handshake and mutually assured each other once more of the delight of having run into a fellow Berlin sophisticate among the Tyrolean rubes. Out of consideration for my incompetent skiing, Lore went down an easy run, slowing the pace of her descent by making one elegant, hip-swinging turn after another. Trying to impress her that I wasn't the total hacker she took me for, I started to speed up recklessly. Pride-before-fall-wise retribution was not long in coming: I took a terrible spill just in front of her. This time my head wasn't buried in the snow. I merely felt a sharp pain in my right ankle.

After I managed to stand up with Lore's help, the ankle hurt even more. There was no longer any question of my controlling the skis. Short of the even greater disgrace of her having to call the ski patrol to come and carry me off on their rescue sled, the only way for me to get down was to put on the pitiful spectacle of gliding tremulously and orthogonally to the slope, reversing direction every once in a while by a kick turn. It took us nearly an hour to reach the valley. The physical pain of my hurt ankle was nothing compared to the spiritual pain of my humiliated self. But Lore didn't seem to mind having wasted an afternoon of prime skiing on me.

"Sit down and take off your right boot and sock. I'll squeeze your foot at different places, and you tell me where squeezing hurts."

The erotic arousal brought on by her hands squeezing my bare skin was dulling the pain so much that I gave my answers more or less at random.

"I think you've sprained a ligament. Walking on your foot is going to do you more good than just sitting down and taking it easy."

So we parked our skis at the lower cable car terminal and went for a stroll (or hobble, in my case) through Kitzbühel, which had survived the war intact. Lore seemed to know every shop, hotel, and restaurant, and, judging by the many passers-by who called out *"Grüss Gott, Frau Doktor T."* as we passed them, she seemed to be known by everyone in town.

Surprised by her being called "Frau Doktor" rather than "Fräulein" and, bearing in mind her expert diagnosis of my sprained ankle, I asked Lore whether she was an M.D. She seemed to find the idea of being taken for a medic hilarious.

"Do I look like an academic bluestocking? Never set foot in a university. They call me 'Frau Doktor' because my husband is a Ph.D. physicist."

"You live with him at the Postkutsche?"

"No. I'm staying there with my mother and my little son. Haven't seen my husband in years and haven't any idea where he is. He's a rocket expert. He was with the secret project at Peenemünde during the war and, like his boss, Wernher von Braun, grabbed by the Amis' 'Operation Paperclip' once the war was over. They probably took him to the States."

This was the only time Lore ever mentioned her husband, and I can't remember that she ever told me his first name. As we passed the parish church of St. Andrew, built in the fifteenth century and done over in Austrian baroque in the eighteenth, a wedding party dressed in Tyrolean costumes came out, got into horse-drawn sleighs, and drove off, sleigh-bells tingling, to the wedding feast. The wedding party made me fantasize marrying Frau Doktor T. in St. Andrew's—not necessarily in Tyrolean costumes—and taking my beautiful bride back with me to the States on the *Noordam*. She being the most stunning woman on board, the captain would ask us to sit at his table. My fantasy didn't take into account, of course, that, even if Lore would have me, the absent Herr Doktor T. might present a bit of a legal obstacle for our immediate marriage in Kitzbühel.

We went into the Grand Hotel, taken over as a French officers' billet and casino, for afternoon tea. The hotel's former splendor—mirrored, crystal-chandeliered, fin-de-siècle-furnished salons—was still evident. Obviously, this was not Lore's first visit to the Grand, since the concierge and many of the waiters greeted her as Frau Doktor T. As we settled down in the salon to our tea, we started calling each other "Lore" and "Günter," albeit still using the formal, third-person plural form of address, *Sie*.

"Tell me, Günter, how did a Berliner get to be an Ami from Chicago?"

By now, more seasoned in my dealings with German women, I was less reticent from the start to disclose to Lore my origins as a Berlin-born Jew. I gave Lore a quick synopsis of my experiences on Kristallnacht and at the Green Frontier, at the end of which she commented,

"It's really horrible what the Nazis did to the Jews. I had no idea about it at the time, of course, and now it makes me ashamed to be a German."

I was relieved that the woman with whom I had just fallen in love was saying acceptable things about the Nazis and the Jews. All the same, unlike Nazi-tainted Hildegard who left me no doubt that she was sincerely, if retroactively, outraged on the Jews' behalf, I wondered whether Lore might not be merely mouthing a politically correct opinion. But then Lore's whole way of talking was much more reserved than Hildegard's extroverted, upfront conversation. I didn't ask Lore how old she was. But I reckoned that she must be in her late twenties, taking into account that she had a son by her long-absent husband. For me, her seniority only added to her appeal. But I did ask her about her situation in Kitzbühel.

"I expect to be kicked out of the Tyrol any day. Can you imagine? Those hypocritical Austrians are claiming that they were the first victims of German aggression! Why don't they apologize to us Germans for saddling us with their favorite native son, Adolf? Didn't they give him a tremendous welcome home in Vienna in 1938 after the Anschluss? I've been able to stay here this long only because of my good connections with the French military."

I suspected that these good connections consisted of a liaison with the commanding general, but, perversely, this suspicion aroused my pride rather than my jealousy. She politely declined my offer to take her home to the Postkutsche:

"With your bum ankle, Günter, you'd better get yourself back to the lodge on the cable car. Here's my phone number; give me a call when you feel up to hitting the slopes again."

I tried to sort out my turbulent emotions on the way up to the Hahnenkamm. I was desperately in love with Lore. Despite unmistakable signs that she was not put off by my ineptitude as a skier and seemed to enjoy my company, I doubted that she could anywhere near return my feelings. And what should I do about Hildegard, whom I had considered marrying, but whom, as I now realized, I would never love as I loved Lore? Obviously, it was all the more urgent that Hildegard go back to Berlin: How could I go on living with her when all my thoughts are now focused on another woman? By the way, how *is* Hildegard doing in our apartment? Shouldn't I find out how her operation came off?

Back at the lodge, I asked the concierge to try to ring the phone I shared with the Thornhills at Kastell Strasse 19. It would have been out of the question to call Durlach via the civilian lines from a Tyrolean mountaintop. But since both the Ehrenbachhöhe and the FIAT billets were hooked up to the Allied telecommunications network, it didn't take long until I had Mrs. Thornhill on the line.

"No, Hildegard isn't at the apartment at the moment; maybe she's over at the Breyers. Yes, she seems fine, at least as far as I can make out."

"Please give Hildegard my love, Mrs. Thornhill, and let her know that I'll try to call her tomorrow at just about this same time."

Next morning, Fritz, the lodge's ski instructor, wrapped my sprained ankle in an elastic bandage, which relieved the pain a lot. But I had to stay off the slopes for at least a day. Marooned high above Kitzbühel, what was I going to do about seeing Lore? Skiing dates were out of the question. In any case, how do you romance a woman on skis? I've got to find some place where I can sit down next to Lore, put my arms around her, and kiss her. Or take her dancing and put my lips on hers while holding her tight on the dance floor. The Grand Hotel would be just such a place, but the last cable car leaves Kitzbühel for the Hahnenkamm at about 5 P.M. So, I'd be stuck in the valley at the end of our romantic evening, unable to get back up to the lodge until next morning.

I called Lore to let her know that I'd spend Saturday night—my last at Kitzbühel—at the Grand, and leave for Karlsruhe on Sunday afternoon. She agreed to have dinner with me at the hotel Saturday night. I was unspeakably excited by the prospect of spending an evening with my dream woman.

Hildegard was home, waiting for my call, when I phoned our apartment from the lodge.

"Everything is just fine at Kastell Strasse 19, Günterlein, except that I'm missing you something awful. I'm going to have my operation tomorrow. There's no need for you to worry. It's really nothing. In any case, the Breyers are a big help. Why don't you call again on Friday? By then I'll be sure to be back in the apartment."

"I'm also very much looking forward to seeing you on Monday, Hildelein."

This was not a totally hypocritical lie. I had begun to feel guilty about my cold-blooded abandonment of Hildegard and wanted to do something—I didn't know just what—to make it up to her on my return. Taking her reassuring message at face value that everything was going well with her, I didn't ask for any details about her medical condition or her state of mind. I didn't worry about Hildegard, except about how to extract myself decently from my entanglement with her and let me devote my full attention to the project of making Lore my own.

When I telephoned Hildegard again on Saturday morning, she reported,

"The operation went well. I'm just taking it easy, lying in bed, catching up on my reading. No need to worry about me, Günterlein. I'll be perfectly OK."

Early Saturday evening, Lore entered the Grand's lobby, wrapped in a black Persian lamb fur, hands stuck in a matching fur muff, infinitely more snazzy than poor Hildegard in her olive-drab overcoat, home-sewn from GI woolen blankets. Under her fur, Lore wore an exquisite silk dress, and, adorned by a golden necklace and golden earrings, she presented a fabulous blend of beauty and elegance. I was speechlessly bewitched by Lore's face and found it hard to believe that this divine apparition had come to join me. As we entered the dining room, I thought of the Prince of Wales and Mrs. Simpson. How had they carried off their affair at the Grand? Had they booked separate rooms? Had they taken their meals together in public in this very hall, or had they dined in private, upstairs in a *chambre séparée*?

The maître d'hôtel greeted Lore, whom he obviously knew, addressing her as "*Gnädige Frau.*" He called me "*Herr Doktor*," knowing that even if I weren't entitled to this honorific, I would be pleased with it.

"I'm not yet quite a Herr Doktor yet, Lore, but I expect to be one before the next year's out. I'm going back to the States within a month, to finish my doctoral thesis research at the University of Illinois. A few days ago, I was still looking forward to going home, but now that I met you, I wish I could stay."

Lore smiled at this first, feeble attempt of mine to declare my feelings for her. I told her about my prewar childhood in Berlin, my emigration to the States, my life with Claire in Chicago, my studies at the University of Illinois, and my present adventures at FIAT, especially my recent visits to our mutual home town.

I didn't ask Lore any questions about herself—not about her family, not about her life in Berlin as a young woman and during the war, not about her son, and not about what, in view of my hope of marrying her, ought to have been Topic A, her relations with her absent rocketeer-husband. From what little she volunteered about herself unasked (as well as from her way of dressing and her demeanor) it was apparent that she came from a moneyed, upper-class social background. I never found out her father's occupation, but from her casual revelation that in her prewar girlhood she had spent some time on the American East Coast, I inferred that he might have been a diplomat or a business tycoon. She left me with the impression that she had never worked a day in her life.

There was a band playing in the dining room, and we danced between courses. I did hold her close to me, but feeling her body next to mine sent me into such a state of rapture that I was paradoxically inhibited from trying my projected amatory-exploratory maneuvers on the dance floor. After the music stopped, we left the Grand, to walk to the Postkutsche. As we crossed the hotel's pitch-dark garden, I was unable to bring myself to grasp the last opportunity for kissing her. We walked in near silence, through the deserted streets of the town, while I said nothing of my love for her. When we got to the Postkutsche, we merely shook hands. I was barely able to sleep. It had been a glorious evening—just being with her and holding her in my arms on the dance floor.

Sunday morning, Lore came to fetch me at the Grand for a walk to the Kitzbüheler Bad, a mineral spa, about half an hour's hike from the hotel. We didn't do much talking, except while we were sitting on a bench at the deserted spa, I said, "I can't remember when I've been happier than on this morning," which was the closest I came to expressing my love for her. "I'm happy to be with you too," she assured me. We had a farewell lunch at the Grand during which I promised that I would come back to see Lore before I left for the States.

On the long train ride back to Karlsruhe, I was in a turmoil. I simply had to get Hildegard back to Berlin as soon as possible. But should I level with her and announce that I had fallen in love with a woman I met in Kitzbühel? It wasn't likely that I could bring myself to be so cruelly honest to kind, wonderful Hildegard, with whom I had tried to fall in love. No, I would have to act as if nothing had happened and pretend that I was desolate over our parting. Once she was back in Berlin, maybe I could break her the news by letter—gradually and, as I supposed, gently.

As for Lore, I had to go and see her again as soon as possible. I doubted that I could get another leave that would provide me with official travel orders authorizing me to enter the French zone of occupied Austria. So I'd have to dope out another travel order scam. Should I try to postpone my departure for the States, even if it meant that I'd have to cross the Atlantic on a troop transport instead of the *Noordam*? That would give me more time to work out how I was going to marry Lore, if she would have me, once the problem of her absent husband had been got out of the way. And then there was her son, whom I had not even met and whom I would have to bring to the States too. Probably, with a mother like Lore, he's a terrific kid.

Back in Durlach, I found Hildegard in tears. A quartermaster team had come to inspect my apartment on Saturday, just after I called, and found her in bed, recovering from the operation. They asked her what in hell did she think she was doing in a U.S. Army billet. Hildegard explained that she had come from Berlin to visit Mr. Stent. He wasn't home right now. She's looking after his place while he is on a trip to Austria, but he'll be back any day now. The officer in charge of the inspection team yelled at her that her staying in this house was totally illegal: Both she and Mr. Stent were going to be in plenty of trouble. She should tell Mr. Stent to report on the double to Major Mountain, the FIAT executive officer, the minute he shows up.

"I'm not afraid for myself, Günterlein," Hildegard sobbed. "What can they do to me that would be anywhere near as bad as what I've gone through during the war and after? I heard that the chow in Ami prisons is better than what you get on your Berlin food rations. But I'm afraid for your sake. Maybe your whole career will be ruined if the punishment goes on your record."

Callously, I accepted at face value her equating my danger with hers and told her not to worry; my career wasn't going to be ruined by this. Actually, that possibility *had* occurred to me too, and the tone in which I dismissed it before Hildegard must have lacked conviction. This was not exactly the kind of homecoming with which I had reckoned.

Next morning, I reported to Major Mountain at his office in the FIAT headquarters building in downtown Karlsruhe. He was a tall, handsome man in his mid-thirties, whom I had seen a few times at the FIAT officers' club. A framed picture of a good-looking, youngish woman with two little children—presumably his family—stood on the desk at which he was seated. The major stared at me coldly, with the look of disdain to which Army officers typically treated civilians decked out in their "simulated" officers' uniforms. He did ask me to sit down, at least, a courtesy that he wouldn't have extended to an enlisted man about to be called on the carpet.

"Keeping a German woman with you in your billet is a damned serious violation of regulations, Mr. Stent. You may be court-martialed. It'll be up to Colonel Osborne to decide whether formal charges will be made against you. Probably the best you can hope for is a stiff fine, and being shipped home immediately, before the expiration of your contract. Anyhow, get that woman out of your billet immediately."

I found some solace in this awful denouement of our brief idyllic life as a young suburban couple, because it seemed to justify me vis-à-vis Hildegard for having pressured her to return to Berlin. I went to see Freddy to ask his advice. He told me that he and most of my colleagues already knew all about the mess I was in. Freddy thought that half the FIAT guys in Durlach had a Kraut woman living with them. To prove his uncompromising vigilance, Major Mountain wanted to make an example of me, a mere civilian rather than a brother-officer. Freddy supposed that Colonel Osborne might be more lenient, especially now that he was about to lose his command anyhow.

I reported to Hildegard what Major Mountain said.

"I'm not surprised, Günterlein, that he ordered you to get rid of me. Unfortunately, I haven't recovered from the operation as fast as I hoped. The doctor says I'm not really well enough yet to handle the trip home. Don't worry though. In a few days, I'll be up to it. The Breyers offered to put me up meanwhile, until I'm in shape to take off for Berlin."

Tremendously relieved by this *deus ex machina* solution of our problem, I moved Hildegard and her few belongings over to the Breyers that same evening. Back, alone at the apartment, I wrote to Lore, to let her know that

I made it to Karlsruhe and that I was thinking of her all the time, mulling over every precious moment we spent together.

Two days later, I was summoned to report to Colonel Osborne. While cooling my heels in his outer office, I was as frightened and nervous as when, eleven years old, I was sitting outside the Rektor's office at the Bismarck Gymnasium, waiting to be punished for having acted up in class. Seated behind a big desk in front of a partially furled Old Glory on a flag stand, mustache neatly clipped, tunic campaign-ribbon-bedecked, lapels field-artillery-emblemed, and epaulets silver-eagled, the colonel cut an imposing figure. Motioning me to sit down, stern but more friendly-mkened than Major Mountain, he came right to the point:

"Did you know, Mr. Stent, that your German girlfriend came to see me yesterday?"

I was astounded. Surely, Hildegard hadn't gone to see the colonel; she couldn't have!

"No, Sir. I had no idea she was going to do any such thing. In fact, I would have tried my best to stop her, had I known she meant to trouble you with this matter."

A feeling of impotent embarrassment welled up in me: How did she have the courage to barge into the FIAT chief's office, when I couldn't have worked up enough nerve to do such a thing in a million years?

"She was here, Mr. Stent, and asked me not to punish you. It had been all her fault, she said. You didn't even ask her to stay with you. She simply came down to Karlsruhe from Berlin and moved into your billet while you were out of town. So if anybody is to be punished, it ought to be her. Mr. Stent is just a nice guy who wanted to help her out. She told me that she has moved out of your apartment and will be going home to Berlin as soon as possible, back to her job at the theater."

"That's correct, Sir. She is staying with a German family in Durlach and plans to get on the first interzonal train to Berlin for which she can get a reservation."

"I pointed out to your girlfriend that what you did was a serious infraction of the rules. You, Mr. Stent, ought to have been aware of these rules. So you, rather than her, are responsible for having broken them. However, in view of the unusual circumstances of this case, I decided not to have you court-martialed. All the same, I can't just let you go scot-free. I'll think it over just what kind of punishment you should be given. There was another thing your girlfriend asked me for. She wanted me to promise that I won't tell you that she came to see me. That, I flatly refused. Damn it, man, you ought to know why you got off so easy!"

Then Osborne spoke to me man to man:

"You're lucky to have such a nice girlfriend, who is not only pretty but a good woman as well. I see from your file that you'll be shipping out within a month; so I'm asking you to promise not to take any more women into your billet until you leave. If you do, I'll throw the book at you. You'll hear more from me about this matter in a few days."

Enormously relieved, I promised the colonel to comply with his orders and thanked him for his consideration. I rushed over to the Breyers to let Hildegard know the outcome of my interview with the colonel.

"I was speechless when he let on that you came to see him, like a mommy pleading for her naughty little boy with the Rektor of the Gymnasium. If I'd known, I certainly wouldn't have let you offer yourself for punishment in my place. But you sure made quite an impression on the colonel, Hildelein!"

Hildegard was beaming: "I'd have never forgiven myself if I had ruined your career."

I was grateful to Hildegard, of course, but I didn't thank her properly for what she had done for me, except giving her a little kiss. I never did hear from the colonel about any punishment, and I don't think any entry was even made in my personnel file.

On March 28, my twenty-third birthday, Hildegard announced, to my great relief, that she now felt sufficiently strong for the trip back to Berlin. So we made plans to go to Frankfurt together on the following day, where, as I had hoped to do two weeks ago, I would put her on the express to Hanover. In Hanover, she'd change to the interzonal train to Berlin. I wasn't able to drive her to Frankfurt, since—with my field work in Leverkusen terminated and my being assigned to the documents branch at headquarters—I had to turn in my jeep.

There was a birthday party for me at the Breyers that evening, for which Frau Breyer had baked a cake and I supplied coffee and booze from the PX. I danced with Hildegard and sang a final *"Je ne suis pas curieux"* into her ear. After the party, I stayed over at the Breyers, the last time Hildegard and I slept together. It was a passionate night, not unlike the one we spent in Berlin on New Year's Eve, making it the happiest of all my twenty-three birthdays. It all seemed like a crazy paradox. Hildegard could give me such great happiness and contentment, and yet I could hardly wait to get rid of her.

Next morning we hitched a ride to Frankfurt on a U.S. Army truck. I had given her a packet of postage stamps and a book with Karl May adventure stories to take to home to her little brother. We got to Frankfurt a couple of hours before the departure of her train, which we spent sitting on a bench on the embankment of the River Main, not far from Central Station. I put my arm around her, but neither of us said very much.

When the time came to get on the train, I put Hildegard in the virtually empty special coach reserved for Allied personnel that was hitched to the train's tail end, ignoring the German railway conductor who was guarding the door to keep out the Krauts. All the other coaches of the express were jam-packed with people. I stowed Hildegard's suitcase in the luggage rack and installed her in a window seat. As we kissed good-bye, Hildegard was crying, and I was not far from tears myself. I stood on the platform, talking to Hildegard, who leaned out of the window of her compartment. When the train pulled out, we waved with our handkerchiefs until we lost sight of each other.

My feelings were all mixed up. I was sad to part with this wonderful woman to whom, in the four months since we first met, I had grown closer than to any other previous girlfriend, including even my sometime fiancée, Dorothy S. Maybe I was making a terrible mistake in letting her go, instead of taking her back with me to the States. I was racked by guilt pangs over my hurry to send my loyal Hildegard back in cold blood to the horrible living conditions in Berlin, instead of trying my best to arrange for her to stay for as long as possible in the moderate comfort I could provide for her in Durlach.

Yet I was glad that Hildegard was gone and that I had been relieved of my day-to-day responsibility for her. No longer fenced in by her presence, my freedom of action had been restored, which had come to seem ever more precious to me since I had fallen in love with Lore. In fact, I had trouble focusing my thoughts on any subject other than Lore.

On my return to Durlach, I called on the Breyers, to let them know that I had put Hildegard safely on the train to Berlin. I casually asked them whether, by any chance, Hildegard had told them, as she had not told me, what her minor operation had actually been for. They said that the operation was an abortion. Hildegard and I had conceived a child on New Year's Eve.

Finessing Mother Nature

BY THE TIME OF MY GRADUATION from Illinois in mid-January 1945, my two dream schools, Caltech and Berkeley, had turned down my application for admission and financial support for graduate study toward a Ph.D. in physical chemistry. So, I resigned myself to staying on at Illinois and accepted the Doc's proposal that I do a Ph.D. under his direction as a research assistant on the War Production Board's synthetic rubber research program.

The research assistantship carried a monthly stipend of $180. Even though I was disappointed that I still wasn't going to make it to California, I felt good about the steady income provided by the stipend. It was going to put an end at last to my precarious hand-to-mouth existence and its constant money worries. There were about ten Ph.D. students working under the Doc on the rubber program. All of them were deferred from the draft for rendering war service as expert civilians, except for me, rejected for military duty as physically unfit.

Synthetic rubber was invented in Germany in World War I, when the Central Powers were cut off from their overseas sources of natural rubber and faced the dire prospect of running out of tires for their military wheels. The first German ersatz rubber was called "Buna." World War I Buna tires were good enough to the keep the Imperial German army trucks rolling, but they were inferior to natural rubber tires, as well as much more expensive to produce. So Buna fell into disuse after the war, once the traditional sources of natural rubber had opened up again.

As Hitler was preparing for *his* war, Germany once more faced the prospect of being cut off from its overseas sources of rubber. I.G. Farben chemists now found that synthetic rubber could be improved by modifying the reaction mixtures in which it was produced, giving rise to a product designated as "Buna-S." Although this was better than Buna, it still didn't measure up to natural rubber. Nevertheless, to satisfy the wartime need for rubber, I.G. Farben built a large Buna-S plant in Auschwitz, as soon as the SS had established its concentration/extermination camp there, a unique opportunity to use a source of cheap labor that could be literally worked to death. My Hebrew and Jewish history teacher at the PRIWAKI, Ludwig Kuttner, was one of the untold number of victims of the Auschwitz Buna-S plant. None of the five I.G. Farben directors in charge of Buna-S production at Auschwitz was ever held accountable for his crimes: They all prospered in postwar West German chemical industry. Even the director personally responsible for the fatal beating at the plant of the Jewish lyricist, Fritz Löhner-Beda, went scot-free. Löhner-Beda's all-time hit parade songs of the Weimar Republic included "*Dein ist mein ganzes Herz*" ("Yours is my heart alone") "*Oh, Donna Clara,*" and "*Ausgerechnet Bananen*" (Yes, we have no bananas).

Soon after Pearl Harbor, the Japanese captured the most important sources of natural rubber in Malaya and the Dutch East Indies, and the Allies were facing the same potential rubber shortage as the Germans. So the U.S. War Production Board took a cue from the Germans and set up production of Buna-S in America. The primary mission of the WPB's synthetic rubber research program at Illinois was to develop procedures that would improve the quality of Buna-S, so that tires made from it would be as good as, or even better than, tires made from natural rubber.

The Doc's crew tried to devise strategies to overcome some of the flaws of Buna-S that impair its elastic properties compared to those of natural rubber, by further modifications of the reaction mixture. For the first few months, I worked with two of the Doc's more senior research assistants on that project. Although we didn't solve the practical problem, we did manage to provide some experimental evidence for a theory the Doc had worked out regarding the relative speeds with which the chemical building blocks enter the chain molecules growing in synthetic reaction mixtures. That work resulted in two scientific publications in the *Journal of the American Chemical Society*, on which my name appeared as the last of four coauthors and the Doc's as the first. Even though I realized that my intellectual contribution to these two papers was virtually nil—I just lent a helping hand to my two colleagues in doing their experiments—I was tremendously proud of having crossed the threshold of published authorship. With my name in print in a premier journal of chemistry read the world over, I felt that I now belonged to the ages.

In the late spring of 1945, the Doc assigned me a project of my own, which addressed another of Buna-S's flaws. Whereas the chain molecules of natural rubber are of uniform length, individual chain molecules of Buna-S vary greatly in length. I was to work out a method by which synthetically produced Buna-S chain molecules could be separated into a series of fractions, each of which would contain Buna-S chain molecules of uniform lengths. I decided to try separation by "thermal diffusion." This method was first developed in 1938 by K. Clusius and G. Dickel. They built a column consisting of two concentric vertical tubes, of which the outer was cooled and the inner heated, and the gap between them communicated with a top and a bottom reservoir. On filling their column with a fluid mixture of two molecular species, they found that top and bottom reservoirs became relatively enriched in the lighter and the heavier of the species, respectively. Unknown to me, and, I suppose, to everyone else in the Illinois chemistry department, at that very time, a virtual forest of Clusius-Dickel thermal diffusion columns had been raised at the Manhattan Project's secret plant at Oak Ridge, for separating ^{235}U from other uranium isotopes in the large amounts needed to fuel the atom bomb.

Perhaps because he didn't know how else he could usefully employ me, the Doc let me try my idea. To see whether there was anything to my brainstorm, I built a column, consisting of two concentric steel tubes separated by a narrow gap, of which the outer tube was cooled by cold water and the inner tube heated by hot water.

The column worked as I had hoped. On filling it with a solution of Buna-S and letting it run for a couple of days, the top reservoir contained Buna-S molecules of shorter chain length than the bottom reservoir. I then modified the column, so that I could sample the solution present also in the narrow gap, at its progressively lower elevations, from top to bottom. As I expected, downward from the top, the gap contained molecules of ever-greater chain lengths. Mathematical analysis of these data then allowed me to provide an entirely novel way of determining the length distribution of chain molecules in the original Buna-S mixture.

I was jubilant: It was my first success as an experimental scientist. I had finessed Mother Nature and made her do my bidding! Perhaps my device would become known as the "Stent column." When people would congratulate me as its eponymous inventor, I would feign modesty and point out that the idea was actually pretty obvious.

I proudly presented my findings at a national meeting of all WPB rubber research groups in the spring of 1946. There was no need for me to feign modesty: My talk aroused only mild interest, mainly because my colleagues thought that my method would never provide a practical way of producing Buna-S molecules of uniform chain length on an industrial scale. Even the Doc didn't bubble over with excitement. Moreover, even if my method *had* proven practical, it would have been moot by the time of the meeting. As the first FIAT industry branch investigators to enter occupied Germany discovered, the Germans had found a way to produce molecules of a more uniform chain length by devising an ingenious modification of the conditions of Buna-S synthesis. When I left Illinois for Germany a few months later, I stopped work on the thermal diffusion column and didn't take it up again on my return in the following year.

My results were never published, and my thermal diffusion separation method for chain molecules of different lengths vanished without trace from the corpus of science. This was a pity, because, unknown to me, biochemists studying proteins and nucleic acids in the forties were in great need of techniques for separating differently sized molecules present in extracts from living cells. My thermal diffusion column could have served that need beautifully. In retrospect, I have no doubt that it would have made quite a stir among biochemists had I gone ahead and adapted it for their purposes. But by the 1950s, when I finally became aware of the opportunity I had missed, much better molecular separation methods had become available.

This disappointing episode proved to be only the first of several instances in my career in which I had hit on a good thing that could have gained me substantial fame—but didn't. It's not uncommon for scientists to retail sour grapes by declaring that one of their brilliant original ideas was stolen by someone who illegitimately claimed credit for it. Such theft of scientific ideas does occur, of course, but not nearly as often as is widely believed. The more banal cause for the failure to get due credit for one's discovery is, as in my case, the personal lack of the qualities needed to have it make an impact. Originality and inventiveness, though necessary, are not sufficient for making a mark in science: One must also have the intuition, stamina, and, above all, the self-confidence necessary to exploit one's inventions and present them as a salable package.

While I was fiddling with my thermal diffusion column in the Doc's lab on May 1, 1945, the news I had been longing to hear all my life came over the radio we had on all the time: Adolf Hitler was dead! Grand Admiral Doenitz, Hitler's designated successor, had broadcast from Hamburg to the German people that the Führer had met a hero's death in Berlin, fighting to his last breath. I was saved! The monster had gone to hell before he got at me with his magic powers, and I had nothing to fear any longer. Or rather, whatever tribulations I might still be facing in my life—never getting over my shipwrecked love for Dorothy, for one, or being prevented by my Jewishness from making a career in chemistry, for another—seemed trivial in comparison with the horrible death Hitler had ordained for me, the fear of which had haunted me for as long as I could remember.

A week later I went downtown to join the crowd celebrating VE Day in the Champaign streets. Everybody was hollering and whooping, women were embracing and kissing strangers, a band was playing in front of the Illinois Central depot, and people were dancing—singly, in couples, or snaking in conga lines. Being too shy to kiss and dance with strange women, I just hollered and whooped, gloriously happy to be delivered at last.

Three months after VE Day, someone ran into our lab, shouting that the United States had just dropped an atom bomb on Hiroshima and that the one bomb released from a single plane had made the entire city go up in smoke. I was no longer interested in the military implications of the news, now that the war in Europe was over. I never had any doubt that the Allies would beat Japan once Germany was defeated. Rather, what came to my mind immediately was that the road for travel into outer space was now open.

Space travel had been a popular subject in the juvenile German science fiction literature I had devoured in my early youth. I had been fascinated by Fritz Lang's interplanetary movie, *Die Frau im Mond (The Lady in the Moon)*, which I saw in Berlin when I was eight. So I was crestfallen when I read an article in a recent issue of *Chemical and Engineering News* that proved conclusively that it would never be possible to build a rocket powered by a chemical fuel that could escape from the Earth's gravitational field:

The weight, f, of any chemical fuel needed to provide the power necessary to lift a given weight, w, into outer space is greater than w. The article concluded, therefore, that space travel would become feasible only if means were found to release some of the enormous energy inhering in atomic nuclei, according to Einstein's formula $E = mc^2$, from a small mass of nuclear fuel whose weight, f, is negligible. And now these means had evidently been found.

As would be frequently the case in my later scientific career, I had bought into a perfectly rational, albeit false, argument. I had prophesied correctly to my fellow students on the day the atom bomb was dropped that we were now at the threshold of the Space Age. But *Sputnik*, which inaugurated space travel twelve years later, was driven by chemical fuels, and no atomically fueled rocket was ever flown.

I became eligible for U.S. citizenship in March 1945, having just come of age and having resided in the States for the required five years. Haughtily, this academic wise guy had disdained looking, even ever-so-briefly, at the booklet that the Immigration and Naturalization Service distributed to help ignorant foreign greaseballs acquire a basic knowledge of American history and civics. At my naturalization hearing at the U.S. District Court in Danville, the judge must have noticed my smirk when he accepted the reply "da Constituzion" given by a Mediterranean graybeard in response to his sole question, "What's the basic law of the United States?" My turn came next.

"I see you're a grad student at Illinois. Maybe I should be more careful, but I'll take a chance on you and skip the literacy test. So tell me how many electors there are in the Electoral College."

I didn't know but shrewdly guessed:

"Same as the number of congressmen in the House of Representatives, Your Honor."

"Wrong! You've got to add the senators in the Senate to the number of representatives in the House. So, here's another easy question for you. What's the subject of the Third Amendment?"

Wise Guy made another shrewd guess:

"Guarantees the right to bear arms, Your Honor."

"Wrong again! The Third Amendment restricts the quartering of troops in private houses."

His Honor seemed pleased to have tripped me up.

"I'll pass you despite your poor performance. But you better study more diligently for your exams at the university than you did for your naturalization test, if you want to get on in academic life."

I felt that the judge's swearing me in as an American citizen was the most important ceremony of my life. An Enemy Alien no more! Now that I had become a certified American, I would shoot right back "Chicago" to the embarrassing question put by strangers "Where are you from?" rather than having to hem and haw about my embarrassing Berlin origins.

"What Is Life?"

NOT LONG AFTER BECOMING A U.S. CITIZEN, I had an encounter in Champaign that would prove seminal for my future scientific research interests. Martha Baylor, a postdoctoral research biologist in charge of the Illinois chemistry department's electron microscope, had invited me to a party she gave in honor of Sol Spiegelman, a visiting seminar speaker from Washington University Medical School in St. Louis. A 32-year-old assistant professor of microbiology, he fixated his interlocutors with a penetrating stare and spoke in staccato sentences, whose patronizing tone suggested that he was bringing enlightenment to us Illinois hayseeds. Provoked by his mannerisms, I got into an argument with him about the applicability of thermodynamics to living matter. He didn't seem to have a firm grasp of thermodynamics, which was my specialty, and I didn't know anything about living matter. But we argued fiercely all the same.

Since Martha had given me to understand that Spiegelman was an up-and-coming star in microbiology, I asked her after the party for some references to his publications. I wanted to find out what the research of a hot-shot in that, to me mysterious, discipline was like. She lent me a reprint of a recent article of Spiegelman's. On the first two pages of that paper I was amazed to find an inelegant derivation of the same standard equations which, in my sophomore year, I had to derive in a few minutes on an exam in the Doc's physical chemistry course.

Next time I saw Martha, I asked her,

"Wouldn't microbiology be a good field for me to get into if it takes so little to have a paper published?"

"Maybe. But if you're really interested in the connection between physical chemistry and biology why don't you get yourself a copy of Erwin Schrödinger's recently published book, *What Is Life?* It's on sale at the Illini bookstore."

I bought the book a few days later and gobbled up all of its ninety pocket-book-size pages nonstop. It had a tremendous effect on me and changed my professional career, as it would change also the careers of many of my coevals trained in the physical sciences during the war. Schrödinger, an Austrian physicist, was the co-inventor (with Werner Heisenberg) of the quantum mechanics that revolutionized modern physics in the 1920s. In *What Is Life?*, which he wrote in the mid-1940s, Schrödinger announced that a new era was dawning for the study of heredity, thanks to some novel ideas put forward by Max Delbrück, whom he identified as "a young German physicist."

Schrödinger argued that since living creatures are large compared to atoms, there is no reason why they should not obey exact physical laws. Yet they do present the physicist with one big problem, namely how they manage to preserve their hereditary information over the generations. For while the genes, which carry that hereditary information, are evidently responsible for the order that an organism manifests, *their* dimensions are not all that large compared to atoms. How, Schrödinger asked, do the tiny genes of a mammal maintain their order while their own atoms, being constantly buffeted by the energetic forces of heat, ought to be jumping all over the place? How, wondered Schrödinger, has the gene specifying the Hapsburg lip passed on for centuries by members of the Austrian imperial family managed to preserve its specific molecular structure, and hence its information content, while being maintained at the human body temperature near 100°F?

Following Delbrück's then ten-year-old proposal that this transgenerational stability of the hereditary information derives from the atoms of the gene molecule staying put in "energy wells," Schrödinger put forward the idea that the gene molecule is *crystal*. Thus its atoms would be kept from jumping all over the place by the strong forces that hold crystals together. He thought that "we may safely assert that there is no alternative to [Delbrück's] molecular explanation of the hereditary substance. If the Delbrück picture should fail, we would have to give up further attempts." Furthermore, "from Delbrück's general picture of the hereditary substance it emerges that living matter, while not eluding the laws of physics as established up to date, is likely to involve hitherto unknown 'other laws of physics,' which, however, once they have been revealed will form just as integral a part of this science as the former."

But Schrödinger added something of tremendous importance to Delbrück's rather vague physical picture of the gene: He proposed that the gene molecule is an *aperiodic* crystal, composed of a long sequence of a few different, over-and-over-repeated basic elements. The exact sequence pattern of these elements would represent a "code," by means of which the hereditary information is encrypted. He illustrated the vast information-encoding possibilities of such a simple scheme by pointing to the three symbols of the Morse code—dot, dash, and space—as its repeated elements. To my knowledge, with this proposal Schrödinger was the first to put forward the concept of the genetic code, one of the most important ideas of twentieth century life science. It became one of the fundamental dogmas of molecular biology.

In posing the question "What is life?" Schrödinger confronted physical scientists, many of whom were suffering from a professional malaise in the immediate postwar period, with a fundamental scientific problem worthy of their mettle. The development of the atom bomb made them doubt the hitherto self-evident humanitarian benefits of their work. At the same time, the triumph of quantum mechanics, which not only accounted for the structure of atoms and their chemical properties but also raised philosophical doubts about the standing of physical laws and the nature of reality at the atomic level, made it seem that the then vanguard discipline of atomic physics was about to enter a jejune phase. So, many physical scientists were eager to direct their efforts toward a new frontier, which, according to Schrödinger, was now ready for some exciting developments. In thus stirring up the passions of its impressionable readers, *What Is Life?* became a kind of *Uncle Tom's Cabin* of the revolution in biology that eventually left molecular biology as its legacy.

When I left for my tour of duty with FIAT, *What Is Life?* was the only piece of scientific literature I took along to Germany, and I read it over and over again. A mere doctoral candidate of twenty-two, I was too green to be suffering from anything as blasé as the professional malaise of my elder colleagues. Yet, I *had* become bored with physical chemistry. My teachers at Illinois had given me the impression that all the important discoveries in physical chemistry had been made by the 1920s. As I understood it, the art of research in my field consisted of finding some little thing that had not as yet been done and then doing it, or devising a novel, more elegant derivation of some well-established theoretical principle. Therefore, the fabulous prospect held out by Schrödinger that by studying genes I might turn up "other laws of physics" so captivated me that I resolved to join the search for the aperiodic crystal of heredity once I had gotten my doctorate. Delbrück, the young German physicist, had probably been drafted into the Wehrmacht and was missing on the eastern front, but perhaps there were people in the States working along these lines whom I could join after my return to the States.

As I was browsing through stale Stateside magazines at the FIAT officers' club in Karlsruhe in the spring of 1947, I ran across an article titled "Tempest in the Cells" in *Time* magazine, which featured a portrait of the man I had met at Martha's party. Spiegelman was smiling, smoking a pipe. *Time* reported that he had called into question a most sacrosanct doctrine in biology. As taught in every biology textbook, orthodox Mendelian genetics maintains that genes are the sole arbiters of heredity. But, according to Spiegelman, this view is incorrect because some cells, such as protozoa, yeasts, and cancer tissue, can change their chemical properties, thus defying their hereditary genes. In Spiegelman's view, for which he claimed to have provided direct support, genes are phosphorus-containing proteins that send out "partial replicas" of themselves—he called them "plasmagenes"— from the cell nucleus to the cytoplasm, and multiply there independently. Thus it is not the genes that govern the cell's global chemical properties, but the balance of their plasmagenes, which can change under abnormal conditions that allow the favored plasmagenes to multiply, "as rabbits once multiplied explosively in Australia after they had been introduced by Europeans." Although I didn't fully understand this news story, it reinforced my resolve to switch into gene research after getting my doctorate. If Spiegelman could get his picture into *Time*, why couldn't I?

Slithering Away

AN ENVELOPE ADDRESSED IN HILDEGARD'S HAND reached me a week after we had parted at Frankfurt Station. I was tremendously relieved to see that it was postmarked "Berlin." Deluding myself, I took Hildegard's writing to me at all for a sign that she wasn't angry with me for abandoning her at the time of her dire need. But as I read her account of the ordeals she had endured on her way to Berlin across the Soviet zone and once she got home, my initial relief quickly changed to dismay and remorse over the suffering my lack of concern for her welfare had caused her.

Günterlein,

Now that I'm back in Berlin, I can't describe my feelings. There is only one thing I know: I had been in Paradise on Earth, but as every fairytale, so also did mine come to an end. Yet I'm happy all the same. You know why? Because we took leave from each other as people ought to. Understand what I mean?

Naturally, as soon as the train left Frankfurt Station and you weren't with me any longer, I was kicked out of the coach reserved for Allied personnel. Since there were no seats anywhere else in the train, I had to stand up all night, all the way from Frankfurt to Hanover. We got to Hanover on Sunday at six in the morning, an hour behind schedule.

The interzonal train to Berlin was going to leave Hanover at one in the afternoon. The boarding passes needed to get on this train were sold out, of course. What could I do? There was no other train to Berlin that day, and I didn't want to hang around Hanover overnight until the following day. So I just had to get on that train, and get on I did. How I managed it I can't tell you. I was unable to eat anything because of the excitement. My nerves were on knife-edge. But at least I didn't have to walk home from Hanover.

Sunday evening I was in Berlin. My mother nearly fainted. She had thought that I was dead. I had no choice other than going back to living at home. It's terrible for me to have to live with my mother.

What's happening with the colonel? I'm still worrying about it all the time. He simply can't be hard on you. After all, he gave me his word that he wouldn't!

As for the theater, it was just as I had feared. They found out, of course, that instead of having been ill, I was out of town. So they sent me a letter saying I'm fired. Even though I couldn't have expected anything else, it did shake me up quite a bit when I had to face the actual fact that I lost my job.

I don't know yet what I'm going to do. I'll try to get my job back at the theater, but only after Easter. At the moment, I'm drifting. I have no desire to talk to anyone. I want to be alone, all alone. But I found a wonderful narcotic: music. It lets you dream of both past and future. Tomorrow I'll go to the Beethovensaal, to listen to HIS *Destiny Symphony* again. Unfortunately, it's not about our destiny; it's merely about everybody's destiny!

It's late, nearly midnight. I'm writing in bed, because that's where I have my peace. Günterlein, do you know what is the most beautiful thing? We're apart and yet together!

Your Hildelein

P.S. Have the hole in your pants darned!!

In my response, I didn't express any of the dismay or remorse her pathetic letter deserved. I replied in a factual tone: I was terribly sorry that she had such trouble getting back to Berlin and even more sorry that she lost her job at the theater. I hoped that she would be able to invent some good explanation for her absence and make them take her back. The colonel hasn't been heard from, and I figure that, thanks to her intervention, the whole affair has been laid to rest, without any repercussions. And I was getting ready to leave for the States.

Even then, I didn't offer any apology for my lack of interest in her having any surgical procedure, let alone one meant to remedy a condition for which I was responsible. Neither did I say that I missed her, nor that my apartment seemed empty without her, or that I understood what she meant by our taking leave from each other as people ought to.

In fact, until I reread her letter when I was in my sixties, I hadn't understood that she was referring to our taking leave without mutual recriminations and that, in her nobility of spirit, she generously overlooked that the civilized behavior at our parting was all by her, and none of it by me. Nor had I appreciated the awful sadness of her postscript—if anything, at the time I had found it slightly annoying—in which she shows that she is still concerned about my welfare and that *she* wants to take care of me. Worst of all, I had simply accepted without demurring her equating her horrendous troubles with my little hassle with the colonel.

Activities were winding down at FIAT headquarters, and people seemed to be working even less hard than they had been before the liquidation of the agency was announced. I reckoned nobody would care, or even notice, if I disappeared from the document section for a few days to visit Lore in Kitzbühel before I went home to the States. I was finally going to declare my love for her and ask her to marry me. Although I didn't have the official papers necessary to enter the French zone of Austria, I figured that compared to getting to Berlin across the Soviet zone without proper travel orders, sneaking into Austria ought to be a cinch.

At the Austrian border, the French MPs came to check the passengers' travel documents. I pulled out a handsome Interallied ID card, decorated in polychrome with crossed Old Glory, Union Jack, and Tricolor banners. It bore the endorsement "*M. Stent est autorisé à traverser la zone française,*" signed by the chief of the French mission to the U.S. military government and stamped with the seal of *La République Française*. As I had hoped, the MPs didn't notice that *la zone française* mentioned in the endorsement was meant to refer only to the French zone of occupied Germany.

The wintersport season was over when I got to Kitzbühel. At the Grand Hotel, which was practically empty, they seemed glad to get my business and didn't ask for any travel orders. I had planned to run right over to the Postkutsche and carry Lore off to dinner. But I had come down with a fever during the trip and noticed to my dismay that a rash was developing on my body. So I merely phoned her to let her know that I was back in Kitzbühel but not feeling too well, besides being dead tired after the long train ride. Would she join me for breakfast at the Grand tomorrow? Lore sounded both very surprised and very pleased to learn that I had actually come to see her, all the way from Karlsruhe.

After passing a feverish night, I became so agitated when Lore showed up at the hotel next morning, wearing a fabulously becoming tweed suit, that all my physical ailments seemed gone. She was so beautiful! Over breakfast, I told her what I had been up to in the three weeks since we parted, leaving out the most important item—the dramatic developments with Hildegard. I presented Lore a few pairs of nylon stockings I had brought along. Unfortunately, the present I would have liked most to bestow on her—the gold chain with my TEP fraternity pin—was now worn by Hildegard and no longer available for recycling.

After breakfast, we went for a walk to the Schwarzsee, a little lake about half way to Kirchberg, the village at which the easy ski run down from the Hahnenkamm wound up. The snow was largely gone from the slopes. Seeing the bare, deserted hillsides on which so many skiers had disported themselves just a short while ago made me maudlin. I grasped her hand, which I held until we got to the Schwarzsee. There I took a picture of her seated on a lakeside bench, with the snow-covered peak of Wild Emperor Mountain towering in the background.

In the evening, we had the Grand's vast dining room virtually to ourselves. There was no longer any band, and hence no dancing. After dinner, we went into the deserted, pitch-dark garden of the hotel. It was a mild spring evening. We sat on a bench and I put my arm around her shoulder. And then we began to kiss passionately. At last!

"Du, Lore-mine, why don't we go up to my room? We'll be more comfortable there."

"I can't risk it, Günter. The hotel staff all know me, and the gossip would be all over Kitzbühel tomorrow."

Because of my lingering illness, especially the rash, I was less insistent on her coming up with me than I might have been otherwise. Ordinarily, I would have figured out some stratagem for our sneaking upstairs unseen. But I didn't know what was ailing me, and whatever it was, I didn't want to pass it on to Lore. So we just stayed put on our garden bench, kissing and petting the night away. Despite this quantum jump in our intimacy, I was unable to come out with it: to tell Lore that I loved her and ask her to marry me.

Late next morning, we met at the cable car station to ride up to the Hahnenkamm. Lore was dressed in the same outfit—ski pants and red cashmere sweater—she had worn as the lovely blonde angel who pulled me out of the snow.

"I'll carve 'L&G,' circled by a heart, into the gondola's wooden handrail, to mark the sacred site of our first meeting."

"Don't, Günter! It's *streng verboten*. The attendant's bound to notice!"

I couldn't help thinking that if I had made this proposal to Hildegard, she would have flirted with the attendant while I was carving away behind his back.

Hand in hand, Lore and I hiked from the upper cable car terminal to the Ehrenbachhöhe. There was still enough snow on the Hahnenkamm ridge for cross-country skiing, and a few die-hard, post season devotees were still at it. We had lunch at the lodge, and spent the afternoon on the deserted sun deck, lacing sunbathing with mulled wine, kissing and caressing.

When I brought Lore back to the Postkutsche at the end of the afternoon, she did not offer, nor did I ask her to introduce me to her mother or her little son. I didn't wrack my brains about why she didn't seem to want them to meet me. My own interest was in Lore herself rather than in her kin, although I was planning to bring the boy to the States with his mother.

In the evening, we had our farewell dinner at the Grand. Lore said,

"I'm thinking of moving back to Germany, Günter. I'm getting fed up with life among the Tyrolean yokels in this cultural backwater."

This would have been the perfect opportunity for me to tell her that I was hoping to take her away from all this to the States, and to ask her to become my wife once her legal situation would allow it. But I didn't. I was tongue-tied, unable to mouth in her presence what I had rehearsed a dozen times in preparation for this evening. So I just gave her the kind of advice I'd have given to anyone.

"Lore, my dear, you'd be crazy to trade the comforts of Kitzbühel for the horrible living conditions in Germany."

The closest I came to declaring my love was "It's really a perverse trick of fate that I met you just as I was about to head home for the States."

To prompt me to open up a little, Lore played dumb and asked, "Why? Didn't you tell me the day we met that you were looking forward to getting back to Illinois as soon as possible?"

"Yes, I did. But that was the way I felt before you pulled me out of the snow on the Kaser slope. Now, I'm totally distressed that I'm sailing home to the States, with little chance that we can meet again until I become a Herr Doktor sometime next year and come back to see you."

Even on this, our last night, Lore wouldn't come up to my room with me. That night, I believed her explanation of fearing the Kitzbühel gossips. But now, in retrospect, it seems more likely that she was unwilling to commit herself to a man who had produced no verbal declaration of his love, nor any assurance that this affair meant more to him than a one-night stand. So I was resigned to postponing the consummation of my love for God knows how long. After dinner, we sat in the hotel bar, discreetly holding hands under the table. Finally, at about midnight, we said good-bye in the garden with a long kiss. I hardly slept during the night, agitated by love and frustrated by this denouement of my search for Fräulein (or, as it turned out, Frau) Richtig. I was also worried about my illness.

Next morning, Lore was waiting at Kitzbühel Station, even though she had not said that she would be coming to see me off. I was so mesmerized by this manifest sign of her caring for me that I was no longer in full touch with mundane reality. As the train pulled in, I embraced Lore for one last time and climbed on board. Lost in reverie, I slumped into a corner seat of a first-class compartment. When the conductor came to check my ID, I discovered that I was on the wrong train. I had boarded the westbound train for Innsbruck rather than, as I had intended, the eastbound train for Salzburg.

Willy-nilly, I was taking the slow route to Karlsruhe, which involved a change at Innsbruck to a *Bummelzug* with a little steam locomotive pulling ancient, two-axled wagons across the Bavarian Alps to Garmisch. The *Bummelzug* was delayed so long at Mittenwald, the Austro–German frontier station, that I was going to miss my connection at Garmisch. By way of a friendly gesture, I offered the conductor a few cigarettes, and at the next station he telephoned ahead to Garmisch, to have the Munich express wait for an important Ami officer he had on board. When we pulled into Garmisch, I ran to get on the waiting express. But the train just stood there and didn't leave. The Garmisch stationmaster had seen only me get on, but not the important Ami officer for whom he was supposed to hold the Munich express. The train left only after I identified myself to the conductor as the important Ami officer they were waiting for.

I spent the night in Munich, too tired to travel all the way to Karlsruhe. Next morning, when I went to catch my train, I saw a city bus with a sign TO DACHAU waiting for passengers in front of Munich Central Station. Dachau!

A name that haunted me ever since I was a kid! It hadn't occurred to me that you can take an ordinary bus to hell, instead of being carted to perdition in a prison van. So, on an impulse I got on the Dachau bus. Karlsruhe can wait. I must visit that infernal place, to see first hand what I, the lucky stiff, had been spared!

Being in Ami uniform, I had no trouble getting into the camp, which was now used as a prison for SS men. They were locked up in the precinct formerly occupied by the concentration camp inmates, behind the old electrified barbed wire fence with its tall watchtowers. Some of the SS monsters were standing just on the other side of the fence and stared at me as I walked past them. It was like going past the tigers' cage in the zoo: scary, but at the same time comforting that those vicious beasts behind bars can't get at you.

A few of the former prisoners' huts had been converted into a museum, in which various aspects of the diabolical camp life were displayed by use of dummies, including the punishing of the inmates by lashing them in a bent-over position or by suspending them from their arms tied behind the back (which had frightened me so when I first heard of it as a child).

These scenes of horror brought back memories of a boat trip to St. Marguerite Island in the Gulf of Juan on the French Riviera I had made with my family when I was nine. We were shown the dungeon in the island's seventeenth-century fortress, in which the Man in the Iron Mask had been imprisoned by Louis XIV. I was horrified by our guide's tale of the jailers' expectation that the unshaven prisoner would be slowly strangled by the growth his own beard within his iron mask. Yet, I had derived some comfort from this story having happened 300 years ago, when, so I believed, people were much more cruel than they are in our own times. Now, at twenty-three, I appreciated the error of my childish belief in moral progress. Whatever may have been the viciousness of the barbarities perpetrated by seventeenth-century French, their scale was insignificant compared to that of the atrocities committed by twentieth-century Germans.

One of the U.S. Army guards at the camp told me that an American military court was holding war crimes trials in the auditorium of the former headquarters buildings of the SS division that had been stationed at Dachau. The defendants were "minor" war criminals charged with specific, individual criminal acts (in contrast to the "major" criminals charged with wholesale, generic crimes, tried by the interallied court at Nuremberg). So I went to watch the trial of half a dozen SS men who had been guards at the Mauthausen concentration camp. They were accused of having brutally beaten to death some American airmen—"gangster bombers" Goebbels used to call them—who had been captured after they parachuted from their disabled aircraft and taken to Mauthausen. The case was heard by a tribunal of three judges, a couple of prosecutors, and a team of interpreters, all in U.S. Army uniform and backed up by Old Glory. The defendants—each prisoner flanked in the dock by a white-helmeted MP—wore SS uniforms stripped of insignia, while their German defense counsels were in civvies.

The defendants behaved respectfully before the court, but they were sullenly unresponsive to the prosecutors' interrogation. They had obviously conspired to adhere to a code of silence and refused to answer any questions directed toward establishing their own guilt or that of their fellow accused. Their German lawyers defended them with surprising vigor and spunk, following a two-pronged exculpatory strategy. First, the lawyers claimed that their clients were misidentified and had never taken part in the atrocities with which they were charged. To that end, the defense cross-examined very aggressively the witnesses produced by the prosecution, seeking to impeach their testimony. Second, in case the evidence *did* prove beyond reasonable doubt the participation of a defendant in an atrocity, his attorney would allege that the accused had merely been following orders of a higher authority and that any refusal to carry out these orders would have amounted to mutiny and entailed the defendant's own death.

The U.S. military personnel conducted itself with dignity and decorum. It made me proud to be an American. There'd be a fat chance that if *we* had lost the war, any Wehrmacht court would have dealt with Allied prisoners accused of war crimes in this civilized way.

I didn't find out the outcome of the particular case I had watched briefly. But I later learned that that by the time the Dachau trials ended in December 1947, a total of 1,672 accused had been tried, of which only 256 were acquitted. The tribunal decreed 426 death sentences, of which 250 were carried out, with the rest commuted to life imprisonment by General Clay.

My brief visit to Dachau had profoundly stirred my soul. It had confronted me more directly with the horrible fate that I had managed to escape than any other experience during my stay in occupied Germany. In view of what millions of Jews had suffered, I felt that I was one of the lucky few who had won the Grand Prize—life itself. Ever since that visit to Dachau, the sense of being one of the lucky German Jews who survived Hitler against all odds buffered my soul against any depression that might have been induced by disappointed hopes and expectations—personal or professional. I had come to believe that whatever other good things might yet happen to me in later life would merely be adventitious fringe benefits to which, having had more than my share of good luck already, I was not really entitled.

Back in Karlsruhe, I went to the Army dispensary, where my mysterious illness was diagnosed as scabies, an infestation of mites on people who live under unhygienic conditions and not uncommon then among GIs stationed in Germany. I had probably picked it up from dirty bed sheets and pillows in the military billets in which I had been staying on my assignments. To get rid of the mites, I had to shave my body hair, take lots of hot baths, and boil all my underwear.

Meanwhile, a letter from Hildegard had arrived in Durlach.

My Günterlein!

Je ne suis pas curieux,
Mais je voudrais savoir . . .

These lines remind me of our most beautiful hours, and that isn't always a good thing. Because then I feel it physically that you are not with me. Do you understand? Are you inwardly lonely as well since I'm gone?

Today is Easter! I thought that my Günter would send me at least a written Easter greeting, because I know that he was thinking of me. But maybe it's still on the way.

Weren't you happy that finally you could get a good night's sleep again? Get some rest; you really need it. After all, I brought enough unrest into your life. I admit it!

It was all a bit comical when they fired me at the theater. I don't feel guilty at all. Is it my fault that my feelings drove me to act in this way? No. I'm trying to imagine what you are thinking while you are reading this. No doubt, you are still of the opinion that I shouldn't have played hooky.

There's still something else, Günterlein, but please don't worry. I am ill, very ill, in fact. My doctor told me that I got up too early in Karlsruhe. I told him everything, and he said that under these conditions I shouldn't have made the trip. I had to stand too long in the train. Now I can't walk at all; I've got so much pain. During the examination my doctor said that I've got to go into the hospital if it gets any worse. I don't dare to talk at all any more here at home. My mother drives me crazy with her questions, even though I was prescribed strict bed rest. She wants to know what is wrong with me. I'm lying to her continuously. What I'm worrying about most is that I won't be able to go to the housing office to see about getting a place to live. Don't worry, though. But I had to tell you about this, because something might happen, and then you know at least what was going on.

Appetite I have none. The food here is really horrible. I hardly smoke anymore, and exchange everything for food, just as I had promised you. I'm looking forward to the sausage from Mrs. Breyer. She had promised it to me, after all. Is that embarrassing for you? No, not if you can help me with it! And I don't have any more aspirin either. That I really need badly, especially right now (as well as Kleenex).

You are probably surprised that I'm writing you such a long letter. But it may be my last one for a while. And it's only to you that I can tell absolutely everything. I don't have anyone besides you. Write me soon, and not so little. It's the only thing I'm waiting for. I hope that my letter will occupy you for a whole evening. Then I'll know at least that you aren't stepping out.

Do you like your work? Still not, eh? You want to get back home! I know. I wish you from all my heart that you'll be back at the university very soon. Main thing is that you are happy. If you're happy, I'm happy.

Do you know that if you were with me, you would bring me flowers on April 12th? You've forgotten, haven't you? No, it's not my birthday: You and I, we'll have known each other for 5 months! It was on the 12th of November, Sir, that you engaged me in a conversation! Tell me, doesn't time pass at a fiendish pace? Christmas, New Year's Eve, and now it's Easter already. Unbelievable. When I was with you, I often wanted to stop the hands of the clock! I wanted to use force to keep the happy time from passing by. But I don't think one can force anything in the Big World, least of all time!! But I'm digressing—I wanted to ask you to look at your diary on the 11th of April and, in your thoughts, live through everything with me another time. I'll be doing the same.

What I liked best in your letter was your bon mot:

> "I am the only decent person! Everybody thinks only of himself, but I am thinking of myself!"

Pure Günter! Could have come only from you. But, by chance, I happen to know that it wasn't you who made it up. That would have been too fitting!

I can't go on. I'm already quite lame in bed. If I weren't so sick I wouldn't have written that much. You are reading this letter much too fast. It's worth a lot more of your attention.

Good Night, my Günterlein.

Your Hildegard

Hildegard's letters—with their gallantry, their humor, their closeness—touched my heart more directly than any I had ever received, and she probably had come to see through me better than anyone. And yet I didn't love her. I was in love with Lore, the embodiment of my Dream Woman fantasy.

My idea of the decent thing to do was to send Hildegard a farewell letter, without letting on that I had fallen in love with someone else. So I wrote her that since I was about to leave for the States, there was little chance of our meeting in the foreseeable future, and it would be best if we tried to forget about each other. In any case, I said, although I very much appreciated that she is a wonderful woman, I hadn't been in love with her. Maybe I just couldn't be in love with *any* woman. Of course, I'd always remember the wonderful times we had together, and I wished her all the very best for the future.

That seemed to me to be the extent of my obligation to her, evidently supposing that it would be less cruel to level with an ill, starving woman who says she has only me, than to let her go on hoping. I also sent Hildegard a little package with some food and aspirin, on which I wrote Claire's apartment in Chicago as my return address.

As I packed my belongings in the Kastell Strasse apartment—my uniform, my spare clothes, *What Is Life?*, and the encyclopedic Langenscheidt dictionaries I had liberated at the Berlin Bartermart, as well as Hildegard's silver 50-Rappen cufflinks, the photograph of her she had sent me from Berlin, the snapshot of us arm-in-arm Neckarside taken in Heidelberg, the programs from the Schiffbauerdamm Theater, and her letters—I was haunted by the memory of living with Hildegard within these walls. She had been a wonderful companion; I had been happy with her here, playing house, getting my first taste of quasi-married life. All the same, my happiness with Hildegard didn't compare with the ecstasy I felt in the company of Lore, the most beautiful woman I had ever held in my arms. I didn't pack down Lore's romantic portrait I took at the Schwarzsee, with the snow-covered Wild Emperor Mountain in the background. I was going to carry it home in my briefcase, so that I could gaze at my beloved's bewitching face *ad libitum* during the trip home.

Although I was far from appreciating how unconscionable had been the insouciance with which I treated Hildegard, I both hoped as well as feared that after having sent her my farewell letter, I would never hear from her again. Yet, when I stepped off my Rotterdam-bound train during a ten-minute stopover at Frankfurt Station and another train was just pulling out on the other side of the platform, I thought I saw Hildegard's sweet face leaning out of a compartment window, her hand waving a handkerchief to me. My love for Lore notwithstanding, something was tying me to Hildegard, something other than what I understood to be love.

At Aachen, the last train stop in Germany, I recalled our ordeal in the station restaurant on Christmas Day 1938, waiting in a terrible fright for the guide who was supposed to sneak us across the Green Frontier and get us out of Hitler's clutches. Now it was April 20, 1947, the late Führer's fifty-eighth birthday: I felt utterly secure at Aachen Station. The son-of-a-bitch was rotting in hell, and I was alive, a homeward-bound officer of the U.S. government.

As I stood on the railing, while the Holland America Line's *MS Noordam* pulled away from the Rotterdam dock and moved down the Rhine toward the North Sea, I was glad to be on my way back to Illinois. Unexpectedly, I had become tired of the dolce vita of a FIAT playboy, with everything at my beck and call that money (black market Reichsmarks and cigarettes) could buy and free to roam the ski resorts of Europe. It wasn't that I wouldn't want to continue the pursuit of pleasure. But now that I had tasted both, I had come to feel that the life of a scientist might actually be preferable to that of an itinerant playboy. Incredible as it would have seemed to me when I left Champaign, I had begun to look on the ambiance of the university as a spiritual need of no lesser importance than skiing and sex. I was eager to get on with my dissertation research, since the Ph.D. was going to provide me with the hunting license for rustling a job in Academia.

Yet I had become aware that though I felt like an American and wanted to shed all vestiges of my German origins, I had remained enough of a European that I might never really feel fully at home in the States. I didn't know exactly why. I liked American people and appreciated that if it hadn't been for them, I would be dead. Maybe I just hadn't gotten used to American women. I wondered whether I shouldn't try to move back to Europe after getting my doctorate and live there with Lore, rather than bringing her to the States. It didn't occur to me that I hardly knew anything about her and that I had no idea what living with her would actually be like.

Mr. Charles Collins, who had hired me, was no longer working at the

Commerce Department's TIIB office when I reported in Washington for my separation from government service. He had been fired when it came to light that the doctoral degree in chemistry he had listed on his Federal Civil Service Form was fraudulently self-conferred. It seemed as if years went by since I was in the TIIB office a mere eight months ago to sign on with FIAT.

This feeling of having been away for a long, long time would overcome me repeatedly, as I was returning to my old haunts. The richness of my experiences in Europe had dilated my perception of time. I was also vaguely aware that I had changed. I sensed that I had become more self-assured—less haunted by the fear of Hitler, and with more money to my name than at any other time since I escaped across the Green Frontier. I had saved most of the $3,000 salary I was paid by the Commerce Department while I was overseas because my living expenses were covered by the military government, and I financed my extracurricular activities largely with cigarettes.

I also felt less frustrated by unfulfilled desires. Until my return to Berlin, I had been forever pining for the future, never savoring the present, always anxiously waiting for the next step that, I hoped, would bring me happiness. As the frightened Krümel in Nazi Berlin, I had put all my hopes for salvation on getting out of Germany. But once I had managed to escape, first to Antwerp and then to London, I had become a hapless refugee, impatiently longing for the day that I would join my Mausi in God's Own Country. Having made it to Chicago and achieved my dream of going to an American school, I could hardly wait for my graduation from Hyde Park High and start going to college. Not long after I had made it in college—as a fraternity man to boot—I put my hopes for the good life on going to graduate school. Although I had started to feel a little more content once I did become a graduate student—in part, paradoxical as it may seem, because my love for Dorothy S. had been requited, albeit ever so briefly—I had still been restive with the present.

Now, on coming home from occupied Germany, I no longer felt that life still lay wholly in the future. Many of the adventures to which I had been looking forward when I left on my trip had actually come my way. Yet the sense of fulfillment, including the miracle that the blonde Berlin shiksa I found in Kitzbühel turned out to be an improvement over my fantasy, had not actually done much to abate my two affective disorders, Jewish self-hatred and empathy deficit.

PART THREE

Stumbling into the Scientific Big Time

The Sophisticate

ON MY RETURN TO CHAMPAIGN FROM GERMANY in May 1947, I stopped in Chicago to visit Claire and her family. She was expecting her second child. Five months later, she gave birth to Barbara, the last-born of my four nephews and nieces. Claire handed me a letter from Hildegard. When I had suggested in my farewell message to Hildegard that it would be best if we tried to forget each other, I was hoping that I would never hear from her again. Yet I must have been hoping also that I *would* hear from her again. Otherwise I would hardly have given Claire's Brompton Avenue apartment as my return address on the last food parcel I sent her before leaving Durlach for the States.

I didn't open Hildegard's letter for a while because I had a premonition that its contents would devastate me. And devastate me they did when I finally worked up enough courage to read her pathetic indictment of my abominable behavior.

> Berlin, once upon a time.
>
> Reply to your farewell letter.
>
> Günter,
>
> I know a fairytale, which is very beautiful; only the ending is ugly: Once upon a time there was a girl, young but too good for this world. One day she got a letter, a letter which . . .
>
> You gave me a piece of advice in your letter. After thinking it over, I decided that I won't lose any jewels of my crown if I accept it. I'll come to terms with reality and not worry about what people call "love."
>
> I'd like to ask you something, though. What kind of life shall I lead? Shall I become a nun, or remain a whore? Since you do know me a little, you certainly ought to be able to give me an answer. Exert your little brain! You know that I'm a demanding person. Although, judging by your letter, one might have thought that you never noticed.

Günter, I'm disappointed, because I didn't think you were a coward. The story you told in your letter I had expected for some time. Why did it have to come in your very last letter from Germany? Did you think I was stupid, or did you want to spare my feelings? I had known for a long time that one of us gave love and the other compassion. I suppose you wrote that letter only because you know that you'll never see me again.

I received the three parcels. Nice idea of yours, because I certainly can use what was in them.

I still have one last request: Forget everything, absolutely everything! I knew that one can't always behave decently in life, but I didn't know that one mustn't toward a man. You've taught me that there's always something more one can learn. But since I'm conscious of my worth and estimate you as a gentleman, I know that you'll honor my request.

Hildegard

Wasn't the fairytale beautiful? It ends with the usual closing sentence: And if they hadn't died . . .

I locked myself into the bathroom, so that Claire wouldn't see that I was crying. The enormity of what I had done to Hildegard struck me at last. I couldn't understand. Why, why hadn't I been able to love her? How could I make it up to her now, when I'm in love with Lore, whom I'm hoping to marry?

The Illinois Graduate School readmitted me as a candidate for the Ph.D. in chemistry, the Doc welcomed me back, and I was reappointed as a research assistant on the synthetic rubber research program. While I was gone, the name of the program's Washington sponsor had been changed from "War Production Board" to "Office of Rubber Reserve." The Doc proposed two sure-fire research projects for my doctoral dissertation, which he thought I ought to be able to complete within a year—before President Truman would cotton on to the idea that cutting off support for synthetic rubber research might be a sensible peacetime economy measure. I didn't find either of these projects super-thrilling. But the idea of completing my doctoral dissertation work in a year had tremendous appeal, especially since I was eager to begin my search for other laws of physics.

One of my new projects was related to a thermodynamic theory the Doc had developed for predicting the vapor pressure of liquids. I was to extend his theory from pure liquids to mixtures of two liquids. This project required neither original thinking on my part nor doing any experiments—only tedious, months-long number-crunching on an old-time office calculator. Doing these calculations for the Doc wasn't going to contribute anything to my development as a scientist. Forty years later, they would have taken just a few hours on a desktop personal computer and have hardly qualified as the subject of a doctoral dissertation.

The other project was a little more substantial. I was to work out another method for sorting, according to their size, long-chain molecules in which different numbers of a small chemical building block are linked end to end. This project thus resembled the size sorting of Buna-S chain molecules for which I had designed the "Stent column" before I left for occupied Germany. The new project differed from the old, however, in that the chemical building block was an acid that can take on a negative electrical charge when dissolved in water. My separation method was going to be based on a clever idea of the Doc's, who reasoned that in longer-chain molecules, a smaller fraction of the incorporated building-block molecules would take on a negative charge. Hence, in view of their lower overall negative charge per building block, longer molecules ought to migrate more slowly toward the positive pole of an electric field.

My task was to build a special electric chamber that would allow the analysis of the lengths of chain molecules present at the positive and at the negative poles of the device after flow of a measured amount of electric current. This second project at least required some originality in the design of physicochemical research apparatus and provided me with additional experience in the conduct of experimental investigations.

I didn't fancy moving back into the TEP house. After my European adventures, I felt too seasoned for fraternity life. So I asked for housing in the university men's residence hall, which had just been reopened for occupancy by civilian students, after having been requisitioned during the war by the Navy. Even though there was a long waiting list, the management made space for me, because they thought that a few graduate students sprinkled among the mainly undergraduate residents could be role models. Two undergraduates shared a triple room with me: John, a journalism major from Chicago, and Sam, a GI veteran economics major from Brooklyn. In keeping with the university-sanctioned residential Jewish–gentile apartheid, my two roommates were Jewish.

A framed enlargement of the photo I had taken of Lore at the Schwarzsee, with the snow-capped Wild Emperor Mountain in the background, stood on my dresser in our triple room. Sam asked:

"Who's the beautiful babe in the picture?"

"She's the woman I left behind in the Tyrol and intend to marry."

He guffawed sardonically. "Don't feed me that bull, buddy! A gorgeous dame like her wouldn't waste her time on a *schlep* like you! You probably cut out her picture from a movie magazine!"

Sam was soon proven wrong by Lore's first letter, which arrived a few days after this colloquy.

Kitzbühel, 1 May 1947

Dear Günter,

I had meant to wait until I receive the first mail from you, but it's taking too long. Are you are back in the States? I'm hoping that you are still in Germany, because then there would be still a chance that you come to visit me here.

I'm unhappy that I got to know you, because you brought such disorder to my placid life; actually less to my life than to my person. I am thinking a lot about you and your kisses. Their taste made me want more (mehr), but there aren't any more because an ocean (Meer) lies between our lips.

Many kisses from

Your Lore

She loves me! But why didn't I tell her on our last night at the Grand Hotel that I love her and let her in on the secret I just shared with Sam? I could have postponed my return to the States by a few weeks and gone back to Kitzbühel to finalize everything. Why had I been in such a big hurry to get back to Champaign, just when I found the woman in search of whom I set out for Europe in the first place? OK. I'll declare myself by mail now. I'll write her a love letter and propose that we get married as soon as her legal situation allows it.

I couldn't do it. I was simply paralyzed. I could manage no more than telling Lore that the disorder she brought into my life was the most wonderful thing that ever happened to me, that I missed her kisses at least as much, if not more, than she missed mine, and that next year, as soon as I was a Herr Doktor, I would recross the *Meer* to get *mehr*. Most of my letter was simply a bland report about my trip home and my work at the university. I enclosed copies of some of the photos taken when we were last together in Kitzbühel, including the treasured portrait of her on the banks of Schwarzsee. Under separate mail, I sent her a little parcel with some coffee and toilet soap.

I also wrote to Hildegard, acknowledging the letter she sent to Chicago, but making no reference to how I had been ravaged by its contents. I said nothing about how painful I found it, how sorry I was for the abominable way in which I had treated her, and how her letter had made me appreciate what a wonderful woman she was.

I made only one oblique allusion to her letter. I denied implicitly her request that, as a gentleman, I should forget absolutely everything, by declaring that I'll never forget her and the wonderful times we spent together, some of which were the happiest in my life. For the remainder of my letter, I just recycled to Hildegard the bland report of my transatlantic travel and Champaign situation that I had just sent off to Lore. I also mailed Hildegard a little parcel.

I lacked even the modest interpretative skills needed to fathom the real meaning intended by Hildegard's devastating indictment, which was not to tell me to go to hell but to try to touch me, in the hope that, all evidence to the contrary notwithstanding, I was not an out-and-out monster. And so I was both surprised and immensely relieved when I received a string of three immediate responses from Hildegard, all written within a week of one another, in none of which she made any references to her horrible fairytale. They were still full of subtle digs, but it was easier for me to ignore her barbs than her earlier outright accusations.

The tone of Hildegard's response to my first Stateside letter to her was so jaunty that it was hard to believe that the same woman had written it as the fairytale indictment, or that it was addressed to the same accused criminal.

> Dear Günter,
>
> Now the gentleman is again in his country. Happy and contented. I was astounded that you were sea-sick on the Atlantic, even for a single day. I thought that you were immune to everything.
>
> I am glad that you have no regrets over finding that most of the guys finished their doctor's degrees while you were gone. Remember, they didn't have all those beautiful experiences that you said you had in Germany!

Fred Thornhill came to Berlin the other day. He took me to dinner at . . . , and they gave us the customary heave-ho. Can you guess where? Think about it! Where you and I had our first meal. Remembrances are the most deadly of all slow poisons!

Seeing how badly off I am, Fred helped me much in a touching way, by giving me cigarettes, soap, etc. Perhaps he didn't even realize how much he relieved me of my worries for a couple of weeks. He also brought me a very nice letter from his wife, Mary.

Perhaps you can do me a favor. The sunglasses that you sent me were stolen. Please send me a new pair. It's possible now to send private parcels to Germany from abroad. You know what else I'm wishing for. A stylish bathing suit. After all, it'll be my birthday soon. Shoes I need as well. Make me a nice parcel. Günterlein, a lipstick. The one you bought me in Heidelberg is finished. And nail polish; you know the colors. And, one of these days, I'd like to taste chocolate again. Please don't get angry over my demands: Remember, I'm allowed to make them because I'm a woman!

Your Hildelein

Hildegard's next two letters were more pathetic:

My dear Günter!

It's nine weeks since we parted in Frankfurt. To survive, I had to sell the black cloth you gave me as farewell present in Karlsruhe. So the new dress will have to remain a dream.

I'm asking you to help me. I can ask you as my friend, can't I? You always wanted to help me, didn't you? Could you send me a parcel every month? Believe me, it's very hard, being unemployed, to live alone and keep house.

The most important items are fat or tallow and meat, such as corned beef (all in cans, of course) and ready-made cake mix (all you need to do is put it in the oven, and, *eins, zwei, drei,* the cake is ready). Also cheese, sardines, cocoa, coffee, tea, a little candy and chocolate. And maybe also a few nuts. So now you have the general idea, at least. I know, Günter, that it sounds like a bit too much. But I need it all so badly.

In addition, maybe you can send a parcel also in-between, with the things you can't pack together with food. Soap, shampoo, Band-Aids, a cigarette lighter, cigarettes, tooth paste, ointment, stockings, possibly yarn for darning in all colors. None of that can be had here. Longingly, I'm wishing for a cornflower-blue suit. But that may be just too expensive. I'm not asking for it.

Günterlein, please don't let me down. Apropos, in my last letter I wrote about shoes. Urgent. Günter, I'm still running around in my clodhoppers. Sport shoes, dark but not brown. With a thick sole, if possible.

Günter, please don't think I'm greedy. I'm just being practical. Don't forget me!

Your Hildelein

My Günter, my everything, my myself!

Why this deep sorrow, when necessity speaks? Is there any way for our love to exist other than through sacrifices, by not demanding everything? Can you help it that you are not wholly mine, I not wholly yours? Love demands all, and for good reason; so it is for me with you, for you with me—only you are so easily forgetting that I have to live also for myself, as well as for you!

Now let's move from the inner to the outer life. Will we see each other again soon? I can't convey to you now the insights I've had into my life during these last few days. My heart is full. I want to tell you so much, but here are moments when I find that language is nothing.

Have a good time—remain my faithful, only sweetheart, my all, as I am yours.

I mailed a couple of parcels to Hildegard, but I didn't send her all the things she asked for, especially not the major items, such as the bathing suit, the shoes, and the cornflower-blue suit. It's hard for me now to reconstruct my then frame of mind that prevented me from showering Hildegard with all the things—and more—on her wish list. What had become of the noble resolve I made after reading her heart-rending response to my farewell letter, that I would do anything in my power to make it up to her?

About a month later, another letter arrived from Lore, which, to my surprise, was franked with German rather than Austrian stamps. She, her mother and her son had suffered a radical downturn of fortune:

Ingolstadt, 8 June 1947

Dear Günter,

We were deported to Germany. It's just as awful here as you had warned me it would be and warned me not to go back if I can help it. Well, I couldn't help it.

The transport was horrible. After we spent six days in a freight car, it was derailed in the Ingolstadt switchyard. We all thought our last hour had come. Thank God, nothing very serious happened to us, except for a few bruises. But many of our fellow-deportees had to be taken to a Munich hospital.

By the time we had to leave, the swimming season in the Kitzbühel lakes was under way. I prefer the beaches on the Baltic or the North Sea. I'm longing to dip into the beautiful, clear, clean ocean almost as much as I'm longing for you.

The longer I'm in this awful place, the more horrible I find it. Maybe I've got so much trouble because I'm not speaking like a Bavarian. They've got nothing but crooks and good-for-nothings here. I fell victim to these gangsters during my very first days, when they stole my suitcase out of my little garret, along with your nylons that I had guarded so carefully. Could you send me a replacement, should your finances allow it? Please forgive my bothering you with this, but everything is so hard to get here. For the time being, I'll stay in Ingolstadt, but I hope that by wintertime I'll be back in my beloved mountains; because I can't imagine a winter without skiing.

Write me soon, because you'd bring happiness to

Your Lore

In my answer, I commiserated with Lore about the cruel turn of fate she had suffered, but said nothing about my wanting to bring her the States, to take her away from all her troubles and make her my wife. I merely provided her with more news bulletins about what was happening to me back at school, and let on that her portrait, prominently displayed on my dresser, had aroused the envious admiration of my roommate. Generous to a fault, I sent her a parcel with a few nylon stockings.

Dear Günter,

I get to be quite besides myself whenever there's mail from you. Maybe it would be better for me if there wasn't any more mail from you (NO, ANYTHING BUT THAT!), because then it all comes back to me, and I become so sad that my dream came to such a quick end, before it had even begun. How often does remorse come too late for things not done! I dreamt of you last night, which is why I've got to write to you.

My main yearning (second only to my longing for you) is for the mountains, which is why I've come down to stay in Garmisch for a couple of weeks. Yesterday I went up the Zugspitze, but the dirty summer snow at the summit depressed me.

I thank you in advance for your gift. Actually, you needn't have sent it, even if I bellyached about the stolen nylons. You oughtn't to have taken it so seriously. You know me! I'm always fussing about trifles. I was always so spoiled! I always got everything I wanted. It does me good not to have all my wishes fulfilled.

I like the photograph you took of me at the Schwarzsee, with the white-topped Wild Emperor in the background. But I prefer the one of me on the ski lodge sun deck, with my black-topped Tame Ami Bear in the foreground. As Bettina von Arnim wrote to Goethe, "If the mails went more quickly, I'd hear from you sooner."

I kiss you.

Your Lore

I wondered whether by her lament that remorse for things not done so often comes too late, Lore meant her refusal to come up with me to my room on our last evening together at the Grand. Heartaches that I gave up so easily that night had certainly consumed me ever since, especially when I realized after my return to the States that she requited my love.

Ingolstadt, 27 August 1947

Dear Günter,

Whenever I'm depressed, which is often since I came back to Germany, I lie down—close my eyes—and dream of our short togetherness, in the evening, dancing at the Grand Hotel, or what was even more beautiful, when we sat, or half lay, on the ski lodge sun deck and kissed to while the time away, until we took the last cable car down the mountain. It was so wonderfully carefree being with you, and the memory helps me to get through all those unpleasant hours.

Last week some friends from Berlin visited me here. They made my mouth water so much that I applied immediately for permission to travel to our hometown. If only you were here. I'd have asked you to come with me. I'm a little scared, but also very excited.

Much love; my thoughts and kisses are hurrying to you across the sea from

Your Lore

Berlin, 12 October 1947

Dear Günter,

I've been in Berlin for three weeks now and still haven't got around to revisiting all the old places where my life was once enacted. Very often, as I stood at people's doors, they looked at me as if I were a ghost. The joy of meeting someone again of whom one had lost track is too beautiful for words. Dearest Günter, when will you stand at my door?

I am not merely toying with the idea of moving back to Berlin: I'm actually running around like mad to get permission to do so. But that turns out to be stupendously difficult. I'm in an awful bind: As a Prussian I'm not wanted in Bavaria, and as one more returnee to house and feed, I'm not wanted in Berlin.

I'm thinking of you. With a kiss, I remain

Your Lore

Ingolstadt, 1 November 1947

Dear Günter!

Now that I was back in Berlin, I am colossally sorry that I was away from my hometown for such a long time. The Berliners are indestructible. Beautiful new shops have grown out of the giant piles of rubble. I wish I could have participated in it all. If only I weren't such an exemplar of ineptitude!

Next spring, I'll move back to Berlin. My mother is going to return to the little house that she owns in Rangsdorf. Unfortunately, it's in the Soviet zone, so that I have little desire to move in with her. I hope to find some place in the West—Grunewald, Dahlem, or Zehlendorf. Then I'll be a Berliner once more and remain one until—well, until the world has something else in mind for me.

When I got back to Ingolstadt from Berlin, a jackass at the city housing office announced that we had to clear out of our hotel rooms within three days and move into the refugee shelter at the Evangelical Mission. According to him, I have no right to live in Ingolstadt. I am a Prussian, and why don't I go back where I come from? (My poor Tame Ami Bear! You're a Prussian too, which is a dirty word nowadays.) I was bowled over. I told him that I hadn't been asked where I wanted to be born, but if I had, I certainly wouldn't have picked Ingolstadt. Well, I was plenty mad—I went to the military government and got their support, thank God—and now I may stay in beautiful, hospitable Ingolstadt as long as I want.

Needless to say, I'll get out of here at the first opportunity. My first choice would be to emigrate to South Africa. But having learnt nothing—only having lived however I wanted to live—with that preparation you can't pioneer a new life all alone in a strange country. So I'm going back to Berlin, but only under the condition that the Red Curtain isn't lowered any further.

Greetings and kisses from

Your Lore

I didn't pity Lore too much that those Bavarian yokels were calling her a Dirty Prussian, which I'd prefer any day over being called a Dirty Jew. But it did send a chill down my spine that racist South Africa—of all places—was Lore's first choice for an overseas country to which to emigrate. Maybe she really meant America as the place where she would prefer to settle and mentioned South Africa only so as not to put me under pressure. What pressure? Wasn't being at Lore's side when she starts a new life in the States exactly what I wanted? She wants to emigrate and is covertly asking for my help. Since she writes as if she were a free woman and never mentions her husband in any of her plans, she is probably divorced by now. Why, for heaven's sake, don't I ask her to marry me? What am I waiting for? I don't think I knew then, or even know now, why I didn't make any move to get what I thought I wanted so badly and seemed to be mine for the asking.

Another letter from Hildegard came in late summer.

My dear Günter!

Man wird bescheiden!
Ein Mensch erhofft sich fromm und still,
dass er einst das kriegt, was er will!
Bis er dann doch dem Wahn erliegt,
Und schliesslich das will, was er kriegt.

One gets to be modest!
A person hopes, devout and still,
That she shall get what she might will!
Until, at last, she stops to fret,
And only wills what she can get.

I'm not working. I've got an acting contract, but it's suspended because I came down with tonsillitis and can only whisper. Anyhow, the Schiffbauerdamm Theater kicked me out. When I showed up on my return from Karlsruhe, they asked, "Did you find your vacation in the West worthwhile? Why didn't you stay a little longer, Dearie?"

I haven't been swimming even once this summer. After all, I can't risk going into the water in the buff! But may I ask you not to send me a bathing suit? It'll be fall soon, and, more importantly, I would very much prefer if you sent me a pair of shoes and something to eat. Do I have to explain to you why? No, I'm sure you can work it out that I need shoes because my clodhoppers are disintegrating, and I need food because I'm hungry all the time.

I had to sell more than half of my food ration coupons to meet my expenses (rent, doctor's bills, and so on). Not much left after that. Neither for eating nor for living. I'm mentioning all this over and over again, so that you can grasp and understand my ruthless demands. For God's sake, you aren't supposed to send me all these things all at once. Of course not, but you should know what I am in desperate need of.

Your Hildegard

Not knowing how to respond to Hildegard in any meaningful way, I just sent her another little parcel with only a very few of the things of which she had written as being in desperate need of. Two months went by until I heard from her again.

Du, Günter, my old friend.

I had turned my back on Berlin because I couldn't stand it any longer. An old pal invited me to come to Hildesheim and stay with his parents. It took me four days to get there. They treated me royally. Best of all, for once I could eat until I was no longer hungry. I spent my birthday, to which I had been looking forward so much, on the train, all alone. Twenty years!! My God, Günter, I'm going on thirty! Did you think of me on my birthday?

Now I'm back. Came in from Munich yesterday. Things were awaiting me in Berlin that put me into such a state that I had to flee to you by way of this letter. You know, wherever I am, you are with me, and it's to you I'm always talking. Everything else is one big nothing!

Your long silence troubles me; before, I was troubled mostly by what was in your letters, when I did hear from you. Günter, there are things in life that one simply cannot explain.

While I was gone, Frau Mama appealed to the Youth Authority, arguing that I'm too young to live on my own. She prevailed. So, I had to move back into our family flat. That was Act II of the tragedy: My mother resorts to force. But in Act III, daughter will be twenty-one, and then . . .

My trip was terrible. From Hildesheim, I had gone to Augsburg to sign an acting contract at a theater there for next spring. A Herr Heinrich, who lives in Paris and whose friend owns the theater, had suggested I meet him in Augsburg. He wasn't of any help: Nothing doing about any contract. I went to Munich with him for a week, and then he went back to Paris and me to Berlin. I didn't stop in Karlsruhe. I'm glad I didn't, because alone . . .

As I gather from your letter, your research is tough going. But hang in there, Günter, you've got nerves of steel! Anybody who was able to stand me for six weeks has got to have them.

I received a parcel, for which I thank you. I was badly in need of the soap, and the lipstick and nail polish are totally along my taste.

I'm thinking of you! I kiss you in my thoughts. I'm waiting.

Your Hildegard

P.S. Here's an order for you: Next time, write by hand; it means so much more to me than something typed.

These concurrent letters from Lore and Hildegard made me aware of the tremendous personality differences between these two flesh-and-blood examples of my abstract ideal of the blonde Berlin woman. When Lore strode into the Grand Hotel in her Persian lamb fur, so much more elegantly got up than poor Hildegard with her home-sewn, olive-drab woolen overcoat, the vast disparity in their social backgrounds and personal fortunes had not escaped my notice, of course. Surely, none of the old friends' doors at which Lore had stood when she revisited Berlin were in Hildegard's proletarian Wedding district, in which Lore had probably never set foot. Her Berlin was upscale Grunewald and Dahlem.

Hildegard, unemployed and down-and-out because she had given up everything just to meet me for a little holiday in the Harz Mountains, was struggling to keep body and soul together for bare survival, begging for the material support that I could have so easily given her but wouldn't, and for the love that I couldn't. Lore, unemployed because she didn't need to earn any money and having never worked a day in her life, seemed to be lacking only nylon stockings and was worried mainly about how she could get back to the mountains for some good skiing in the winter and to the seashore for some good swimming in the summer. And I had sent her the stockings.

Yet despite the differences in their situations, both Lore and Hildegard were having a hard time, each in her own way. Both were needy, feeling deprived, and looking to me for the help that I couldn't bring myself to give either of them.

A woman identifying herself as the wife of Heinrich Selver, the former director of the Jewish school I attended in Berlin, called me in Champaign in June 1947 to invite me to a reunion of former staff and students of the PRIWAKI living in the Chicago area. I was thrilled by the prospect of seeing some of my old teachers and schoolmates again. The reunion was held at Marks Nathan Hall, a home for Jewish waifs in Chicago of which Selver was now the director. He seemed hardly changed during the ten years I hadn't seen him, as was Erwin Jospe, our music teacher, who had also turned up and was now the musical director of a fancy Reform temple on the North Side.

The eight former students who came to the reunion included two girls from the lower grades to whom I hadn't paid any attention at the PRIWAKI because we considered the younger kids as little insects, unworthy of our notice, except when they annoyed us. Both had developed into comely women now definitely worthy of my notice. One of them, Steffi, was the daughter of Mrs. Guttmann, the chief cook presiding over the school's kitchen whose tasty hot lunch in June 1947 and moved down to New York City, where she had found a thick cauliflower soup, mock hare—roast of ground beef, potato salad, fruit gelatins—I preferred to the food I got at home. Mr. Guttmann père had been the lead-baritone of the Charlottenburg City Opera. Thanks to a special intervention with the Nazi minister of culture on his behalf by the general manager, he was not fired until 1935. As one of those fervent German–Jewish patriots, Mr. Guttmann refused to leave the Fatherland, and his wife and daughter were stuck with him in Berlin. By the time he died of a stroke in 1941, it was too late for Steffi and her mother to get out. So they went underground, using false identity papers. Mrs. Guttmann was taken on as a housekeeper on the country seat of a noble Prussian family. When her mistress discovered her true identity, she kept her concealed from the Gestapo on the estate until the end of the war.

Steffi went into hiding in Bavaria, where she was hired as a draftswoman at the Messerschmidt airplane factory. She took on the identity of the young widow of a fallen Wehrmacht officer and simulated a fanatic Nazi harpy who intimidated her Aryan coworkers. When Steffi was liberated by the U.S. Army's sweep through Bavaria in the very last days of the War, her sudden release from years of ever-present fear of discovery, imprisonment, and death precipitated a nervous breakdown. But her cheerful and witty presence at the reunion suggested that she had recovered from her wartime trauma.

The other noteworthy former insect was Edda Schürmann, who, although she had left Berlin when she was only eleven, had turned into one of those vivacious, sassy Berolinas who appealed to me so much. Her father had been a wealthy garment manufacturer in Berlin, whose household included a butler, a domestic grandeur unique even among our affluent PRI-WAKI crowd. Mr. Schürmann had managed to reestablish his business in London, where he became just as wealthy as he had formerly been in Berlin. During the blitz, Mrs. Schürmann took Edda and her elder sister to Chicago, where the girls went to high school on the North Side. After the war, Mrs. Schürmann rejoined her husband in London, but the sisters stayed on alone.

Here I had been vainly combing the City all those years for an ideal girlfriend, while, the whole time, she was right here under my schnozzle—a childhood schoolmate, to boot! Yet come to think of it, being a German-Jewish girl, Edda might not have held out any oomph for me if we had met in Chicago as teenagers. Being romantically attracted to her now must have been one of the early symptoms of the abatement of my Jewish self-hatred, set off by my triumphant return to the Fatherland.

Ready to throw overboard my old erotic prejudices on behalf of Edda, I took her for a nightcap at the Congress Hotel's Glass Hat bar after the reunion, and we finished off the evening at her apartment. She played for me a set of precious, pre-Nazi German 78-rpm records with songs from the *Dreigroschenoper*, sung by Lotte Lenya, Kurt Gerron, Willy Trenk-Trebitsch, and other members of the original cast. The Schürmann family had taken the records with them on their flight to England before the war. These songs overwhelmed me with Berlin nostalgia when I had recently heard them at a showing on the Illinois campus of the sole surviving print of Georg Pabst's 1931 movie adaptation of the Brecht–Weill operetta. I made Edda play them over and over again.

We got on famously and gabbed through the night. But Edda gave me to understand that she didn't feel attracted to Jewish men. Anyway, she was engaged to a former Air Force noncom, a handsome goy, who worked as a tie salesman at the Carson Pirie department store. I could just picture the jerk: a big, dumb hulk with a vacuous kisser.

Good news reached me in Champaign not long after my return from Germany. Contrary to my lugubrious premonition, the protagonist of Schrödinger's *What Is Life?*, the young German physicist, Max Delbrück, had not fallen on the eastern front after all. He was alive, and in America!

But there was also bad news. Delbrück was teaching at Vanderbilt University in Nashville, Tennessee. After my wartime visit to Claire and Bob in Mississippi I had little stomach for moving south of the Mason-Dixon Line for my postdoctoral studies. No doubt, working with Delbrück in Nashville would have given me a leg up in the search for other laws of physics. Yet forfeiting once again my dream of living in California seemed too heavy a price to pay for that opportunity.

By late summer, better—no, super-tremendous—news reached me. Delbrück had moved from Vanderbilt to Caltech, where he was appointed professor of biophysics. I wrote to him right away, to ask whether there was any possibility of my working under his direction in Pasadena as a postdoc, after I received my Ph.D. in physical chemistry at Illinois next summer. I explained that my research experience was in the physical chemistry of high polymers, but I had recently become interested in biophysics.

Delbrück's reply came within a few days. He couldn't encourage me to apply for a Caltech postdoctoral fellowship, because the research going on in his laboratory was concerned with bacteriophage. There was no physics in any of this, and hence not the type of work in which I seemed to be interested. Therefore, in fellowship appointments for which he could speak, he would give preference to candidates who wanted to work on research problems that were very close to what was already under way. But if I came to Caltech on an outside fellowship, he would be more inclined to have me work on problems unrelated to his current interests.

So much for my getting started on the discovery of other laws of physics in California! Who's going to give me an "outside fellowship" for such a hare-brained project, especially since Delbrück himself seems to have lost interest in it? I'll just have to reconcile myself to carrying on with dull, old-time physical chemistry after my doctorate, probably not even in California but somewhere in the boonies, like New Jersey or Delaware.

A week or two after the Delbrück debacle, Ernest Grunwald, the son of the German–Jewish friends of Gunther Steinberg's mother in Los Angeles, visited me in Champaign for a weekend. Ernest had just got his doctorate in organic chemistry from UCLA and taken a research job with the Portland Cement Association in Chicago. While I took him for a walking tour of the Illinois campus, he asked me what I was going to do after I got my Ph.D. next summer.

"I'd hoped to get into biophysics research with Max Delbrück at Caltech. But he just brushed me off. He wrote he wouldn't take me on unless I came to Pasadena on an outside fellowship. How am I going to get an outside fellowship, for Christ's sake?"

"Have you heard about the new type of National Research Council Postdoctoral Fellowship, bankrolled by the Merck Chemical Company?"

"No. What's it all about?"

"The announcement I read in *Science* said the purpose of the new Merck NRC Fellowship is, quote, 'to provide special training and experience to young men and women who have demonstrated a marked ability in research in chemical or biological science and wish to broaden their fields of investigational activity.' That's you, isn't it?"

"No, it isn't. I haven't demonstrated any marked ability in research in chemical or biological science."

"How much demonstrated ability in research can they ask of a graduate student who is going to get his Ph.D. only next year? Don't be a drip, Gunther! Go ahead and apply! What you've got to lose?"

I didn't think that there was any possibility that the NRC would give me one of its prized postdoctoral fellowships. Very few were awarded each year, and they were reserved for the young superstars-presumptive of American science. Former NRC Fellows included my first academic hero, Linus Pauling (1925–1926); the inventor of the cyclotron, Ernest O. Lawrence (1927–1928); and the founder of molecular virology, Wendell Stanley (1930–1931), all of them future Nobel Laureates. Maybe one of the new, special Merck NRC fellowships might be easier to get, but still . . .

But Ernest was right. I had nothing to lose. So I wrote to Delbrück again, to let him know that I was thinking of applying for an NRC Merck postdoctoral fellowship and to ask him what sort of problem I could be working on in his laboratory if I were awarded one. He replied that my question was rushing him a little bit, since he was not now in a position to state in any detail the type of problems he was going to work on next year. He was thinking of doing some experiments on phototaxis in purple bacteria, which might be a good opening for the study of excitatory processes.

Delbrück's second reply was no more friendly than his first, but at least it didn't seem to amount to an outright rejection. I didn't know the meaning of "phototaxis" or "excitatory processes," had never heard of "purple bacteria," and was totally in the dark about what all this might have to do with genes and the paradox that would lead me to the discovery of other laws of physics. Never mind! There was no point worrying about such fine details now. There'll be plenty of time to find out what Delbrück's proposed project is all about, in the unlikely event that my fellowship request would be granted.

So I applied to the NRC in Washington for award of a Merck fellowship. All I could do by way of documenting my "demonstrated marked ability in research in chemical or biological science" was to provide a brief summary of my modest research accomplishments in the Doc's laboratory. As for my wish to broaden my field of investigational activity, I declared that I hoped to apply my knowledge of physical chemistry to the study of biophysical problems, with special emphasis on the investigation of life processes from the point of view of thermodynamics and reaction kinetics. My immediate plan was to study at the California Institute of Technology under the direction of Professor Max Delbrück, investigating the general nature of excitatory processes. I had the good sense not to let on in my application that I had a hidden agenda: looking for other laws of physics.

My Hero

RITA HAD GRADUATED FROM VASSAR in June 1947 and moved down to New York City, where she had found a glamorous career-girl position, an editorial assistantship on the *Scientific American*. That magazine, dating back to the late nineteenth century, had been a long-decrepit popular science monthly when Dennis Flanagan, a former *Life* editor, took charge of it after the war and tuned it into the world's premier serial of its genre. In December 1947, the magazine sent Rita to Chicago, to cover the annual meeting of the American Association for the Advancement of Science. It was going to be held at the Stevens Hotel on Michigan Avenue over the Christmas holidays and was expected to be the biggest scientific meeting ever held in the States.

I was looking forward to the AAAS meeting and my first participation in a scientific congress, fantasizing that one of these days, I would be a main speaker at such an event, admired and applauded by the colleagues crowding the lecture hall. My reverie was hazy about the topic of my brilliant presentation. It could have hardly concerned my present studies on the vapor pressure of liquid mixtures or on the relative speed of migration of negatively charged chain molecules, with which, as they say in Berlin, you couldn't lure a dog away from a warm stove. No, I'd have to be acclaimed for the discovery of as yet unimaginable other laws of physics.

I invited Rita to come down to Champaign from Chicago on the weekend before the AAAS meeting, to avail herself of my expert help to get an exclusive scoop for her magazine on the stupendous developments at the World's Greatest Synthetic Rubber Research Center. The university being on holiday, I was able to book Rita into one of the grand bedrooms on the top floor of the Illini Union, normally reserved for VIP visitors. We dinner-danced at my old standby for maxi-dates, the Hotel Urbana Lincoln, and then, at the Illini Union, on the sixth anniversary—almost to the day—of our first meeting as high school kids at the B'nai B'rith dance at Temple Sholom, slept together for the first time.

That night was fraught with very different meanings for each of us. For Rita, it was the removal of the last barrier to our intimacy, whereas for me, it was merely a natural development of a close, albeit loveless, friendship of long standing. Rita talked about our getting married and how everybody was saying that we were obviously meant for each other. I pretended not to understand, and she dropped the subject. She left Champaign in the morning and wrote to me as soon as she got back to Chicago. The romantic sweetness of her brief note was unique among the many letters she had sent me over the years.

> Gunther, it was a wonderful week-end. After I got back from Champaign, it hardly seemed as though I had been there only little more than 24 hours. It seemed much longer than that—or maybe just timeless—it passed too quickly. It was a memorable moment—the kind that fiction writers are always trying to record—and never quite catch. Enough slush!

Rita and I went to a few talks together at the Stevens Hotel. The one I liked most later won the $1000 prize for the best lecture of the whole AAAS meeting. It was the presentation by the thirty-year-old University of Chicago physical chemist Harrison Brown of thermodynamic arguments showing that meteorites derive from the breakup of an ancient planet. I thought that if the other-laws-of-physics project didn't pan out, I might try to follow in Brown's footsteps and apply physical chemistry to astronomy, the only science subject that had interested me in high school.

As a representative of the press, Rita was asked to lots of receptions and cocktail parties thrown for the nabobs of science, to which none of its foot soldiers, let alone graduate students, were invited. She took me along, and I thus got my first taste of scientific high-life. There were very few women at these affairs, and even fewer attractive ones. Rita was mobbed by senior statesmen old enough to be her grandfathers. Obviously, she had a great future in science journalism.

It was a heady experience for me, too. Rita introduced me to her boss, and we met such luminaries as James B. Conant (the AAAS president), Harlow Shapley (the astronomer), and Philip Frank (the philosopher). We also ran into Sol Spiegelman, who obviously didn't remember meeting me at Martha Baylor's a couple of years ago. Instead of resuming our earlier argument about thermodynamics and living systems, he was raving flirtatiously to Rita about plasmagenes.

New Year's Eve, I was back in Champaign. I spent it with a few of my fellow graduate students of the Doc's and their wives and girlfriends. I remembered last year, when I had been so happy with Hildegard, first at her theater and then in our love nest at the Gossler. I suddenly missed her. If only I could have had her with me now! My yearnings for Hildegard were strictly at my convenience, namely when I was lonely—while I continued to ignore her mental and material welfare. Far from having sent her a parcel at Christmas time with some of the things she needed so badly, I had not even troubled to send her my wishes for a happy 1948. Yet I heard from her.

> My dear Günter,
>
> My wishes for the New Year come a little late; but better sending them late than not sending them at all, as you didn't! Did you remember last New Year's Eve at least a little? It was a beautiful day: my première and you. This time I went into the New Year all alone. I went for a walk around midnight, looked for Jupiter and brought a little order into my soul.

Why don't you write? Günter, I don't understand you. Are you mad at me because I haven't written for such a long time?

Did you get married? There's nothing special to report about me, except that I'm still in love. But this disease is going to come back again and again. I can't find any antidote.

Your Hilde

I wrote to Hildegard that I too had felt lonely on New Year's Eve, even though I spent it in the company of some fellow students, that I had thought of her all evening and of how happy we had been together a year ago, and that I would never forget that night in all my living days. No, I said, I'm nowhere close to getting married. I made no response to Hildegard's closing remarks, which, in their simple, direct pathos had made me want to love her in return.

Early in March, Rita came back to Champaign, this time on a bona fide, albeit not wholly fortuitous and probably self-initiated, assignment from the *Scientific American*. She was to gather material for a story on Professor W. C. Rose's identification of eight of the twenty kinds of amino acids present in natural proteins as indispensable ingredients of the human diet. This study, whose urine-bottle-carrying, feces-collecting student-guinea-pigs in my chemistry classes had first given me the idea that biochemistry is a disgusting subject, was now nearing its completion. Although Rose's laboriously collected results were of little fundamental biological importance, they were of great practical significance for nutritional science. This was the angle Rita was supposed to emphasize in her on-site collection of background material for the story.

Rita stayed in Champaign only one night, which we spent together at the Inman Hotel. Feeling that we were sufficiently grown-up, I simply rented a double room for Mr. and Mrs., instead of going through my old college-boy routine of sneaking upstairs from the basement. It didn't occur to me that the registration charade at the Inman must have been fraught with bitter meaning for Rita, since she so badly *wanted* to be Mr. and Mrs. So I treated our pulling a fast one on the hotel as a big joke. If she didn't share in this mockery, I was unaware.

After several months of waiting to hear about the fate of my application, I had written off the NRC Merck fellowship and applied for instructorships in chemistry at Yale, Washington, Florida, and Duquesne. After all, I needed a job if I was going to marry Lore and bring her to the States. And then, just a couple of days after Rita's visit to Champaign, I received a telegram asking me to come to New York for an interview with the Merck fellowship board. First-class rail travel and accommodations at the Hotel Pennsylvania would be paid by the NRC.

My sky-high exultation over this fabulous news subsided as soon as I began to think about the interview. It wouldn't take more than one or two incisive questions by the board to show that I knew nothing about the excitatory processes on which I was proposing to carry out advanced postdoctoral research. I had never even bothered to find out what "phototaxis of purple bacteria" was all about, the phenomenon whose study under Max Delbrück was supposed to lead me to novel insights about life processes from the point of view of thermodynamics and reaction kinetics. Maybe I could cram some biology in the few days left before the interview? No, there wasn't any point in even trying. How, starting from a base of total ignorance, could I pick up enough savvy in such a short time to fool the board? Rita would be back in New York by the time I got there for my interview, and promised on-site moral support.

My rival finalist-candidates for the Merck fellowship were waiting outside the board's meeting room at the Hotel Pennsylvania. Had I been aware of their credentials—they included one future Nobel laureate, Carleton Gajdusek, and three future members of the National Academy of Sciences, I probably would have packed up and gone home. Age twenty-three, I was also the youngest. All the others, except Gajdusek, who was twenty-four but looked sixteen, were in their late twenties or early thirties. There were no women candidates.

The Merck fellowship board, whose chairman was the president of the National Academy of Sciences and ex-officio high priest of American science, comprised six formidably distinguished, awe-inspiring senior scientists. Yet I was so pitifully ignorant that I had previously heard of only one of my six illustrious interviewers, namely the geneticist and future Noble laureate, George Beadle, chairman of the Caltech biology division. I had been an involuntary attendee of Beadle's lecture on his recently developed "one-gene-one-enzyme theory," which he gave as the keynote speaker at my induction into the scientific honor society of Sigma Xi in my last undergraduate semester. Listening to Beadle, I had been amazed that the mysterious Mendelian gene, of whose existence I had only just learned in Will's philosophy course, now seemed to be amenable to study by chemical methods.

The board members were deployed in Last Supper formation behind a long table covered by a green felt cloth. Two senior staffers of the NRC and a stenographer sat at either end of the table. I was put in a chair on the other side of the table, facing my inquisitors. My fears that it wouldn't take the board long to discover that I knew virtually nothing about the research proposal for which I was seeking support were not unfounded. The interview opened with a few preliminary questions by the chairman about my personal history and background. Since my biographical information was already lying on the table before the board members I reckoned that he was asking these questions only to put me at ease. Presumably I knew the answers to them. But it didn't put me at ease at all to have to own up that I was a Berlin-born, German–Jewish refugee. Anyway, trouble started right away with the chairman's next question.

"So you want to go into biology; what do you plan to do?"

"I want to study excitatory processes to test whether the Second Law of Thermodynamics applies to living systems."

"How are you going to do it?"

"I'm going to study the phototaxis of purple bacteria."

"How? And how's this going to tell you something about the applicability of the Second Law?"

"I'm not exactly sure. Professor Delbrück suggested that this would be good experimental material for my project."

On this answer, the board members frowned and shook their heads in disbelief. Their reaction evoked in me, not for the first time in my academic career, the refrain of a tragico-comic rhyme by the German versifier-cartoonist, Wilhelm Busch, about a failed scholar flunking an oral exam, which haunted me ever since I first came across it as a school boy:

> *Über die Antwort des Kandidaten Jobses*
> *Geschah ein allgemeines Schütteln des Kopfes.*
> Upon this answer of Candidate Jobs
> Ensued a general shaking of nobs.

After I proffered a few more obviously unsatisfactory responses to the questions of other board members, one of them finally asked me sarcastically,

"Then, if I understood you correctly, your proposed postdoctoral studies in biology at Caltech would have to be at the—(pause and emphasis)—undergraduate level. Isn't that so?"

"Yes, Sir. I guess so."

This response concluded the interview, and I was dismissed summarily. As I shuffled out, humiliated as never before in my life, I noticed, to my astonishment, that Beadle winked at me, as if we shared some private joke.

Having blown my main chance, I went upstairs to my room at the Pennsylvania, hoping that a hot bath and a nap would calm my nerves before dinner with Rita. I picked her up at the apartment on West End Avenue in which she was sharing a room with a would-be career girl from Wisconsin. Their landlady was a bleached blonde of about 40, who, according to Rita, didn't approve of her sublessees engaging in free love because she was selling hers. We ate at a restaurant chosen by Rita, who evidently knew her way around Gotham's checkered red-and-white-table cloth and candle-in-Chianti-bottle bistros.

Rita asked, "How'd the interview go?"

"There were six famous scientists who quizzed me, including the president of the National Academy of Sciences. He's the board's chairman."

"Well, how'd it *go*?"

"It didn't take very long. I guess they were short on time. They had a lot of candidates to interview."

"Come on, Gunther, don't play dumb! I mean how well did you *do*?"

I realized that she could tell from my downbeat mood that something had gone wrong. So I stopped pretending and availed myself of the opportunity to unburden my soul to my long-term girlfriend.

"I blew it, but totally, Rita! It was awful, awful. At the end, one of 'em even humiliated me by asking whether my so-called postdoctoral studies at Caltech wouldn't have to be at the *undergraduate level*. Hell will freeze over before I become an NRC Merck fellow."

Rita didn't believe me: "Aren't you a little paranoid, Gunther? How can you be so sure that they aren't going to give you the fellowship? The interview was probably just a formality. Most likely, they'd already decided beforehand to give you the award. They just wanted to have some fun making the candidates jump through the hoop."

"Think about it, Kiddo," I answered. "If it were really the board's policy to make an award to any interviewee who turns out to have a pulse-beat, they could save the NRC a lot of dough by cutting out the interviews altogether. No, Ma'am, there's no chance that they'll give me the fellowship."

Although I overtly rejected Rita's antidefeatism as nonsensical, covertly it did raise a tiny glimmer of irrational hope in my bosom that maybe all might not be lost after all. "Is it possible," I asked myself, "that Beadle winked at me to let me know that it was a done deal? No, it's not possible! Rita doesn't know what she's talking about!"

After dinner, we went to a loft on 6th Avenue, where Lili Cassel, who had been the premier artist among my PRIWAKI classmates, had her studio. I hadn't seen Lili since 1938. Struck by Lili's angelic face, I was dumbfounded that my callow love-sick yearning had been wholly focused on Ursula and that I had paid no heed to Lili's beauty during our school days. Lili had already made a name for herself in New York as a freelance calligrapher, book jacket designer, and illustrator of children's books. I had never been in an artist's studio before, but it wasn't too far from my movie-tutored imagination. The place was in picturesque disarray, with books and papers all over the place, and a large drafting table strewn with Lili's work in progress. Lili disappointed me in only one respect: Her German accent was even thicker than mine. Lili, whose perfect English elocution had always been held up to us for emulation by our teachers!

There were also two other PRIWAKI classmates at Lili's studio whom I hadn't seen since before the war. One was Gisela Rosenfeld, who had recently come to the States from Belgium, where she survived the Nazi occupation by hiding with false papers on the outskirts of Brussels. She was hoping to make a living in New York as a graphic artist. The other classmate was Gerd Rawitscher, our tough star athlete, my rival for Ursula's affections, fellow member of the quadrumvirate of bullies at the top of the class' pecking order, and my partner in the disorderly conduct that brought about our joint expulsion from the school by Monsieur Jacquot. Gerd had fled from Berlin to a *hachshara*—a Zionist agricultural retraining farm—in Holland, although he had no more desire to settle in Palestine than any of the other kids. In 1942, he escaped from one of the collecting stations from which the Nazis were shipping Jews from Holland to the extermination camps in Poland and was hidden by a Dutch family living in the center of Amsterdam.

The formerly boisterous Gerd was subdued and hardly spoke all evening. I supposed that it was his terrifying wartime experience—locked in a closet for three years while constantly afraid of discovery by the Gestapo—that was responsible for the transformation of his formerly ebullient personality. After deliverance from his closet-hideaway on the liberation of the Netherlands, Gerd had married Inge, a lively, half-Jewish Berlin girl, who was also at Lili's studio. Like Gerd, Inge had survived the war by hiding in Holland. They had recently come to New York, and he had Americanized his name to Gary Raven.

When I told my classmates about my bombing the fellowship interview, they all insisted that they won't be fooled by my phony modesty. Gisela said,

"Obviously, the board thought that you're a hot-shot. Why else would they have invited you to make this all-expenses-paid-trip to New York?"

"Because I fooled 'em with my glib *schmooze* on the application form, that's why!" I replied. "You're just a naïve greenhorn, fresh off the boat, Gisela. You don't yet appreciate how far you can travel on mere *chutzpah* in the States."

I returned to Champaign by way of Indianapolis rather than Chicago, because I didn't want to face Claire after this fiasco. Thoroughly discouraged, I was now hoping that one of the instructorships for which I had applied would pan out, if not at the first-class Yale, then at some second-, third-, or fourth-class school. Three days after I got back, a telegram arrived at Noyes Laboratory that said that I had been awarded a Merck fellowship. Someone was playing a joke on me, of course. Anybody can go to Western Union and send a telegram purporting to be from the NRC, can't he? But then a letter came from the NRC, confirming the telegram and informing me that my tax-free stipend would amount to $3,000 per year. I, the arch-four-flusher, King of the Phonies, had really won an NRC fellowship! How, after my disastrous interview, could the Merck fellowship board have possibly selected *me* as one of only seven awardees among a total of 46 postdoctoral applicants?

As I eventually learned when I got to Pasadena, Delbrück's appointment to the Caltech biology division had not been uncontroversial. After taking over the division chairmanship from Thomas Hunt Morgan, Beadle had engineered Delbrück's return to Caltech over the opposition of most of his faculty. Beadle, I conjectured, must have reckoned that it would help vindicate his disputed recruitment if an NRC fellow joined Delbrück's laboratory, even if it were an ignorant jerk. Beadle had probably turned his not inconsiderable charm and persuasive powers on the formidable fellowship board to sway them in my favor. Even though Beadle and I later became good friends, I never asked him about this conjecture, and, though probably true, it remains unproved.

I wrote to Delbrück.

> I suppose you probably don't remember our earlier correspondence, but in the meantime the NRC has awarded me an outside fellowship for work with you. Will you have a place for me in your laboratory?

Delbrück had known all along about my fellowship award through Beadle, probably even before I did. In his reply, he asked me to meet him in Chicago in early May at the house of a friend, so that we could talk about what I might do with him at Caltech. He would be stopping in the City for a day on his way from Pasadena to Indiana University, where he was "going to visit Luria." I had never heard that name before, but savvy friends at Illinois explained to me that Delbrück was referring to his collaborator, Salvador Luria, with whom he had published an awfully important paper in 1943.

Little more than a month had passed since Rita had visited me in Champaign on her *Scientific American* assignment, when she called me from New York to tell me that she thought she was pregnant. I was petrified. Not considering the possibility of marrying her, I ventured the opinion that it was probably a false alarm and said that I trust that everything would work out satisfactorily.

New York, March 31, 1948

> I'm glad that you trust that everything will work out satisfactorily. Unfortunately, I can't share your opinion. So far nothing has turned out satisfactorily. A test should reveal all by Saturday. But I have a sneaking suspicion it will be positive. In the meantime, my cousin has been looking around for

someone to help me out on that event, but he hasn't been too successful there either.—So you can see, I am not sleeping in a bed of roses these days. In fact, I'm not feeling so hot at all.

Goodnight, Gunther, and please pray for me.

New York, April 7, 1948

I wish I had something definite to tell you, but unfortunately I don't. The test said no, but my cousin doesn't seem to trust it. He's giving me a series of shots, which should bring about certain physiological reactions that simply haven't appeared So far the shots haven't done any good, and I can say nothing. My nerves are somewhat frayed around the edges, as you may well guess, but there is nothing to do right now, except wait. Will let you know if anything happens, but am losing heart rapidly at this point.

New York, April 9, 1948

All right, you can relax now. These shots worked, with a vengeance. I feel as though I were about ready to die, but am glad, of course. God, these last few weeks have been pure hell! But my cousin was very nice to me ... moral support and all.

Rita finally understood that I was an irresponsible womanizing cad, who wouldn't lift a finger when he got her into trouble.

Like me, Delbrück had left Berlin for America shortly before the war. But unlike me, he was not a Jewish refugee. He came to the States to continue his scientific training. Other European scientist-emigrants who settled in the States before the war were driven here by the cosmic geopolitical forces that unleashed the war. But Delbrück was driven to abandon Germany for love. When war broke out, he stayed on to marry his fiancée, a Californian college girl, rather than going home to Berlin and report for military duty.

Born in Berlin in 1906, he was the scion of a well-known Prussian family, whose members included bankers, clerics, and savants, among them his father, Hans, professor of history in the University of Berlin. His maternal great-grandfather was Justus von Liebig, one of the nineteenth century founders of organic chemistry, and his maternal uncle was the Harnack Haus's eponymous Adolf von Harnack, founder and first president of the

Kaiser-Wilhelm-Gesellschaft for the Promotion of the Sciences. Delbrück disliked his uncle—as did his father—for being one of Kaiser Willie's reactionary toadies.

Delbrück studied physics in Göttingen from 1926 to 1929, at the time of the excitement that had attended the development of quantum mechanics by Schrödinger and Heisenberg. After flunking his doctoral examination, he was given a second chance, eventually passed, and was granted his Ph.D. in 1930. He then went to Copenhagen as a postdoctoral fellow in Niels Bohr's Institute of Theoretical Physics, where he worked on a problem concerning the decay of radioactive atoms.

Delbrück returned to Berlin in 1932, to take a job in the Kaiser-Wilhelm-Institute (KWI) for Chemistry in Dahlem as an assistant to Otto Hahn and Lise Meitner, who would discover atomic fission in 1938. In later years, Delbrück liked to point out that he had played a crucial role in world history. Hahn and Meitner were chemists, and if Delbrück, as their house physicist, had not given them bad advice about the interpretation of their experimental results, they might have discovered atomic fission a few years earlier —maybe in time for the Germans to have developed an atomic bomb before the end of the war.

While working for Hahn and Meitner, Delbrück's interest had already turned from physics to biology, after hearing Niels Bohr's 1932 lecture "Light and Life." In that lecture, Bohr had laid out the philosophical implications for living systems that the new quantum physics had brought to the conception of natural law. He proposed that the ultimate understanding of life itself will require the discovery of some novel principle of matter, analogous in its radical character to the "uncertainty principle" of quantum mechanics. This lecture was the fountainhead of the beguiling idea that the study of living processes bids fair to turn up "other laws of physics."

To get in touch with biology while working for Hahn and Meitner in Dahlem, Delbrück joined a discussion group led by the Russian fruit-fly geneticist, Nicolai Timoféef-Ressovsky, who was working at the KWI for Brain Research. The central focus of this group was on the then utterly mysterious physical nature of the gene. Their discussions resulted in a paper Delbrück published jointly with Timoféef and Karl Zimmer, entitled "On the Nature of Gene Mutation and Gene Structure." This paper remained virtually unknown until Schrödinger drew attention to it a decade later in his *What Is*

Life? (in which he made no reference to Bohr as the original source of the idea of "other laws of physics").

Hoping to make a career as a university professor in spite of his unpromising start, Delbrück signed up for the *Dozentenlager*. This was a political indoctrination camp, which the Nazis required aspiring *Dozenten*, or assistant professors, to attend before their habilitation, or license to lecture at a university, could be granted. One day, after his class had solemnly intoned the sacred Nazi anthem, the "Horst Wessel Lied," Delbrück pointed out to the instructor that the last two lines of the first stanza

> *Kameraden die Rot Front und Reaktion erschossen,*
> *Marschier'n im Geist in uns'ren Reihen mit.*

are semantically ambiguous. Since the relative pronoun *"die"* can mean either "whom" or "who," two alternative readings are possible:

(1) Comrades *whom* Reds and reactionaries shot march along in spirit in our ranks.
(2) Comrades *who* shot Reds and reactionaries march along in spirit in our ranks.

So, Delbrück asked, were our dear, defunct SA comrades victims, or were they killers of Reds and reactionaries? The flabbergasted camp director reported Delbrück to the National Socialist German Union of *Dozenten*, which, in turn, blacklisted him in a circular letter found in the Berlin University archives after the war:

> National Socialist German Union of Dozenten
>
> Berlin, 25 November 1936
>
> To His Magnificence, the Rektor of the University of Berlin
>
> Re: Application of Dr. Max Delbrück for Habilitation in Theoretical Physics.
>
> Since Dr. Delbrück does not yet satisfy the political criteria that have to be met by a National Socialist University teacher, I request that you reject his application for a *Dozentur* in Berlin. I could agree to his appointment to a *Dozentur* at some other Institution only if he would find himself there in a salubrious National Socialist environment.
>
> Heil Hitler!
> W. Holtz

Delbrück was lucky to have merely ruined his university career by making fun of Nazi hagiology. A few years later, his joke would have cost him his neck. Here was another difference between Delbrück and me. Raised as a member of the German intellectual aristocracy and totally self-confident, Delbrück felt free to be a nonconformist. I, by contrast, as a self-hating German Jew, wanted to blend in with the wallpaper, to avoid drawing any attention to myself. Had I, a super conformist, been in Delbrück's place, I would have been a model student of the *Dozentenlager*, enthusiastically joining in with the singing of the "Horst Wessel Lied."

Delbrück left Hahn and Meitner in 1937, just as they were about to discover atomic fission. He went to the States as a postdoctoral fellow supported by the Rockefeller Foundation, to gain professional competence in genetics. His intent was to join Thomas Hunt Morgan's stellar fruit-fly genetics group at Caltech, but he soon realized that fruit-fly genetics was not for him. Its jargon was too arcane and its corpus of data too complex for mastery during his intended one-year stint in Pasadena.

Yet he managed to find an experimental material at Caltech that he thought more suitable than the fruit fly for his fellowship studies. This was a virus that infects and multiplies in bacteria, with which another postdoctoral fellow was working in the basement of the Caltech biology building. Delbrück realized that the virus-infected bacterium is an ideal experimental material for studying biological self-replication. He had been looking for just such material, because the gene's capacity for self-replication was its most mysterious aspect.

Delbrück thought that the mechanism of bacterial virus replication was a problem that he might be able to clear up before it was time for him to head home for Berlin. He made very good progress on this project during his first fellowship year in Pasadena, but by its end, he hadn't quite managed to solve the problem he had posed for himself. So his Rockefeller fellowship was extended for a second year, which expired just as the war broke out. As it turned out, it took another twenty-five years to solve the problem of gene self-replication, and the solution was provided, not by Delbrück himself, but by his disciples.

On May 2, 1948, I set out for the City to meet Delbrück. Throughout the three-hour drive up in the decrepit 1935 Plymouth I had bought after my return from Germany I feared that the old jalopy would have one of its chronic breakdowns. In between worrying about getting stuck on the highway and missing my appointment, I agonized about what might happen if I did *not* miss it. What if Delbrück found out during our talk that I didn't know anything about biology beyond what I had picked up from Schrödinger's little book? Might he not decide that his standards were not as low as those of the evidently witless Merck fellowship board? I also tried to imagine what Delbrück would be like: short, heavy-set, bald, wearing horn-rimmed glasses, and speaking English with a heavy German accent.

The Plymouth seemed to appreciate the importance of our mission and faithfully carried me to the Ellis Avenue address on the South Side that Delbrück had specified. It happened to be just around the corner from where I had lived with Claire and Bob in my Hyde Park High School days. When I rang the bell, there was Delbrück, exactly as I had pictured him! Except that the man who opened the door was not he but his friend and host, Hans Gaffron, professor of biochemistry at the University of Chicago. Delbrück, then forty-two, was actually tall, thin, with a full head of clipped brown hair, unbespectacled, and spoke English with only a slight, and not particularly German, accent.

What I had expected to be a brief, one-on-one interview with Delbrück —the briefer the better—turned out to be a luncheon party given by the Gaffrons for him and an elderly couple visiting from England, Mr. and Mrs. Ernst Pringsheim They had been waiting for me, and we sat down to eat as soon as I arrived. Delbrück asked me,

"D'you still speak your mother tongue?"

"Yes, Sir, I certainly do."

So they went back to talking in German. Their conversation was at a stratospheric intellectual level, fluttering between literary, artistic, political, and scientific topics. I felt like an interloper, having burst in on this distinguished company of old friends. At first I didn't dare to speak up, fearing that I had little to contribute that would interest my elders. But then the conversation turned to Berlin. Delbrück and I were born and raised there. Gaffron, though born in Peru, had gotten his doctorate in Berlin in the twenties, and worked at a KWI in Dahlem until he emigrated to the States in the mid-1930s. Pringsheim, an eminent microbiologist nearing retirement age, had been professor of biochemistry at the University of Berlin in the 1920s, before moving to the University of Prague, whence he fled to England after the Nazi invasion of Czechoslovakia.

Since I happened to have been the first of this party to have gone back to Berlin after the war, I could contribute something to the conversation, after all. My chance to show off came when someone mentioned the Harnack Haus.

"It was in perfect condition when I dined there in November 1946. It's been converted into a fancy U.S. Army officers' club."

I didn't trouble to mention that I had been thrown out of the Harnack Haus when I tried to dine there with a Fräulein I had just picked up.

"You were in Berlin in 1946? What were you doing there?"

"Yes, Ma'am, I sure was. I worked as a scientific document analyst for the U.S. military government."

"What happened to the State Opera?"

"Its house Unter den Linden is a total ruin, but they're putting on performances in the Admiralspalast. The Philharmonic is playing in the Titania Palast cinema in Steglitz."

When the conversation turned to Bertolt Brecht and his current troubles with the House Un-American Activities Committee in Washington, Mrs. Pringsheim reminisced, "I'll never forget the opening night of the *Dreigroschenoper* that Ernst and I attended at the Schiffbauerdamm Theater in 1928."

"The Schiffbauerdamm is alive and well, Mrs. Pringsheim. It's one of the main Berlin stages these days. In fact, I happen to have a girlfriend who's an actress at the Schiffbauerdamm."

I also didn't trouble to mention that my girlfriend lost her job at the theater more than a year ago, when she came to live with me in Karlsruhe.

I had never been at such a conversationally dazzling meal. This was the kind of sophisticated crowd with which I wanted to keep company in the future! I was enchanted by Max Delbrück: lightning-quick on the uptake, funny, and amazingly well informed on a wide range of subjects. He seemed to know everybody, especially the all-time greats of quantum physics, onward from Max Planck (who lived next door to his parents' villa in Grunewald and in whose garden little Max Delbrück had "bought" cherries "in the English way" back in the days of World War I), through Niels Bohr to Werner Heisenberg, Wolfgang Pauli, and Paul Dirac. Delbrück entertained us with a Niels Bohr anecdote, illustrating Bohr's typically Danish love of paradoxes.

"Bohr showed Heisenberg around his country house north of Copenhagen, and Heisenberg noticed a horseshoe mounted over the door. Heisenberg said, 'Niels, I'm surprised that you, of all people, believe in this silly superstition!' Bohr answered, 'No, Werner, I don't really believe in it. But I've heard that some superstitions bring good luck even to people who don't believe in them.'"

When the conversation turned to the then very fashionable philosophy of existentialism and someone asserted that Kierkegaard's novella, *En Dansk Students Eventyr (A Danish Student's Adventure),* was a seminal contribution to its origins, Delbrück interrupted,

"Nonsense! *En Dansk Students Eventyr* was written by Poul Møller, Kierkegaard's senior by twenty years."

I was happy and excited over the prospect of going to work with him. As it turned out, this was one of the very few occasions in my lifelong association with Delbrück, first as a disciple and then as a friend, on which we spoke to each other in German. He believed that everyone should speak English on American soil, regardless of national origin or linguistic competence. But, like Hildegard and Lore, he always called me "Günter," since to fellow Berliners, the archaic, Wagneresque form "Gunther" just didn't seem to fit a lad my age.

When the time finally came to discuss my future projects, Delbrück didn't quiz me. As I later found out, he generally assumed that his interlocutors or lecture audiences were wholly ignorant of the subject under discussion. He used to say that he was asking of his conversation partners no more than they be infinitely intelligent. Delbrück didn't mention phototaxis of purple bacteria or sensory excitation at all.

"I take it that you want to work on phage?"

"Yes, Sir, that's exactly what I want to work on. But could you refresh my memory as to just what 'phage' is actually all about?"

"No need for that now. You'll find out what it's all about soon enough at Cold Spring Harbor Laboratory on Long Island. You're going to spend the summer there and take the phage course. In early September, we'll all head out West, to Pasadena."

At the end of the afternoon, I drove Delbrück to the Loop, where he caught the train for Bloomington. I felt that in chauffeuring the protagonist of Schrödinger's *What Is Life?* I was on my way to becoming the Number One Sidekick of the Big Number One. We parted with a cordial "*Auf Wiedersehen in Cold Spring Harbor!*"

Evidently it didn't bother him that I was such an ignoramus. Super-satisfied with the outcome of this first meeting with my new boss, I drove back to Champaign. On our return trip, the ancient Plymouth seemed to realize that it was no longer critical for my professional future. Its worn-out master cylinder would no longer hold hydraulic fluid for more than a few depressions of the brake pedal. So we crawled home brakeless, without any means of stopping other than throwing the gear into reverse. But in spite of being in dire danger of losing life, limb, and property, I felt relaxed. A few weeks ago, I turned twenty-four, and within a few weeks I'd be entitled to style myself "Doctor Stent" and then escape from Illinois at last, to start an exciting new life in California. As I would soon learn, however, styling yourself "Doctor" was considered vulgar on the Big Time Scene into which I was about to stumble, and addressing a colleague by that honorific was often a sarcastic preface to insult.

Stepping Out with Mother Nature

AFTER HAVING BEEN VERY SLOW to answer their letters, I heard from both Hildegard and Lore not long after I had been notified of the award of the Merck fellowship. Hildegard's came first:

Hello to a birthday child!

Does there exist a more conceited man than you? No, impossible. Just one question, Günter, may I reckon with the possibility that in one of your lucid moments, you will reach for your pen, turn your mind back one year, and write to me, possibly at Easter, at Whitsun, or on my birthday?

With me, everything is as usual. There is still no word from my father. I read, I grow, I thrive, sometimes with a little more, sometimes with a little fewer calories. May I give you a tip? Give me a little pleasure again one of these days and sweeten my birthday in advance. Of my thanks, Sir, you would be assured.

Do you remember your last birthday, March 28, 1947? Our climb up the

Turmberg to the observatory tower in the afternoon? Your birthday party at the Breyers in the evening? *Je ne suis pas curieux?* And then, you and me, we were all alone for the last time. Was it more beautiful this year?

Your Hildegard

Instead of replying, I thought, but did not write,

Yes, dearest Hildegard,

The final night we spent together on my 23rd birthday was beautiful. I spent the 24th in the company of fellow-graduate students. I appreciated the effort they made in my honor, but it was not beautiful at all, at least not in the way it had been beautiful with you.—But why are you calling me "conceited"? Maybe you mean it as a euphemism for "swinish," in which case I couldn't agree more. To assuage my guilt feelings, I am sending you a little parcel.

Then came Lore's letter, whose envelope astounded me. It was franked with Austrian stamps and bore the imprint of Kitzbühel's Hotel Postkutsche, as had her first letters to me. While poor Hildegard was struggling for survival in Berlin, Lore had managed to solve her problem how to get in some spring skiing before the 1947–1948 season was over.

Kitzbühel. Hotel Postkutsche. 20 February 48

Dear Günter,

You unfaithful tomato, you, as we say in Berlin! Why hasn't lazy bones been heard from for such a long time? Seems like an eternity to me. Were you snowed under by a blizzard?

You'll be surprised to get a letter from me from Kitzbühel—our Kitzbühel. Thanks to my persistence I succeeded in getting a passport, which made me very happy. Now I'm spending happy hours skiing down fabulous runs in marvelous weather. Wouldn't you like to be of the party?

Kitzbühel has changed very much since we were deported. Prices went up to dizzy heights. Everybody complains. Nobody has any money. That's why there are so few people here and the hotels are almost empty, which can't be said of the resort hotels in Germany, which are bursting at their seams. It's hard to predict how all this is going to work out in the long run.

Before coming here, I spent six weeks in Garmisch and Mittenwald. Unfortunately, the weather was very mild; so there wasn't much skiing to speak of. The only compensation was the carnival, which I celebrated as I hadn't done for years.

Now, get your pen and write me your latest news!

Fondest greetings from

Your Lore

Within a few days, there arrived another, much less upbeat letter from Lore. It was franked with German stamps.

Ingolstadt, 9 March 1948

Dear Günter,

I came home half dead from my long rest and recreation trip to Mittenwald and Kitzbühel. I'm getting daily heart injections to revitalize me. I expected too much of myself, and unfortunately it didn't do me any good. So I became quite run-down and now have my hands full to restore my old zip. The food situation is worse than a year ago. As ordinary citizens we get only 75g fat per month and 300g meat. And from what one can get besides that, such as 62g cheese, one can't live.

I still haven't moved back to Berlin for good, because I still haven't received the last stamp. Moreover, at this time there is little enthusiasm for moving to the East. But the situation changes daily, like the weather. My mother wants to leave for Berlin as soon as possible, but she'll have to postpone her move because of my physical condition.

My son is going to the local *Gymnasium*, and since he is a good student, nothing is hard for him. But in order to be an even better student, he has a private tutor twice weekly. I'm afraid that the children aren't going to learn too much from those ancient, ossified *Pauker* whom they seemed to have dug up in antique or curiosity shops. And I believe that a lot of knowledge will be needed for making a living in the future. Don't you think that's also true in your own case? On what kind of new research problem have you just started?

The winter didn't mean well with us. There was lots of rain. It was only during the ten days in Kitzbühel that I was able to make some terrific runs. Since I thought of you often as we were schussing downhill, I sent you greeting from Kitzbühel.

I remain with a kiss

Your Lore

I wrote to Lore that I would take my final doctoral examination on June 14 and that I had been awarded a postdoctoral fellowship. This fellowship would allow me to quit boring physical chemistry and become a biologist.

Unfortunately, to prepare myself for my new line of work, I would have to take a summer course at Cold Spring Harbor on Long Island. So I wouldn't be able to visit her in Germany in July, as I had hoped. But I was determined to come to see her in the summer of 1949, after the end of my first fellowship year.

Ingolstadt, 1 June 1948

Dear Günter,

Your letter reached me in Mittenwald, where I spent an extended, three-week Whitsun holiday. Unfortunately, horrid Ingolstadt remains my enforced permanent place of residence, and since I hate that place for many reasons, I try to get away from it as often as I can. I was twice in Austria again for a short time, in Seefeld. As long as we still use the old Reichsmark money, I can allow myself a little excursion from time to time; but I don't know what's going to happen when they introduce a new currency.

I congratulate you on your postdoctoral research fellowship. I hope you'll like it in California. I remember that it can get pretty hot on Long Island in the summer, but on the shore one can endure the heat better.

I'll be thinking of you on the 14th of June and knocking on wood on your behalf so much that I may chop up my entire furniture. But your taking it easy after becoming a Herr Doktor and just hanging around idly has got to be out of the question. After all, mankind isn't living only to eat, but is eating in order to live, and that costs money. So, my little-big would-be lazybones, you'll have to behave yourself and get down to earning some money. OK.?

I'm sorry that because of your (heavy) work nothing is going to come out of your trip to Europe this year, but then the year 1949 is coming. It'll be only a year.

My mother was in Berlin during the critical time, when all of us here feared that the borders to the East would be closed overnight. Thank God this nasty situation has blown over. Otherwise, the mess continues here and no clear prospects have come into sight. Waiting—waiting for—waiting for everything—life has become an eternal waiting-only. Until—

Affectionately,

Your Lore

My tedious, year-long number crunching had shown that the thermodynamic theory the Doc had developed for predicting the vapor pressure of pure liquids can, in fact, be successfully extended to mixtures of two liquids. My calculations allowed me to identify the conditions under which the vapor pressure of either of the liquids in the mixture deviates significantly from the simple formula known as Raoult's law. I had also managed to design and build the special electric chamber that permitted the analysis of the lengths of negatively charged chain molecules present in the anode (positive terminal), middle, and cathode (negative terminal) compartments after the flow of a measured amount of electric current. By use of this chamber I was able to confirm experimentally the Doc's theoretical prediction that shorter-chain molecules will migrate faster toward the anode than longer-chain molecules.

So the Doc suggested that I get going on writing up my results for the dissertation. It was my first major writing project—a much more substantial literary undertaking than the little baccalaureate honors thesis I had done for him in 1945. Contrary to the opinion almost universally shared among my graduate student colleagues that thesis writing is a big pain in the butt—the worst part of getting a Ph.D.—I enjoyed it immensely. Thus becoming aware of my bizarre preference for writing about my scientific findings over making them in the first place gave me my first inkling that there seemed to be something oddball about my motivation for doing science, quite aside from my hoping to use it as a way to professional fame and glory for impressing women.

According to conventional wisdom, scientists are driven by a thirst for understanding the natural world. Their reward for slaving away in the lab day and night is supposed to be a "joy of discovery"—the elation that attends reaching a novel insight into nature, as well as satisfying one's innate curiosity. This can't be the whole story, of course. As is well known to professionals, the satisfaction of a scientist's curiosity about nature provided by a competitor's discovery is cause, more often than not, for disappointment rather than joy.

I had chosen science as my vocation even though my curiosity about the workings of nature was not all that fervid, anyhow. What attracted me to science was the lifestyle that came with it. I found it hugely satisfying to work in a lab, to make good conversation—scientific or otherwise—with intelligent colleagues, to travel all over the world, and to have friends everywhere. Whenever I did manage to find something really new—which didn't happen very often in my career—I too felt elated, of course. But on those rare occasions, I derived my main satisfaction, not from a joy of discovery, but from a joy of telling. What excited me most was the thought that, thanks to my discovery, I will really have something to say next time I meet a colleague or go to a conference.

That is why publishing papers was what I liked best about science. No sooner had I started out on a research project, than I was thinking about the paper I would write about it. Long before I had found anything worth reporting, I was already composing the opening paragraph of the report in which I would present my yet-to-be-made discovery. Publishing gave me such great pleasure because I thought of it as a way to get a conversation started. In fact, compared to the joy of telling, the joy of discovery played such a minor role in my motivational makeup that—at least so I believe—I wouldn't have done science if I had been Robinson Crusoe. Out of my colleagues' earshot and with no women around to impress by my fame and glory, I wouldn't have made experiments, even if there happened to be a fully equipped lab on the island, with Man Friday available as a research assistant.

While I was writing my dissertation, I heard from Edda that she was no longer engaged to her tie salesman, who had reenlisted in the Air Force. According to her sister, she was well rid of the jerk and ought to show more interest in me. Edda complained that we always want what we can't get. Evidently lacking Hildegard's philosophical depth, she didn't draw the inference that therefore we ought to want only what we *can* get.

To test whether Edda was ready to take her sister's advice, I suggested that, on Memorial Day, we take a trip to the Indiana Dunes lining the Lake Michigan shore. It was a sunny day, and we found a place to picnic on the sandy beach. Lake Michigan was still too cold for swimming, but we ran, hand in hand, along the shore through the shallow water, spritzing it madly. When we sat down again, dripping wet, on our beach blanket, I put my arm around Edda and kissed her. She didn't resist, but as soon as I let go, she shook her head.

"I'm a lost cause! Would you mind my staying just your good buddy?"

Although my social and intellectual qualifications as a swain of a nice Jewish girl from Berlin—an imminent Herr Doktor coreligionist—were obviously superior to those of her ex-fiancé, I evidently lacked sex appeal, probably, I thought, *because* of those very desirable qualifications. So I wrote off Edda for the second time.

In accord with ancient academic tradition, the time and place of my final doctoral examination and thesis defense was announced in the official university weekly calendar. This announcement was strictly ritual, since no one except the candidate, the dissertation supervisor, and the examiners was expected to be in attendance. The Ph.D. candidate would present a brief oral summary of the dissertation to the examiners, who, since they had already read and informally approved it, never asked questions. No candidate in Noyes Laboratory had ever flunked the final examination, within living memory, at least.

To everyone's surprise, there were three Japanese sitting in the back of the seminar room when the Doc, my three examiners, and I came in. The Doc asked,

"Did you gentlemen perhaps get lost on your way to some other event on the campus?"

"Excuse us, please, Sir, but is this not the room where a Ph.D. candidate in physical chemistry is going to defend his dissertation? My colleagues and I are Japanese physical chemists on a study tour of the United States. We are interested in learning how this critical step in the maintenance of high academic standards in our discipline—a Ph. D. candidate's thesis defense—is handled in American universities."

The Doc decided that there were no grounds for excluding these strange aliens—they must have been one of the first scientists to have come over from Japan after the war—from what was, after all, a publicly advertised event. So he would let me present a summary of that part of my dissertation that was concerned with the vapor pressure of mixtures of liquids. I was not allowed to discuss the other part, devoted to the differential rate of migration of short- and long-chain molecules in an electric field. Those results, obtained through contract research for the Office of Rubber Reserve, were classified, pending clearance from Washington.

So my final doctoral examination was even more perfunctory than the usual quickie affair. After I had spoken for about half an hour, the Doc interrupted my spiel and said, "Congratulations, Gunther. I mean *Doctor* Stent!" He and the three examiners signed the title page of my thesis, and that was that.

Despite the lack of facial display of emotion by Japanese, well known to me from Hollywood wartime movies, I could tell that my auditors were amazed by the offhand nature of these proceedings. Maybe they were wondering how, in view of the lack of rigor in the training of its recruits, American physical science had managed to build the atom bomb.

With its hideous engine knock, my old Plymouth would never make it to the East Coast. Luckily, I managed to find a buyer who took the wreck off my hands without troubling to take it out on the highway for a test drive at high speed. Used-car buyers were not too fussy in those days, since automobile production was still far from satisfying the enormous demand for new cars that had built up during the war.

The owner of the garage to which I had taken the Plymouth several times for major surgery told me that I was in luck. He had a 1942 Ford two-door sedan for sale, the last model built before production of civilian cars was stopped after Pearl Harbor. He told me that this was as young a used automobile as I could hope to find. It was a bare-bones, "Victory" version of the 1941 model, with a painted, rather than chrome-plated, radiator grille and lacking the standard decorative chrome stripping.

Despite its youthful vintage, the prize car looked decrepit. The maroon paint job had lost all its former sheen, the safety glass window on the driver's side had been replaced by a piece of Lucite, and the upholstery was in tatters. The five-digit mileage register on the speedometer read 13,489, which meant that the car had at least 113,489, if not 213,489, miles on its hide. My benefactor promised to put the car into tip-top mechanical shape, although he wouldn't dress up its looks.

The lucky find turned out to be the Mother of All Lemons, running the gamut of automotive pathologies. Just as my nightmarish dread of Nazi persecution had finally begun to subside, it became replaced by another neurotic obsession. Whenever I was at the wheel of my Victory Ford I was afraid that it would suddenly and unaccountably stop in its tracks. Although nerve-wracking, owning this worn-out relic of the prewar motor-car era was actually a formative experience. Unable to afford having it routinely repaired, I had to become a mechanic to keep it running. Over the two years that I drove it, I gradually replaced almost all of its mechanical and electrical parts (and had to have a professional install a rebuilt engine and transmission). In the end, only the chassis and body were left of the original chariot.

No sooner had I been hooded as Doctor of Philosophy by the dean of the graduate school at the solemn commencement exercises in the Illini auditorium and marched out of the hall to the concert band's rendering of Elgar's "Pomp and Circumstance March No. 1," than I got into my Victory Ford, and, New York-bound, quit Champaign for good.

The Good Scientific Life

ON TURNING OFF INTO COLD SPRING HARBOR LABORATORY from Route 25A, which leads out to Long Island's North Shore from New York City, I entered a charmed world—a cozy medley of small laboratory buildings and dormitories scattered over an enchanting tree-studded plot bordering the Cold Spring Harbor inlet of Long Island Sound, thirty-five miles from Manhattan. Some buildings dated back to the first half of the nineteenth century when Cold Spring Harbor was a major whaling port. Others had been built at the turn of the twentieth century when rich New Yorkers were building fancy dachas on the North Shore. Sensing the atmospheric electricity of a highbrow life, I instantly fell in love with the place.

The lab assigned me a single room on the second floor of Blackford Hall. Blackford was one of the first reinforced-concrete structures built on the Eastern Seaboard and, as far as I could tell, had had no maintenance work in the forty years since its erection. Blackford was the lab's main social center. Its ground floor housed the kitchen and cafeteria on the west side and a lounge on the east side. Even though I was a hardened six-year veteran of institutional food, I found the three daily meals served in the cafeteria as awful as they were skimpy. The crummy lounge was furnished with dilapidated sofas and easy chairs that the Salvation Army would have refused to accept as a charitable donation. It was used as a parlor, as a seminar room, and, incredible as it seemed, as the venue of the world-famous annual Cold Spring Harbor Symposium on Quantitative Biology.

The second floor of Blackford was divided into a couple of dozen cell-like single rooms for senior (postdoctoral) gentleman summer visitors, who did their ablutions in a communal, minimalist toilet/shower facility. Blackford also had a basement, whose layout suggested that it once served some social functions. It was now under water most of the time.

The lab was founded in 1890, as a marine biology field station. By the turn of the century, it had become a popular summer camp for high school and college biology teachers in need of refresher courses. The first crucial step towards the emergence of Cold Spring Harbor as the breeding ground of molecular biology was taken in the early thirties, when the first annual Symposium on Quantitative Biology was held there. The symposium was put on for the avowed purpose of stimulating a dialogue between biologists and the physicists, chemists, and mathematicians whose ideas would eventually revolutionize life science research. The second crucial step came in the early forties, with the appointment of Milislav Demerec as the lab's director. Demerec had been a fruit fly geneticist, but after taking on the directorship, he shifted the lab's principal research focus to the genetics of bacteria and their viruses.

During World War I, the British bacteriologist F. W. Twort discovered an infective agent that kills bacteria. He published this finding in a brief note that remained unnoticed for several years. Meanwhile, the French–Canadian bacteriologist Felix d'Hérelle, working at the Institut Pasteur in Paris, announced *his*—maybe independent, maybe not—discovery of a very similar agent. He called it "bacteriophage" ("eater of bacteria"), a name that later usage shortened to "phage." D'Hérelle's announcement produced a sensation in the world of medical microbiology. He claimed that phages rather than antibodies in the blood serum are the chief mediator of natural immunity against bacterial diseases and should provide the means for a universal prophylaxis and therapy. This undreamt-of possibility of ridding humanity of some of its most dreaded ills fired the popular imagination, an aspect of the medical Zeitgeist of the 1920s that is preserved in Sinclair Lewis's novel *Arrowsmith*.

Arrowsmith had fascinated me when I read it as an assigned book at Hyde Park High. Yet I had forgotten—or, more likely, was never interested in —its scientific gist. All I remembered was that I looked to Martin Arrowsmith, the novel's protagonist, as a role model for his fun life as a fraternity man at a midwestern state university, his romantic adventures in exotic places as a scientist, and (ever-mindful of my father's matrimonial advice) his marrying a wealthy widow. I had managed to accomplish the first two of these goals. I *had* been a fraternity man at a midwestern state university, and I *had* had plenty of romantic adventures in exotic, war-ravaged Europe as a scientist. But it seemed uncanny that now I was about to emulate him in a way I had never imagined. Hoping that it would lead me to the discovery of other laws of physics, I was going to work on phage, the very biological material to which the fictive Dr. Arrowsmith had devoted his scientific career.

In spite of twenty years of frantic research efforts, phages never became a successful medical tool. And once antibiotics turned out to be far more effective in the control of bacterial diseases than even the most fervent enthusiasts of phage therapy had ever dared to hope for their panacea, the bizarre phage therapy chapter in the history of medicine came to a close. Yet despite holding mistaken notions about its medical potential, d'Hérelle managed to recognize phages for what we know them to be today: viruses that are parasites of their bacterial host cells.

Within two or three years of his original discovery, d'Hérelle had invented the *plaque* method of assaying the number of infective virus particles in a given volume of solution. Thanks to his plaque assay method, he was able to outline the life cycle of phage. The free virus particle attaches itself to the surface of its bacterial host. It then penetrates the cell, where it reproduces itself to generate an issue of many progeny particles. The progeny particles are ready to infect further bacteria when the infected host cell bursts open and releases them. It was d'Hérelle's plaque assay method and the possibilities it offered for carrying out quantitatively precise experiments with a rapidly self-reproducing organism that attracted Delbrück to working on phages when, as a young postdoctoral fellow, he first came across them in the basement of the Caltech biology building in 1937. He realized that the central question of life—"How does like beget like?"—could be distilled into the question "How does the parental phage particle manage to replicate itself inside the bacterial host to give rise to its multitudinous brood?"

D'Hérelle's view of phage as a bacterial virus found few followers among his contemporary bacteriologists, not least because his flamboyant personality seems to have precluded his being considered *un serieux* by his colleagues. The seeming scientific validity of some of his claims appeared to have caused his detractors to redouble their efforts to show him wrong. It was not for suggesting the (ultimately fruitless) medical applications that d'Hérelle usually came under attack. They were thought to be of such self-evident merit that he deserved little credit for them. Instead, it was for his credo that the phage is a virus that his detractors showed their greatest disdain. Twort's obscure brief note was exhumed by d'Hérelle's detractors to prove that his contribution to the discovery of the so-called bacteriophage was limited to coining its catchy moniker.

As I had already found out as a graduate student, the high intrinsic merit of a scientific discovery is not sufficient to assure its acceptance. My successful adaptation of the Clusius-Dickel thermal diffusion column for separating chain molecules of different lengths had aroused no interest in the synthetic rubber research community because I lacked the ability to dress it up as a salable package. Now, on learning of the put-down of d'Hérelle's correct identification of phage as a bacterial virus by his bacteriologist colleagues, I realized that the vendibility of the package also depends on the personal demeanor of the discoverer. If one is too timid and too dull, as was Twort, the package is likely to be ignored. But if one is too aggressive and too flamboyant, as was d'Hérelle, the package is likely to arouse resentment among the peers, who consign it to the trash can of history by managing to find fatal flaws in the package, as is always possible in biomedical science. So I arrived at the deep sociological insight that making a successful career in science depends on, among other things, acting neither too timidly nor too aggressively, as well as on being neither too dull nor too flamboyant.

In 1940, at a meeting of the American Physical Society in Philadelphia, Delbrück happened to run into Salvador Luria (then thirty-six years old), who had only recently arrived from war-torn Europe. A Turin-born Italian Jew, Luria had left Italy in the late 1930s, when, by way of showing Axis-solidarity, Mussolini started to implement anti-Jewish measures on the German model. Delbrück found that Luria, too, appreciated that the phage-infected bacterium offered an ideal experimental system for studying genetic self-replication and that he and Luria both wanted to address the same problem. The encounter of the German Delbrück with the Italian Luria brought into being the American Phage Group, whose members were united by the common goal of solving this little enigma and, on Demerec's invitation, made the Cold Spring Harbor Lab their communal summer resort.

The group's first recruit was Alfred Hershey, then age thirty-four, an assistant professor of bacteriology at Washington University, St. Louis. Its further growth was greatly accelerated by Delbrück's setting up the annual summer phage course at Cold Spring Harbor in 1945. The purpose of the course was frankly missionary: to spread the new gospel among physicists and chemists, a purpose that was not exactly hindered by the appearance in that same year of Schrödinger's *What Is Life?* Within a few years, the Group came not only to dominate phage research but also to exert a decisive influence on the then nascent, yet unnamed discipline of molecular biology. For their pioneering contributions to that revolution in twentieth-century biology, Delbrück, Luria, and Hershey were awarded the 1969 Nobel prize in physiology or medicine.

On registering for the phage course, I discovered that the summer of 1948 was going to be the first time that Delbrück was *not* going to teach it. He was about to take off for a conference in Paris and for his own first postwar return to Berlin. It was as inconceivable for Max (as everybody at Cold Spring Harbor called Delbrück) as it was for me to travel from the States to Europe without paying a visit to our native city, with which both of us were obsessed.

Before he left for Paris, Max introduced me to two people from Luria's lab. One, Renato Dulbecco, age thirty-four, Luria's postdoctoral associate, had recently arrived from Italy, where he and Luria had been fellow students at the University of Turin. During the war, Dulbecco had served as a medical officer in the Italian Eighth Army, which Mussolini sent to the Russian front. Dulbecco's outfit was surrounded by the Soviets, just as were its Big-Brother German troops at Stalingrad. Dulbecco was now in Cold Spring Harbor rather than in a Siberian POW camp—or in a mass grave on the banks of the River Don—only because he happened to be wounded and sent home for medical treatment just before the Soviets tightened their noose.

The other associate of Luria was James Watson, a twenty-year-old graduate student, who would receive the 1962 Nobel prize for codiscovering the DNA double helix with Francis Crick. Watson was born in Chicago and had attended Hyde Park High's rival, South Shore High. Now he was working for his doctorate at Indiana, doing research on the effects of X-rays on phages.

Jim Watson seemed like a strange guy. His goggle-eyed face was in constant motion, ever-changing from one grimace to another, and his peculiar way of speaking was not one I recognized as being native to Chicago's South Side. (As I discovered in the following year, when I met the Canadian microbiologist, Roger Stanier, Jim had been affecting the Vancouver Island accent of Stanier, whom he had admired at Indiana.) Jim's tall body was emaciated, despite his enormous appetite; he bought a double set of meal tickets, so that he could go through the Blackford Hall chow line twice.

Like me, Jim had been fascinated by Schrödinger's *What Is Life?* He had chosen Luria as his mentor because he had heard that Luria was a collaborator of Delbrück, the "young German physicist" protagonist of Schrödinger's tract. Jim first met Max at Indiana University on the day following my own first meeting with him in Chicago. Jim too assumed, as he later wrote in his memoirs, "that as a German with his reputation [Delbrück] must be balding and overweight."

I liked Dulbecco right away. Despite being my senior by ten years, he treated me as an equal. Well informed about a broad range of scientific and cultural subjects, he was a wonderful conversation partner. I was less enthusiastic about Jim. Despite his being my junior by four years—a mere graduate student—he treated me as an equal, acting as if his opinions were just as good as mine. But before long, I had come to terms with the sobering fact that whenever we disagreed about some scientific proposition, his opinions were almost always better than mine. And so we became lifelong friends.

Max's substitute instructor in the phage course was Mark Adams, professor of bacteriology at New York University. He taught us the standard laboratory procedures and experiments of the Phage Group, as they had developed since Max revolutionized phage research. I found experimenting with live bacteria and viruses much more thrilling than making the kinds of physicochemical measurements on dead macromolecules I had done at Illinois. What excited me most was the rapid pace with which significant biological findings could be made with phage and definitive answers to clearly posed either/or questions be obtained. I had no doubt that this was the kind of research that I wanted to do in the future.

By the time the course was over, I felt I had become an expert phagologist. I had also imbibed the conceit of the Phage Group that there was no point in paying any attention to the work of our predecessors or of contemporaries external to the "Church," as the French microbiologist, André Lwoff, referred to the coterie of Pope Max's disciples. Reading publications lacking the Church's imprimatur was worse than a waste of time: The unsubstantiated claims based on poorly designed experiments presented by such confused heathen outsiders would just put wrong ideas in the True Believer's head.

Max had established the tradition of a final examination and graduation ceremony for the end of the phage course. Held after dinner in the Blackford Hall lounge, it was preceded, as well as accompanied throughout, by heavy consumption of Red Rotgut, a wine a.k.a. "Mussolini's Revenge." The graduates, who had to wear academic robes of their design—a pirate, devil, or angel getup would do—stepped forward one by one, to be examined by the Lord High Inquisitor. Finally, each graduate was handed a certificate conferring a personally customized degree, such as "Master of Pipette Breaking."

Normally, Max played the role of the Lord High Inquisitor. The straight-laced Mark Adams seemed an unlikely substitute, and so we guessed that Luria would probably be the absent Max's understudy. Just before the start of the ceremony, we graduates were informed that Max had made it back from Europe just in time for the commencement. Costumed in a white bathrobe over my swimming trunks and crowned by a laurel wreath, I was already well into Mussolini's Revenge. The lights were turned out, and the lounge was plunged into darkness broken only by the flicker of candles. A figure wrapped in a white bed sheet walked in and announced that he was Max's ghost, come to preside over the exercises. With my visual acuity severely diminished by Red Rotgut and darkness, I thought that the mysterious figure really was Max, pretending to be his own ghost. It turned out to be merely the lowly graduate student, Jim Watson, who conferred on me, the exalted doctor of philosophy, the degree of "Master Mechanic," in recognition of my having spent as much time under my jalopy as in the lab.

When Max did come back from Europe, I met his wife, Manny. She was even more different from my image of the dreary, homely Frau Professor wearing steel-rimmed granny specs than had been the real Max from the balding, portly Herr Professor. No wonder Max had abandoned the Fatherland for her! She was good-looking, amusing and clever, easy to talk to, ever ready to play a joke on someone. She was always planning projects: picnics, swimming parties, tennis games, trips to restaurants. Her body posture and movements remained—and until her death in her late seventies—those of a college girl.

Besides the Phage Group votaries and the lab's small year-round resident staff, other people with biological or biochemical interests were also summering at Cold Spring Harbor. Many of the summer residents, as well as some short-term visitors, presented seminars on their current work (most of which were beyond my ken), followed by vigorous, and sometimes acrimonious, discussions. Although there was never a shortage of powerful and opinionated personalities among the audience, Max always dominated the colloquy, unless he had swaggered out in mid-seminar with a look of disgust. No matter how distinguished and self-confident the lecturers or discussants might be, none hassled Max.

Max himself gave a seminar in which he presented his contribution to the conference that he had just attended in Paris. There, Tracy Sonneborn's idea of "plasmagenes," whose alleged discovery by Sol Spiegelman was featured in the 1947 *Time* magazine article I had read at the FIAT officer's club in Karlsruhe, had been a major topic of discussion. Max said that he had nothing against plasmagenes as an explanation of how the surface of single-celled protozoa can change from one hereditarily stable feature to another without undergoing any gene mutation in the cell nucleus. But he pointed out that Sonneborn's findings could be explained just as well in an entirely different way, without invoking any genetic mechanisms at all.

The plasmagene hyothesis was the first example I encountered of an important generalization that distinguishes biology from physics and chemistry. Whenever two reasonable alternative hypotheses are put forward to explain some biological phenomenon, *both* will eventually be found to be applicable to one or another living creature. Thus, some types of cells turned out to contain the plasmagenes envisaged by Sonneborn, while others, including those that Spiegelman had incorrectly alleged to contain them, actually undergo stable changes in their physiological properties by the nongenetic mechanism Max proposed.

One of the seminars I heard was particularly memorable, because it dealt with a discovery that seemed to be important but that nevertheless had not made an impact on contemporary research. It was given by Rollin Hotchkiss, a biochemist at the Rockefeller Institute for Medical Research. Hotchkiss spoke on the hereditary transformation of bacteria by DNA discovered by Oswald Avery in the early 1940s. This discovery clearly pointed to DNA as the genetic material. Yet there had been very little mention of DNA as the genetic material in any of the other seminars, even though the physical and chemical nature of the gene was of capital interest not only to the members of the Phage Group, but also to most of the other people at Cold Spring Harbor. It was only four years later, after the Phage Group's own Alfred Hershey and his young assistant, Martha Chase, showed that DNA is the genetic material of the phage that DNA is suddenly in, as far as thinking about the nature of the gene was concerned. Within a few months there arose speculations about the nature of the genetic code, and Jim Watson and Francis Crick were inspired to set out to discover the structure of DNA.

I brooded about the reason for the ten-year delay in the impact of Avery's discovery. Finally I concluded that when Avery made it, his discovery had been premature, in the sense that its implications could not yet be connected with generally accepted, or canonical, knowledge by a series of simple logical steps. In the summer of 1948, DNA was still generally held to be a monotonously uniform chemical molecule that is always the same, regardless of its biological source. This belief made it impossible to fathom how DNA *could* be the carrier of hereditary information. But by 1952, at the time of the Hershey-Chase experiment, it had been shown that DNA is *not* monotonous and that its composition *does* vary according to its biological source. Now one could readily imagine that genes are inscribed in DNA as a specific sequence of its four kinds of iterated chemical building blocks, whose long string makes up the giant DNA molecule. In other words, DNA turned out to be Schrödinger's aperiodic crystal of heredity, composed of a succession of a small number of different elements.

This general refusal to accept the obvious implications of Avery's experimental results until they could be reconciled with the chemical structure of DNA later reminded me of the Doc's mistrust of the data I had gathered for my senior honors thesis at Illinois. He was suspicious of them because they didn't fit the bond strengths he had predicted on theoretical grounds. Like the Doc, the disbelievers in Avery's data evidently acted in accord with the maxim put forward by the astronomer Arthur Eddington that it is unwise to put too much confidence in observational data until they have been confirmed by theory.

The Cold Spring Harbor Lab's marvelous scientifico-intellectual ambiance was not its only allure. A bevy of attractive female laboratory assistants held out splendid prospects for summer romance. Getting to know these splendid prospects was as easy as falling off a log, since etiquette at the lab called for joining others at the dining tables rather than eating alone. Moreover, there was a continuous cadence of social activities. Their diurnal rhythm was reset from day to day to accommodate the swim at high tide on the lab's private Sandspit beach on an inlet of Long Island Sound. There were tennis and softball games, mussel and clam collections for cookouts at the Sandspit, and square dances on the lawn fronting the unmarried women's Carnegie Dorm, a run-down Victorian mansion built in 1884. Square dancing was big at Cold Spring Harbor because one of the staff members was a semiprofessional caller. The hard day's fun would usually be topped off with a walk of the singles crowd to Cold Spring Harbor Village for a nightcap at its waterfront tavern, Neptune's Cave.

At a late-evening party during my second weekend at the lab, I found myself sitting next to a Japanese girl. I had noticed her in the cafeteria but not considered her one of those splendid prospects. Asian women, of whom I had hardly met any—there were none at Hyde Park High and very few at Illinois—had never attracted me. And so I was wondering how I could move away from her politely, to position myself next to a more enticing quarry. But then I became aware of an erotic signal in her eyes, and stayed put. Her name was Taeko Y., a Nisei from Seattle. I asked her what she was doing at Cold Spring Harbor.

"I'm on the all-year staff of the lab. Dr. Annie Macnamara's technician. They tell me you're Max Delbrück's new postdoc. Taking the phage course?"

"Yes. How'd you fare during the war?"

"After Pearl Harbor, my whole family was locked up in the relocation camp at Tulare. Even though my brother, my sister, and I are native-born American citizens. In 1944, they let me out of the camp, after I was admitted to Oberlin College. I just graduated from Oberlin in June. Are you German?"

"I used to be. I was born in Berlin, but I grew up in Chicago. I was naturalized after the war."

"And how'd *you* fare during the war? You were an enemy alien, weren't you?"

"I was. But they let me run around free. In fact, I did war work for the government as a grad student at Illinois. Synthetic rubber research. I just got my Ph.D. in physical chemistry."

Around midnight, someone announced that it was high tide and a great opportunity to swim by the full moon. Taeko and I were the only people to act on that information. We took a couple of bath towels and set out into the sultry summer night for the fifteen-minute stroll from the lab to the Sandspit. Silently holding hands, we were sung to by a loud chorus of chirping cicadas in the bushes and trees lining the deserted road. About halfway to the Sandspit, I stopped to kiss her. When we got to the beach, we undressed, spread the towels on the sand, and made love on the moonlit beach. Then we went for our swim.

The party was over when we got back to Carnegie Dorm. Taeko took me up to her room, where I spent the rest of the night. I had found my summer romance. For the remainder of my stay at the lab, we always sat next to each other in the cafeteria, went swimming on the Sandspit together, and took part as a pair in all the lab's social events.

I introduced Taeko to my PRIWAKI classmates living in New York—taking her along to parties and dinners at their homes. She got on very well with my childhood chums from Berlin, especially with Gert and Inge Raven. I am sure that Taeko found these German Jews so congenial, because, as I soon discovered to my surprise, there was a peculiar symmetry between Taeko's psyche and mine. Once Inge asked her whether she liked kabuki.

"I'm not interested in that stuffy Japanese culture and its ridiculous traditions. Why should I be? After all, I'm a native American, not a Japanese."

She too was a self-hater! She was just as ashamed of being Japanese as I was of being Jewish. Hadn't I once thought that I was not a Jew but a native German? And now I had become Taeko's blonde Berlin shiksa!

Taeko created an aura of harmony about her person that I had never met before. Our relation remained wholly free of tension, from beginning to end. That is not to say that Taeko was without any opinions of her own, always deferring to mine. On the contrary, she was very opinionated, especially about people (not to speak about Japanese) and highly intelligent. She reminded me of the judo class that I took for the obligatory physical education credit in my freshman year at Illinois. In judo, you oppose force, not by counterforce, but by skillful yielding, until you have exposed your opponent's weaknesses. Then, seizing on them, you can bring your opponent down to the mat before he or she even realizes what's happening. This summer romance with Taeko was a wonderful thing for me, even though, far from being in love with her, I wasn't even *trying* to fall in love this time.

One evening in August, Taeko and I drove to Manhattan to take in one of the first German postwar films, *The Murderers Are Among Us*. It was made in the Russian sector of Berlin in 1946, set in the gigantic pile of ruins of the war-wrecked city. The character portrayed by Hildegard Knef (then twenty-one) in her first movie role is the girlfriend of a surgeon driven to drink by his remembrances of gruesome wartime experiences as a Wehrmacht medical officer. He is particularly obsessed by having witnessed mass executions of civilian hostages on the eastern front ordered by his commanding officer, a captain who, now demobilized from the army, is leading a contented, guilt-free life in postwar Berlin as a prosperous family man. The surgeon feels he has to kill the captain to punish him for his crimes, but his girlfriend convinces him that the captain must be turned over to the authorities investigating Nazi atrocities. The captain is indicted; he is tried; and the movie ends with the defendant shouting "I'm not guilty!" I'm not guilty!" as the camera pans over photomontages of Nazi atrocities.

The Murderers Are Among Us was not a commercial success in Germany. No other German films would be made for many years thereafter that treated the theme of personal guilt of ordinary citizens in the Nazi state. The few films that did deal with the Nazi period at all invariably showed that the German people too had been victims of that awful man, Adolf Hitler, and his gang of bullies.

The movie confronted me again with my shamefully ambivalent feelings about the collective guilt of the Germans for the Nazi atrocities. Of course, I thought that the captain, and probably also the surgeon, were guilty. But what about the Berlin girl played by Hildegard Knef, whom I identified with my own Hildegard? Of what was she guilty? After the movie, I couldn't get my Hildegard out of my mind. So far, I had not replied to the letter she had sent for my birthday five months earlier. I hadn't thought much about Hildegard in Cold Spring Harbor because I was so content there. Now I wrote to thank her for her birthday wishes.

That summer of 1948, I discovered my brother Ronnie as a companion and friend. I had always considered him a grown-up, too superannuated to share my interests. But now, at his thirty-four to my twenty-four, we were near-contemporaries, both footloose in New York. Ronnie was sent over by his London firm, Empiria Products, to manage its recently opened American branch office. If the firm decided that the New York operations were successful, Gabi and the girls would join him in New York in the fall.

Ronnie and I got together almost every weekend. Sometimes he came out to Cold Spring Harbor and stayed overnight with me in my Blackford Hall cell. Other times I drove into Manhattan and slept on the floor of his tiny room at the Tudor, a second-class hotel on 42nd Street near the United Nations headquarters. We would do things together with Taeko as a threesome, such as going to Broadway plays and open-air concerts in Lewisohn Stadium.

Rita was still living in Manhattan that summer, still working at the *Scientific American*. I had made no effort to contact her, not only because I was spending all my time with Taeko, but also because I was ashamed to face her. But Rita got in touch with me at Cold Spring Harbor. She asked me whether it was OK if she wrote a little puff piece about me for a series called "Our New Generation" in the German–Jewish émigré weekly, *Aufbau*. We met in Manhattan for dinner in a Chinese restaurant, during which she interviewed me. She was smartly dressed—very different from Taeko's plain togs—and as good-looking as ever. Our meeting was very cordial, but wholly professional. Her article, which wildly exaggerated my modest attainments, appeared in the *Aufbau* a week or so later.

Twelve years later, to commemorate its twenty-fifth anniversary, the *Aufbau* ran an update on the young *vedettes* whom it had featured in the summer of 1948. The biographees who, like me, had apparently lived up to the *Aufbau*'s expectations, included Max Frankel, of the *New York Times*, Judith Malina, of the "Living Theater," and Henry Kissinger, of . . . , well, the mightiest maven of us all.

I wrote to Lore soon after I got to Cold Spring Harbor—before I met Taeko—but I didn't repeat my pledge that I would definitely come to see her in the summer of 1949, let alone declare my love for her. Lore's reply took more than a month to reach Cold Spring Harbor, arriving as I was getting ready to leave for California.

Ingolstadt, 23 July 1948

My freshly baked Herr Doktor!

Congratulations! Got used already to your new title? It has to be a lofty feeling. How much longer are you intending to still play the Eternal Student? I would have thought the time for that is over. Well, as long as you are having fun and your bankroll permits it, why not? Especially since you seem to have made your nest in the most exclusive academic circles.

I too have changed, after the currency reform obliged me to say bye-bye to the expensive life in hotels. After endless struggles with the authorities I got two rooms of my own, at last. We are living at the edge of town, with a view over two little ponds, on which our Christmas roast is cooling her belly for the time being. It's pretty nice, yet I miss still the mountains.

Yes, I'm happy I didn't go back to Berlin for good. The conditions there are certainly not pleasant. The black-marketeers are not too badly off, but most of the Berliners are in bad shape. The worst is that they don't have any water or electricity. You can be glad that you don't have to live there as a German.

The currency reform has pared my cash reserves to the bone. Occasionally it's not clear to myself how and on what I am living. Yet I manage. I've got to manage. The world isn't going to come to a stop for my sake. And I'm not going to get myself gray hairs on that account. We'd lived on our capital. Now it's worth very little. It would go a little better for us if the banks released some of our money, but unfortunately they are taking their time. They paid out only DM40 per person and DM250 per family. I hope they are going to shell out some more before too long. Now, when I see all those nice things for sale in the shop windows again, I've got to look at every five-mark bill ten times before I dare to spend it on anything.

Write me again soon, if you can find the time, Mr Eternal Student. With affectionate greetings, I am

Your Lore

Whereas every previous letter I received from Lore had raised goose pimples of rapture, this one felt like a cold shower. Her sardonic remarks about my playing the Eternal Student made me suspect that when she congratulated me on my fellowship award in her preceding letter, she had not grasped that with a prestigious postdoctoral fellowship you're merely on the Autobahn to the big time, but you're not there yet. She had probably thought that, being a Herr Doktor at last, I had managed to get a career job at Caltech that would support a family. Perhaps she expected that I would now come to Germany to take her back with me to the States. But when she read in my last letter that, thanks to the fellowship, I was starting afresh in a new field, facing several further journeyman years before I could settle down somewhere to a steady job, she realized that I had no such intentions.

Incredible as it now seems to me, I didn't answer Lore's letter. All I can think of by way of an explanation of my atrocious behavior toward the dream woman I loved, whom I had wanted to marry and who was obviously willing to become my wife, is that the feeling of contentment I derived that summer from my romance with Taeko—whom I didn't love—quenched the torch I was carrying for Lore. I never heard from Lore again.

One Sunday, I drove down to Princeton from Cold Spring Harbor, to visit Martha Baylor, the electron microscopist who had first introduced me to Schrödinger's *What Is Life?* in the spring of 1946. She had made the crucial intervention in my life that led to my becoming Max Delbrück's disciple and brought me to Cold Spring Harbor. Meanwhile, Martha had moved from Illinois to Princeton, where she was working at a branch laboratory dedicated to the study of animal and plant viruses of Manhattan's Rockefeller Institute for Medical Research. She was taking electron microscope pictures of viruses for Wendell Stanley, who had received the 1946 Nobel prize in chemistry for his crystallization of tobacco mosaic virus. Martha told me that the Princeton branch laboratory would soon be closed, and that Stanley would move to Berkeley to become the founding chairman of a new department of biochemistry at the University of California at Berkeley. A virus laboratory would be built specially for him at Berkeley, under the "One Nobel Prize, One Building" rule then common in academia. By providing me with this news, Martha made a second crucial intervention in my life. She gave me the idea that after finishing my postdoctoral *Wanderjahre*, I would ask Stanley for a job at Berkeley, my long-term dream place.

There was a final, summer's end party at the lab on the night before my departure for California, including square dancing, at which Taeko and I— the Nisei maiden from Seattle and the Jewish *bocher* from Berlin—made a proficient hoe-down couple. When we took leave from each other, I avoided mentioning the subject of love, mouthing only general platitudes that I could just as well have whispered in Jim Watson's or Renato Dulbecco's ear:

"Taeko, dear. I was so happy spending the summer in your company. I'm going to miss you one hell of a lot in Pasadena."

She didn't point out that there was actually a very simple way—called "marriage"—by which I could continue to enjoy her company in Pasadena on a long-term basis. I merely promised to write often and come to see her at the first opportunity I would have to re-cross the continent.

The summer of 1948 marked a third critical turning point in my life. The first had come on New Year's Eve 1938, when I fled from Germany and escaped Nazi persecution. The second came in the spring of 1940, when I arrived in Chicago and turned from an uprooted, refugee into an American student. And now the third had come when I fell in with a scientific elite at Cold Spring Harbor. I had acquired a new set of lifelong friends, many of whom would eventually join the all-time-greats in the pantheon of science. Having adventitiously stumbled into the vanguard of the then nascent but yet unnamed discipline of molecular biology, I had risen from the workaday world of parochial American chemists to the international scientific big time.

The Even Better Scientific Life

CALTECH LIVED UP TO MY VISION of a palm-tree-studded academic Nirvana: A double tier of adobe-colored California-Mission-style laboratory buildings, facing a subtropically landscaped central mall, stretching for half a mile between two Pasadena streets, set off against the sun-lit San Gabriel Mountains peaked by 10,000 foot Mt. Baldy and peopled with brilliant minds, such as Pauling, Beadle, and Max.

The three-story Kerckhoff Laboratory of Biological Sciences, whose mall side was adorned with an arcade topped by little domes, lay at the northwest corner of the campus. Its hallway floors were covered with ceramic tile, the ceilings hung with wrought-iron chandeliers, and the walls lined with decorative fish tanks. By comparison, the red-brick Noyes Chemical Laboratory seemed like a bleak factory.

Max was waiting for me in his office in Kerckhoff, lounging in the yellow-leather easy chair that was the centerpiece of his otherwise frugal working mobiliary. Max's disciples would follow their master's example in eschewing fancy offices. In the 1980s, a few years after Max's death, I was provided with a deluxe departmental chairman's suite. I would have loved to spend all my time in it, but piety obliged me to keep mainly to my own austere faculty office, with a reproduction of an expressionist portrait of Max as its sole decoration—and, of course, a leather easy chair like his.

Max and Manny took me for lunch to the Athenaeum, the Caltech Faculty Club. If I had not been already convinced that I landed in Paradise, this would have done it. Modeled after a seignorial Spanish or Italian villa, the Athenaeum has a baronial lounge, with wood-paneled walls, a tall fireplace, and a Persian rug said to be one of the largest in Christendom. I thought this was the classiest living space in which I had ever set foot, not excepting the sumptuous salons of the Cristallo in Cortina and of the Grand in Kitzbühel. Filipino waiters in natty uniforms served us in the elegant dining hall, light years distant from the crummy Blackford Hall cafeteria where I had last eaten in the Delbrücks' company.

Aside from a small desk set against the wall, I was not going to have any working space of my own in Max's laboratory. The sole, double-sided laboratory bench would be used by anyone doing phage experiments—at first only Max and me, but before long, also by other disciples, who turned up one by one. The lab accommodated also a couple of helpers, who washed and sterilized the glassware. There was none of the kind of fancy apparatus with which the Doc's labs at Illinois were studded: A little tabletop serological centrifuge, a couple of hot-water baths, a home-built light box for viewing petri plates for plaque-counting, and a body temperature ($37^{\circ}C$) incubator were about the size of it. Soon after my arrival, a refrigerated water bath on wheels was installed; I would spend much of the next two years bent over it doing my experiments.

The research project on which Max put me was one of the few he could have picked for which my training as a physical chemist happened to have eminently qualified me. One of the favorite phage types studied by Max's Phage Group, strain T4, was known to fail to attach to its bacterial host cell unless it has been previously "activated" by contact with the amino acid tryptophan. Max suspected (or maybe hoped) that the hitherto known facts about this activation process were not compatible with ordinary physicochemical principles. So maybe there was a paradox hidden here that might lead us to one of those "other laws of physics."

But, much as I was hoping to run into a biological system manifesting such an "other law," I feared this was not one of them. I thought I wouldn't have much trouble coming up with an explanation of the seemingly bizarre tryptophan activation phenomenon within the framework of conventional, house-and-garden theories of physical chemistry. Max informed me that I'd have a partner in my research project.

"His name is Élie Wollman. He's a French M.D. bacteriologist from the Institut Pasteur in Paris. Working there in the department of microbial physiology headed by André Lwoff. Met him in Paris this summer; very nice guy. The Rockefeller Foundation is sending him for a two-year stay with us. Wollman and you are going to complement each other like liverwurst and rye bread. He's got the bacteriology, which you don't know from beans, and you've got the math and physics, of which he's largely innocent. Together, you'll make the perfect phageology sandwich."

Max had asked me to fetch Élie and his wife, Odile, at Pasadena's Green Hotel on the morning after their arrival by train from the East Coast. I had no trouble picking them out among the tourists in the hotel lobby, who were clad according to the relaxed Southern Californian standards of matitudinal attire. Élie had on a natty double-breasted, pin-striped suit, while Odile wore a hat and had a fox stole draped over her shoulders. (Before long, they did assimilate their style of dress a little to Pasadena custom, but they never went native. Élie always showed up in the lab wearing a necktie).

Odile spoke French with a lovely Provençal accent. She was from Albi, near Toulouse. where she had met Élie during the war and helped him hide under a false identity. They had only recently been married, and, to my amazement, still addressed each other as "*vous*" rather than "*tu*."

I helped the Wollmans find a tiny apartment within walking distance of Caltech. Élie often took me home for dinner after work for meals that would have a lasting formative influence on my eating habits. It wasn't merely that the meals Odile served were tastier than I had ever eaten in a private home, not to speak of the sordid institutional cuisine to which I had become accustomed in the previous six years. But the Wollmans treated every meal as if it were a special occasion, an impression augmented by Élie's always playing the role of the *sommelier*, serving an apéritif before we sat down at the table and then keeping the glasses filled with wine throughout the meal.

Although Élie was my senior by only seven years and Odile was just about my own age, they took to parenting me. I responded to their parenting by developing a filial attachment to them, for which—with Claire and Bob, my original foster parents no longer within range—they filled a strong spiritual need.

One Saturday in the late fall, the Caltech biology division held a picnic in the foothills of the San Gabriel Mountains, to which Odile, Élie, and I had gone as a threesome. Chairman George Beadle had not only organized the affair, but also took on the role of its *maître de plaisir*, directing the canonical picnic amusements, such as the softball game and the sack races for the children of the faculty and staff, and barbecuing the hamburgers over an open fire. Beadle's concern for the well-being of his minions, his attention to detail, and his luminous goodwill were a big part of the explanation of how he managed to forge the biology division, whose faculty consisted of persons no less contentious, opinionated, and pig-headed than those in any other first-rate academic department, into a harmonious entity. During my six-year stint as a chemistry student at Illinois, there had never been a chemistry department picnic. The very idea of Roger Adams, the formidable chemistry department head at Illinois, barbecuing hamburgers for his underlings was as unimaginable for me as was for Élie the notion of Monsieur Tréfouël, the director of the Institut Pasteur, socializing with his staff, let alone cooking for them.

Not long after we arrived at the picnic, the Wollmans struck up a conversation with Donna G., a secretary working in Beadle's administrative office. I had spoken to her a few times about some business matters, and had not overlooked her charms. She was an attractive, well-got-up, personable blonde, whom I reckoned to be in her late twenties. I had wondered what kind of husband or boyfriend she was going to bring to the picnic. No doubt, it would be one of those handsome, muscle-bound California-beach-boy hulks pictured in the body-building ads of pulp magazines. But she had come alone, which didn't necessarily mean, of course, that there was no Mr. Donna. Probably he was a pilot, flying to New York that day, or a detective working on an urgent case. She stuck to the Wollmans all afternoon, and so did I, not solely motivated by my filial obligation to steer my new foster parents from the Old World through the unfamiliar rite of the American picnic. As the party was ending, Donna proposed that we come to her place for a predinner drink, an invitation which I instantly accepted on behalf of the three of us.

Donna was living in a three-room cottage on Fair Oaks Avenue in Pasadena. My quick scrutiny of her quarters indicated that if there was a Mr. Donna, he was not staying at her place. I grew more and more agitated by the fantasy of becoming our hostess's lover. The predinner drinks extended into a light supper of sandwiches, and the Wollmans and I finally left in the late evening.

When I called Donna next day to ask her out for dinner, I still found it hard to believe that such an attractive woman wasn't already bespoke. I took her to Robert's French Restaurant (whose appellation "French" Odile declared a slander of her nation's cuisine). Donna and I got on very well.

I asked her, "What's it like to work for Dr. Beadle?"

"I love my job. He's a swell boss. Very kind and considerate."

"Yeah. He doesn't act at all like the big shot he is. They say he is in line for a Nobel prize. Élie Wollman is amazed that Dr. Beadle drives an old Model A Ford coupe with a rumble seat, instead of being chauffeured to work in a big limo, like Monsieur le Directeur of the Institut Pasteur. Of course, Élie doesn't know that a Model A in good condition is worth almost as much as a new Cadillac."

I felt that there was a good chance to make her mine. But when I brought Donna home, I didn't have the nerve to ask her to let me into her cottage. I merely made a date to take her to J. B. Priestley's new play, *An Inspector Calls*, at the Pasadena Playhouse and gave her a little good night kiss at the doorstep.

After a couple of more dates at the movies, Donna invited me to dinner at her place. She had prepared an Italian-style meal—antipasto, *cottoletta di vitello*, and pasta, washed down with Chianti. After dinner, we sat on the couch in her living room and began to kiss and pet. Before long, she asked,

"Want to move to the bedroom? We'll be more comfortable there."

"Do I ever!"

It was marvelous. She was a very skilled lover. I began virtually living at Donna's, went to the lab only during business hours, and spent almost no time in my own room. Yet as far as I was concerned, our relation was strictly carnal. The very idea of falling in love with Donna did not occur to me, nor did I give much thought to what her feelings might be toward me, except that she seemed to enjoy sleeping with me as much as I did sleeping with her. As far as I remember, the word "love" never came up in our conversation.

I didn't find out very much about Donna, except that there *was* a Mr. Donna, after all. She was married to a sailor, who was stationed near Paso Robles, about 150 miles north of L.A. There was something the matter with him, although I couldn't make out what. Maybe he was seriously ill, physically or mentally, or maybe he was locked up in the Navy brig. I never asked for or learned his name.

Four weeks after the Biology picnic, the Delbrücks gave a Saturday night party. Since I hadn't as yet quite cottoned on to Max's ultra-anticonventional standards, I thought that it wouldn't be proper for me to bring my married mistress to my mentor's social. But I felt I *had* to ask Donna whether she'd like to come with me to the party, about which she'd be sure to hear.

"I don't think that'd be such a hot idea. There'd be a lot of gossip around the office if I went to the Delbrücks with you. Remember, I've got a husband in Paso Robles! I'd love to come with you, of course, but you'd better go to the party alone."

"OK. I'll come over on Sunday morning for breakfast."

After the Delbrück party, I went home to my room, to sleep alone in my own bed for a change. Early next morning, I drove to Donna's cottage for our usual prebreakfast lovemaking. I rang the bell, but Donna didn't answer. The front door was unlocked, and so I entered, calling out her name as I went from room to room. When I got to the bedroom, I found her lying on the bed in her seductive negligée. I tried to awake her, first by shouting and then by shaking her, but it was to no avail. I was dumbfounded: Why won't she wake up? How could she be so tired? What's the matter with her? Is she ill? Then, as I happened to look at the night table next to her bed, I noticed an open, empty medication vial. I suddenly remembered how, ten years earlier, Ronnie had found the corpse of Aunt Katharina v. Ameringen in her bed in London, with an empty vial of sleeping pills at her side. I grabbed the vial: Its label read "NEMBUTAL."

What should I do? Telephone for help, obviously. What number do you dial? How do you call an ambulance? Anyway, it probably would take too long for help to show up. So I carried Donna in her negligée out to my car —hoping that the neighbors wouldn't watch this dramatic scene from their Sunday breakfast tables—and raced to the Pasadena police station. I ran in and stammered to the duty sergeant,

"I've got a dying woman in my car!"

"There's nothing *I* can do about it, Mister. Take her to the emergency entrance of the Huntington Hospital, a couple of miles up Colorado Boulevard. I'll phone the hospital to let 'em know that you're on your way."

The orderlies lifted Donna out of my car, put her on a stretcher, and wheeled her into the hospital. I handed a nurse the empty Nembutal vial, which, with a residue of presence of mind, I had taken along. Then I was sent to the admissions office, where they were going to ask me some questions. First they wanted to have lots of information about the patient, most of which, except for her name, address, place of employment, and religion (she had told me that she was Catholic), I couldn't provide. I didn't even know the date and place of her birth, let alone the names of her next of kin, her social security number, or type of medical insurance. Then they wanted to know the details of what had happened, and all I could tell them was how, coming to join Mrs. G. for Sunday breakfast, I had found her about an hour ago, lying unconscious on her bed.

Finally they asked me about myself and my relation to the patient. Now I began to fear that this sudden lugubrious turn of what had begun as such carefree, easy romance would have fateful consequences for me. I pictured Beadle as a latter-day Colonel Osborne, who would summon me to his office to inform me that driving a member of his staff to suicide, after establishing an illicit, adulterous relation with her, was incompatible with the standards of conduct expected of postdoctoral fellows. Hence he was terminating my brief tenure at Caltech. He would notify the Merck fellowship board of my dereliction, of course, and I could kiss my fellowship good-bye.

I considered bolting from the hospital without telling them who I was; but I realized that this would only make matters worse. They would call in the police, who would have no trouble tracing me through my license plate. So I gave them my name and address, identifying myself as a coworker and friend of Mrs. G.'s. All that time, I didn't give any thought to the fateful consequences that Donna's illicit, adulterous relation with me—dramatized by her suicide attempt—would have for her job at Caltech, not to speak of her marriage.

After I had waited for about an hour, a nurse told me that I brought in Donna just in time.

"On pumping out Mrs. G.'s stomach we found that most of the Nembutal hadn't been absorbed yet into her circulation. She'll probably regain consciousness in a few hours. But we're not going to release Mrs. G. after she comes to. She'll be taken to L.A. General downtown for a couple of days' observation. Here's the number to call at L.A. General tomorrow for information about her."

I reckoned that I had to have played some role in Donna's decision to kill herself, since she evidently swallowed the sleeping pills shortly before she expected me to show up and thus must have intended me to be the person who would find her dying. But I wasn't aware of having done anything awful enough to her to drive her to this desperate act—nothing nearly as awful as what I did to Hildegard or Rita when I acted as if I had no responsibility for their pregnancies.

I was badly in need of spiritual succor and went to seek it from Élie and Odile. Élie asked, "What's the matter, Gunther? You look awful!"

"Donna's tried to kill herself with an overdose of sleeping pills. I found her unconscious in her bed this morning and took her to Huntington Hospital. They expect that she'll regain consciousness. Once she wakes up, they'll take her to the psycho ward at L.A. General, probably this afternoon."

Odile asked, "How'd you happen to come to her house this morning?"

"I guess you didn't know that we became lovers after the biology picnic. Actually, it was wonderful sleeping with Donna every night. And then, all of a sudden, she tries to kill herself! I thought we were both happy with each other."

"Actually," Élie said, "it did occur to us that you and Donna might have become lovers, after you suddenly dropped out of circulation. But that such a self-assured, attractive woman like Donna would try to kill herself is really amazing!

"What d'you think I could have done to her that drove her to suicide?"

Élie trie to reassure me: "Whatever Donna's reasons were for trying to kill herself, they couldn't have come from anything you did or didn't do in those three short weeks of your affair. At most, you could have been the straw that broke the back of an emotionally already heavily overloaded camel."

"Might as well say good-bye to you, dear Odile and Élie. This is the end of my postdoctoral stay at Caltech. Obviously, Beadle's going to kick me out once he finds out."

"Don't worry!" Élie said. "Beadle won't throw you out. Why don't you join us for Sunday lunch?

I did feel a little better after their reassurances. And I couldn't help noticing that Odile was impressed by the discovery that her young American foster son, whom she had believed to be interested only in playing with his test tubes in the *labo*—as he had pretended, by way of a backhanded compliment, to Hildegard on the evening they first met—was actually involved in a tragic *affaire de coeur*.

On Monday, I went to L.A. General to see Donna, who, I had been told over the phone, was out of danger. She was in the psycho ward, lying in a large room housing a dozen or more patients. It was a ghastly, noisy scene, with screaming and babbling coming from many of the beds. Donna was awake, staring silently at the ceiling. Her wrists were tied to the bed frame with leather straps. She looked horrible; her normally carefully coiffed hair was disheveled; she wore no makeup, and seemed to have aged by about ten years. She acknowledged my presence with a nod. Acting as if I was just making an ordinary hospital visit to a friend who was recovering from a tonsillectomy.

"How d'you feel, Donna, dear? I hope you'll be able to go home soon. Is there anything I can do for you?"

I didn't ask any substantive questions, especially not why she had tried to kill herself or whether I was to blame. She neither said anything nor responded to any of my questions. I left after sitting at her bedside for half an hour, telling her that I would be back tomorrow afternoon.

Tuesday noon, I phoned the psycho ward, and was told that Mrs. G. had been discharged an hour ago. She never returned to work at the biology division office, and her fellow secretaries told me that she had resigned for medical reasons. I thought that they looked at me in a funny way, probably because they did know that I had been dating her. Beadle, who, as Donna's employer, must have been privy to the suicide attempt story and been told that another of his employees, a Mr. Stent, had brought Mrs. G. to the Huntington Hospital, not only didn't play Colonel Osborne but he never said a word to me about it, although I had the definite feeling that he knew. And if he did know, he couldn't have told Max. Because if he had, Max would have certainly insisted that I tell him all about my dealings with Donna—to the last detail.

There was no answer when I called Donna's cottage during the next few days, and finally an operator came on the line informing me that the number had been disconnected. I never saw Donna again. The memory of finding Donna seemingly lifeless in her bed and then seeing her, suddenly aged and tied down, in the psycho ward haunted me for many months and cast a pall over my relations with women during the whole remainder of my stay at Caltech.

Delbrück's Salon

IN JANUARY 1949, THE DELBRÜCKS moved into their newly completed house. Its bizarre design was Max's—in New-Age-Speak—"statement" of his anti-conventional sentiments. An unpainted, L-shaped wooden bungalow lining the edges of a large corner lot, it was neither traditional nor modern; maybe it was premature postmodern. It clashed brutally with the ubiquitous Ibero-Californian, white-washed, two- or three-story, red-tile-roofed houses in an upper-middle-class Pasadena neighborhood near the Caltech campus. Half a century later, the bungalow had become partially hidden by trees and shrubs and no longer presented quite as grating an aspect as it did when there was no vegetation to assuage its stark facade.

More than half of the total floor space was given over to a large, stone-floored living room, sparsely furnished, except for an ancient, ornately carved rosewood Steinway grand. No doubt predating the opening of the Panama Canal, the piano had probably rounded Cape Horn. The rest of the interior was parceled out over two tiny bedrooms, a small study, a mini-bathroom, and a cramped kitchen, which lacked a dishwashing machine, in line with the Delbrückian credo: "The family that washes dishes together, stays together." A wall-mounted small table in the living room served as a dining area. I suppose that the niggardly area dedicated to private, as compared to public, space reflected the Delbrücks' gregarious bent and lack of concern for privacy. The principal living area was the large backyard garden, while the interior's main function was to provide shelter at night and during inclement weather. The kitchen opened out on a paved patio with a picnic table, at which the Delbrücks ate all their meals.

In striking contrast to your typical European intellectual's lair, the Delbrück house had almost no books. Even the few built-in bookshelves in the small study were practically empty. Although widely read, Max didn't believe in owning books. Once having read a book, he usually gave it away to someone he thought might (or *ought* to) enjoy reading it. One noteworthy exception was a shelf mounted above the water closet in the bathroom, which held a beautifully bound, definitive multivolume edition of Goethe's

Gesammelte Werke. I was too shy to ask Max whether he was making regular use of this treasure, but I sometimes did.

The walls of the living room were hung with strange paintings that Max considered, in his words, the "crowning glory" of the house. They were the work of Jeanne Mammen, a German painter whom the twenty-year-old Max had met at a party in Berlin in the early 1930s. He and Mammen, who was his senior by sixteen years, became lifelong friends (platonic, no doubt, since he described her physical appearance as "small, homely, and unremarkable"). When Max left Berlin in 1937, he took along some of Mammen's pictures to Pasadena, hoping to make her work known in California.

One could walk in on Max and Manny any day, at almost any time, although sometimes at one's peril. Once, after a Sunday brunch in the Delbrück's garden on which a few of us had dropped in, Max ordered all present to get down on their knees and weed the lawn, pulling out crabgrass, blade by blade. The scene reminded me of a picture I had recently seen in *Life* magazine, which showed an Alabama chain gang clearing the roadside of litter under the steely eyes of an armed prison guard.

At the slightest provocation, Max and Manny would invite people for a meal or a party—usually including a musicale, in which, on Max's command, every woman, man, and child present had to participate. He brooked no such lame excuses as not being able to play an instrument. Any would-be deadbeat had to sing or join the percussion section.

Once, Max selected a Bach chorale, transcribed for voice, recorder, flute, and piano. For lack of a trained vocalist among the guests, Max ordered a young visiting physicist from Cologne to intone the German lyrics. The man blanched and stammered,

"I don't know how to sing, Herr Professor Delbrück."

"Nonsense! Here's the music with the words. Sing!"

After they had played a few measures, Max put down his recorder and stopped the music. He looked at the poor physicist in utter disgust: "And you call *that* singing?"

I received three letters from Hildegard during the spring of 1949.

My dear Günter,

I send you my best wishes for the New Year: good health, much money, and professional success, which is exactly what I'm wishing for myself too, since there doesn't seem to be anyone wishing it for me. In the New Year, I'm going to make up for all the things I missed in the Old Year. Fat chance! I'm tired, endlessly tired. I haven't got the willpower to do anything any more.

Where were you Christmas? What are you going to do New Year's Eve? This year I'm alone; that is, I want it that way. I can't stand being together with someone on New Year's Eve who doesn't mean anything to me. Are you still thinking about our New Year's Eve from time to time? It was something really special, wasn't it?

Why do you write so seldom? Are you engaged? Write soon, Günterlein, and when, at midnight, you clink your glass of champagne, think of me for once, briefly but really!!!

Hildegard

It's difficult for me to understand now, as I am rereading this pathetic letter after all these years, how I could have been so stone-hearted as not to answer it. Despite my not responding to her New Year's wishes, Hildegard wrote again six weeks later.

Dear Günter,

What's going to become of us all? The crisis here in Berlin is getting worse all the time. They turned many theaters into cinemas. And one's got to try to change East Marks for West Marks, because anything worth having can be had only for West Marks. Nobody dares to go into the East Sector any longer, for fear of not being able to come back. But you can't keep a Berliner down.

What's new with you? Married? It's slowly getting to be about time. Maybe that's the only way you will be really happy. Are you going to let that pass by you?

Krone des Lebens,	Crown of Life,
Glück ohne Ruh,	Bliss in heart,
Liebe bist Du.	Love, thou art.

Goethe

I still can't make up my mind about my career. The movies are always a one-shot deal. Theater is very difficult at the moment. I went to see Kurt Seifert (he's running the Berliner Theater). He would take me immediately, but since he's putting on only modern plays and there's a shortage of costumes, the actors have to wear their own clothes, of which I haven't any.

The weather is very strange today. Neither cold nor warm; not raining yet wet; like my letter, lukewarm. I don't know what's the matter with me. It's been like that the whole week. Maybe that's why I'm writing to you; actually it's your turn. Take care of yourself and try to read between the lines. Because what I wanted to write was something totally different. Don't let me wait so long and write me a beautiful letter. It's been a long time since I've had a nice one from you.

Your Hildegard

I *hadn't* read between Hildegard's lines and didn't really know what totally different things she had wanted to write. All I could think of was that she expects me not to abandon her in her misery and help her as one friend would help another, especially when that help is so easy for me to give and so desperately needed by her.

Overcome with guilt feelings, I did comply with Hildegard's request not to let her wait long. I replied by return of post. But I didn't send her the beautiful letter she asked for. Instead, I merely composed one of my standard, cold-turkey activity reports: I'm working on the tryptophan activation of the kind of viruses I had learned about at Cold Spring Harbor last summer; I made lots of new friends in Pasadena; I'm going on lots of camping trips to the desert; I drove to Mexico with two of my new friends over the Christmas holidays, where we climbed Mt. Popocatepetl, higher than Mt. Blanc; I am neither engaged nor married, and spent a dull New Year's Eve with my two friends in a Mexican motel room, without any clinking of champagne glasses; I have not forgotten, and never will forget, the New Year's Eve I spent with her, the most wonderful of my life. Not a word about the Crown of Life.

Easter was yesterday. Today is April 19, 1949

Günterlein,

How old did you become? I forgot to keep count. It's still the same that I always wish you on your birthday: the best isn't good enough.

No news from you for nearly a year. I thought you were going under, and now I learn that, on the contrary, you climbed a 4,000-meter mountain. An ordinary mortal can't keep up with you.

Otherwise there is no news. Berlin has become an enigma. The things that are going on in the Soviet zone are really sinister. One kidnapping after another. The poor people.

Some nice things have been happening to me lately. But—knock on wood—nothing is definite as yet. I was introduced to the director of the French Film-Union, Mr. Saint-Saëns. Made my application and learned from him personally that he wants to send me to Baden-Baden for a tryout for dubbing.

I'm working at the Palast-Kabarett at the moment. I'm very satisfied, have little time, and finally don't get around to thinking. As for the earnings, it's a little less rosy. You can buy anything in Berlin; but even the necessities of life are unaffordable for many people.

As an old friend, you'll be able to help me a little. What I need urgently and is still most expensive here are a pair of black summer sandals, size 38, a beautiful summer material, blue or green, and a pair of black leather gloves, size 7. They cost up to 30 West Marks here. Günterlein, will you be so kind? They would be a belated Christmas present! Or a premature birthday present. Are you horrified? It's not that often that I have tortured you with such things.

Once I'm in Baden-Baden, I'll write to you right away how it was. Du, and if it works out, maybe we could meet in the West sometime, after all. The dubbing is done over there.

A little kiss for birthday and Easter.

                    Your Hildegard

I neither sent Hildegard the things for which she asked me nor replied to her letter. She never wrote to me again.

Forty years later, on rereading Hildegard's last two letters, I asked a savvy friend whether she, by any chance, could, as I could not, read between the lines and knew what that totally different something was Hildegard had really wanted to write. It's obvious, my friend said, what she meant was

> Despite all your glowing reports about your exciting experiences in Cold Spring Harbor and Pasadena, Günterlein, I know damn well that, letting the Crown of Life pass by you, you aren't—can't be—happy. You big idiot you! Come back to Berlin to get me, the woman who loves you desperately, and we'll live happily together ever after!

Although I wasn't in love with Taeko, I missed her harmonious companionship, except during my brief affair with Donna. And after that affair's dire denouement, I missed Taeko even more. We had kept up a lively correspondence since we parted in September 1948. In fact, most of the letters I had written from Caltech were addressed to Taeko. But ever oblivious of the effect my impersonal letters created and so totally self-absorbed as to make me think that just telling my news was enough, I sent her my usual matter-of-fact bulletins: the research I was doing in Max's lab, the new friends I had made in Pasadena, the camping trips on which I had gone, the Delbrück parties. But never "I miss you." My detailed account of one of these trips provoked her to comment,

> Gun, you write so well. Your letters, when they are impersonal like this travelogue, sound exactly like you, you wonderful stupe!

Taeko usually responded in kind, providing me with lab gossip and activity reports on what she was up to in Cold Spring Harbor, but sometimes her emotions broke through. It was so lonely at the lab. Wouldn't I please put her to bed and kiss her good night? Then, after I ignored her attempt at prompting me to say something sweet, she would ask me to forgive her for getting sentimental:

> I've resolved that I'm not going to bother you with my thoughts, but I'm finding it harder than hell to keep my pen from writing what I spend most of my time thinking. Chalk it up to my desire to make like Nature Boy says, 'The only thing worthwhile is to love and be loved in return.'" I want to be loved, even if merely by a gesture. So will you please come and kiss away my tears?
>
> My sister is planning to get married to a "Caucasian," with or without parental permission. Mom and Dad are unlikely to take this well. I had always figured that I would be relieved of the obligation of marrying a Nisei, by virtue of Sis marrying one. Now, I'll be trapped by obligations toward my parents, to whom I owe so much. But I have so little time; parents don't live forever! I want to live my life my own way—and I want to see you and be around where you are.
>
> Maybe it would be so much simpler if I could just go off and get a job any old place, start from scratch, make new friends, while profiting from the lessons I've learned so far. But knowing myself, I'd make the same mistake again—I wouldn't wait until I knew whether a guy's intentions were honorable or not before I fell in love and got all mixed up.
>
> If I had any sense at all, I'd tear up this letter and throw it away. But I can't—it somehow tries to say what I think and feel—and you might as well get a blast of that side of me once in a while—the side that makes me wonder where I've lost all my pride. I can't quite quell the feeling deep inside that premarital relations are wrong, and I can excuse it only on the ground that I am very much in love.

But that wasn't true the first time I slept with you. I wasn't in love, and I knew you weren't in love with me, but somehow it seemed so natural after a while. When I tried to make you say what you thought of me one night, it was because I needed something to grasp at. Sleeping with you would have been all right if you loved me, but I couldn't have stood myself if I thought that I was merely an outlet for your sex needs and nothing else. So I had to invent things to ease my conscience.

I didn't like the idea of thinking of myself as a whore, but sometimes I wonder where the line is. What's right? What's wrong? You made it all seem so natural: Sex is a natural thing, and an expressing of that sex is a normal function and I agreed and lived with it, but people speak so differently! Is it all talk, and do most of them live like I do? Or do they really live chaste lives?

I know you respected the girls you've lived with. Did any of the others fall in love with you too? And how did you get rid of them?

I just finished reading *Tales of the South Pacific* before I started this letter, which partly explains my mood. It brought home to me in every story what the average American subconsciously thinks of mixed marriages. White people are steeped in a tradition of white supremacy, and it takes a good mind and a lot of willpower to destroy the seeds of prejudice in a man's mind—most people don't have the mind or the incentive for it.

Be good, darling.

Taeko

The easiest thing for me to deal with in her letter was her insinuation that my intentions toward her were not honorable because I was a racist, thinking of her as just a gook lay. In my reply, I explained to Taeko that, while it is not totally untrue that I am a racist, my racism is directed against my own race: I *want* to enter a mixed marriage, rather than marrying my own Jewish kind. I thought I didn't harp on her being obviously the same kind of self-hating racist as I was, what with her unwillingness to marry a Nisei—sight unseen—not even to honor the obligation she acknowledged as owing her parents.

To substantiate this self-hating quirk of mine, I told Taeko more about my dealings with Rita, whom, during our Cold Spring Harbor summer, I had casually mentioned as an ex-girl friend of mine. She knew that Rita was living in New York and that I'd gone to meet her in Manhattan in connection with the "New Generation" *Aufbau* article that Rita was doing. But I hadn't told Taeko, as I did now, that everybody, not only Rita herself but also our families and friends, thought that she was the perfect wife for me and that I tended to agree, except for one thing: I didn't love her, and her being Jewish must have had something to do with it.

The account I gave Taeko was incomplete in at least two important details. For one, I didn't reveal my unconscionable failure to help Rita when she thought that I had made her pregnant on her visit to Champaign. For another, I didn't mention that, my anti-Semitic self-hatred notwithstanding, I *had* been in love with a Jewish girl during my senior year in college, whom I would have married if she hadn't broken our engagement and married a non-Jew, in violation of her obligations to her parents, by the way.

But it was harder for me to explain to Taeko my lack of honorable intentions toward other girls I slept with and, indeed, respected and whose exotic racial background was actually the stuff of my dreams. I hadn't figured out the reasons for myself. I gave her a brief synopsis of my involvement with Hildegard, telling how I tried very hard to fall in love with her but never succeeded, and drove her to desperation by my abominable behavior. I didn't mention the most important detail from Hildegard's story, however. Namely that I did fall in love with another, strikingly beautiful woman from Berlin, while Hildegard was living with me in Germany, and, though that other woman returned my love, I made no moves to marry her, either. And I closed my answer to Taeko by putting in writing for the first time that I did miss her.

> Gun,
>
> I feel as though I understand you almost as well as I understand parts of myself. All the things you say only confirm what I sort felt of subconsciously all along, and being on this side of the fence I can understand so well what Rita and Hildegard must build their dreams around. My second sense always tells me to play it safe: Never say anything so that those irreversible decisions are never forced, or even brought to mention, which is fine if you have a lifetime ahead of you. But, impatient as I am, once in a while, all the things I think and dream and wonder about bubble out, and in the end I think I'm better off for giving expression to these doubts and wonders.
>
> Actually, faithfulness in the moral sense doesn't really matter so much. It's whether you can make the person you're with at any one particular time feel as though she is the only one, and that you can do. I think you are an opportunist: The person who's around is the one who is important. I have this feeling that if I could contrive to be where you are, then everything would be

all right, that for me time was the essential thing. If only I had another month with you last summer!.

I rather expected to get an answer to my letter from you—and yet, and yet, I hoped you would ignore it and go on as we have been doing. It's my favorite policy too, to avoid things as long as possible. I wanted you to do it for me; slide over it and give me more rope to go and hang myself on. You did, sort of. I didn't think you would commit yourself; I would have been very surprised if you had.

So spent my time day-dreaming, hanging on to whatever sliver of hope there was, just as I suppose Rita and Hildegard will go on doing as long as you are not tied to someone else. I'm afraid you're going to be stuck with my loving you and wanting you until I find someone else to love. And until that day comes (if it ever does) I'm going to go on fighting for a stronger foothold in your world. I'm going to go on dreaming about eventual success, writing to you and living for your letters. Just go on missing me even once in a while and I'll be relatively happy—and love me when you're with me—I won't ask for any more, though I'll go right on hoping.

I miss you.

Taeko

I was utterly distressed by Taeko's two letters, especially since they reached me at about the same time as the final letters from Hildegard. How did I manage to cause so much misery to the women for whom I bore the fondest affection, even though I did not love them? I didn't think I had acted under false pretenses. I had never said, or even hinted that I loved them, except that I had not concealed that I was happy in their company. I had made no demands on them, especially not for exclusivity or fidelity, since I considered them sovereign individuals, who had given themselves to me for more or less the same reasons that I had desired them and didn't want return demands.

I saw that Taeko had put her finger on at least some of my characterological deficits that had led me to cause, however unwittingly, such havoc among people dear to me, and that were probably also not unconnected with my failure to form a stable union with Dorothy and Lore, the two women I actually did love. But, on the whole, I remained as much of an enigma to myself as to the women to whom I had caused so much distress.

In my response to Taeko's second letter, I didn't enter into its substance, not least because I didn't know what to say. But this time, I supplemented my standard news bulletin with a telegraphically ordered bouquet, hoping, as I indicated in the ultra-brief accompanying telegraphic message (without which, so the Pasadena florist told me, people don't telegraph flowers), to mitigate her distress. Unfortunately, my message was misinterpreted by the Long Island florist as a condolence and delivered to the Cold Spring Harbor Lab on a black-bordered "Deepest Sympathy" card. Taeko happened to be bedridden with a bad cold when the flowers arrived, and she interpreted what she thought was a floral expression of my compassion with her illness as an instance of transcontinental lover-to-lover telepathy.

After her emotional flare-up, our correspondence settled back to its previous sentimentally low-key pitch. She wrote that she found winter life at Cold Spring Harbor "unbearably horrible" and that she had concluded "that you've got to have interesting men around to make life worthwhile for a woman. Whether the men are attached or not doesn't matter as much as whether they're worth knowing and talking to."

Although I didn't think much about it at the time, in later years I began to wonder why all these wonderful women—Rita, Hildegard, Taeko—persisted in loving me, when they knew that I didn't love them in return. It couldn't have been that I had so many other lovable qualities that they were willing to forgo the humane treatment that I didn't bother to offer them and that they had every right to expect from a decent man to whom they had given themselves. Perhaps it was that they, like me, were yearning for the unobtainable dream lover, and emotional unobtainability was certainly a quality of which I had an overabundance.

In the early spring of 1949, Max was puzzled by a picture that he had seen in a recent issue of *Life* magazine. It was an electron-microscopic image of a chromosome, whose caption put *Life* readers on notice that they were looking at the first portrait ever taken of a gene. Max wondered, as the Socrates of Plato's *Meno* would have, how one could recognize the portrait of a gene if one had never seen one before and didn't know what it was supposed to look like.

Max decided that I would drive with him to downtown L.A., where we would call on Drs. Baker and Pease at the University of Southern California, credited by *Life* with this first snapshot of a gene. He asked Marko Zalokar to come with us, a chromosome expert postdoc from Yugoslavia, recently arrived at Caltech after getting his doctorate in zoology from the University of Geneva. Max hoped that Marko could ask Baker and Pease some tough technical questions, while he put them through the Socratic third degree. My role wasn't too clear, but maybe I was supposed to provide backup, in case Baker and Pease became violent and tried to throw us out physically.

Marko told Max that he was in luck. Marko would bring along Jean Jacques Weigle, his former professor at the University of Geneva, who happened to be visiting him. In the early forties, Weigle had taken part in the development of the first Swiss electron microscope and was now a supreme expert in the very tricky interpretation of electron-microscope images. According to Marko, a recent heart attack had caused Weigle to retire from the University of Geneva, but not from mountain climbing or skiing. He was passing through Pasadena on his search for a place to settle in the States—possibly a high-altitude cosmic ray observation station in the Rockies or the Sierra Nevada—where he could combine his interests in mountaineering and science.

Weigle turned out to be a stocky, lively man, who spoke wonderfully expressive, strongly French-accented English, not wholly free of obscenities. He was wearing a spiffy, cornflower-blue Harris tweed jacket and was roaming the States in a matching powder-blue, 1948 Cadillac convertible. We drove downtown—top down—in Weigle's gorgeous automobile, my first ride in the Original Dream Boat of postwar America, which was more exciting for me than having a look at the first portrait of the gene.

It didn't take Jean long—by the time we got to USC, we were on a first-name basis—to ascertain that the gene portrait hullabaloo was mostly scientific media hype (which existed in the 1940s just as it does now in the 1990s). All that Baker and Pease had done was to develop a novel method of cutting ultra-thin sections of chromosomes for examination under the electron microscope. Jean complimented them for that as an important technical breakthrough. But under Max's questioning, they admitted that they had no way of knowing what features, if any, in the chromosome photograph corresponded to the portrait of a gene. They tried to put the blame for the hype on *Life*'s overenthusiastic editors. Since Max wasn't interested in ultra-thin sectioning and didn't want to waste any more time on people he considered four-flushers, we left USC after little more than half an hour, on somewhat less than cordial terms with our hosts.

Although we learned very little about seeing genes on our investigative mission to USC, that excursion was to have substantial personal consequences for us. Jean had meant to visit Pasadena for only a couple of days, but he stayed on for four months. He gave up his plan of moving to a high-altitude cosmic ray observation station and decided to join the Phage Group instead, at a mere 700 feet above sea level. He returned to Pasadena in the fall and became a permanent member of the Caltech biology division.

Max and I each began a lifelong friendship with Jean, despite Jean's rabid, typically *Suisse Romande,* antipathy toward Germans. He turned out to be one of the few people who was not awed by Max, which endeared him to supremely self-confident Max as a true peer. And Jean was willing to forgive—though not to forget—that I too was German-born. Even after I had reached full professorial rank at a major university, he still continued to treat me like a little Heinie Wisenheimer with half-baked ideas, which I didn't mind because he was my avuncular senior by more than twenty years.

Jean Jacques Weigle was born in Geneva, where he obtained his doctorate in physics in the early twenties, after he was already renowned in Swiss alpinist circles as a ski mountaineer *de grande classe.* There was a Central European elegance in Jean's style of thought, experimentation, and lecturing—indeed of living—with which I never before been in touch. By comparison, the academics I had known at Illinois, and, more lately, even at Cold Spring Harbor, seemed dowdy. They tended to be suspicious of rhetorical elegance, considering it as an indicator of intellectual superficiality and insincerity. And they eschewed stylish clothes, fancy cars, and beautiful furniture as tokens of moral corruption. But when I saw Jean in his spiffy Harris tweed jacket driving a matching Cadillac, I vowed that one of these days, I, too, would own such a jacket and matching convertible. Within a mere five years, I had the blue Harris tweed, but I needed fourteen years to get the Cadillac, and even then I had to settle for a nonmatching white convertible. It's a 1963 model, which I still keep in running condition and which makes me the laughing stock of my Berkeley colleagues. The jacket was worn out long ago, and since Harris tweeds seem to have gone out of style, I haven't been able to replace it.

On weekends when there were no Delbrück-ordained camping trips, we usually went skiing in the San Gabriel mountains north of Pasadena. Neither the runs nor the snow conditions were outstanding, but it was thrilling for me to get into my car in the morning and be on the slope in an hour or so. Jean, the renowned ski mountaineer *de grande classe,* was a

Schusskanone, whose downhill *wedeln* was a marvel to behold. Jean's exquisite style (not only of skiing but also of living) reminded me of my beloved Lore, Queen of the Hahnenkamm slopes, with whom I had ceased corresponding. I agonized again, as I had many times before, why I hadn't done anything to bring her to the States. It would have been such a wonderful life with her in Pasadena: skiing, camping, party going, dining at the Athenaeum. Max would have flirted with her, the most beautiful woman at Caltech.

At about the time when Jean first showed up in Pasadena, Élie received exciting news from Paris. His *patron*, André Lwoff, had just finished some experiments that proved beyond doubt that there exist bacterial strains, termed "lysogenic," whose cultures keep on producing phages while seemingly enjoying the best of health. That there are such lysogenic bacteria had long been claimed by some bacteriologists. Most bacteriologists, however, dismissed lysogenic bacteria as artifacts, created by incompetent, self-deluding experimentalists. After all, the existence of such lysogenic bacteria seemed incompatible with d'Hérelle's meanwhile generally accepted view of phages as killer viruses that inexorably destroy the host cells on which they grow. This dismissive view was shared also by Max and was accepted as dogma by Phage Group devotees like me. To avoid corrupting the innocent minds of the Group's neophytes, lysogenic bacteria were not even mentioned in the Cold Spring Harbor phage course.

What Lwoff had now shown was that each lysogenic bacterium propagates a covert hereditary entity he called "prophage," which is not itself an intact, infective phage particle. The covert prophage endows the bacterium —and all its offspring—with the ability to give rise to overt, infective phages at some future time, without intervention by any exogenous, infective phage particles. Even hypercritical Max had to accept Lwoff's results as definitive. Once one cottoned on to the notion of the covert, noninfective prophage as an agent responsible for the long-term coexistence of virus and host cell, it didn't require much imagination to appreciate that a new age was dawning for understanding the role and survival of viruses in nature.

Delbrück's Conclave

THERE WAS GOING TO BE A PHAGE GROUP MEETING, to be run by Luria at Indiana University in the second week of April 1949. Almost all of the dozen or so of the Church's communicants were going to Bloomington, including Élie and me. The National Research Council agreed to pay for my trip from Pasadena, but when Max asked the Rockefeller Foundation for reimbursement of Élie's ticket, they informed him that they didn't send their fellows to private *Kaffeeklatsches*. So Max talked Beadle into shelling out some of his discretionary monies to cover Élie's expenses. Jean Weigle was going to come to the meeting too, driving back east in his dream boat convertible and stopping over in Bloomington on his way home to Geneva. He was going to leave the Caddy in the East and reclaim it in the fall, on his return to Pasadena. Another neophyte was expected to show up at the meeting: Wolf Weidel, a young German biochemist, who would be on his way to Pasadena from Tübingen. Max was bringing him to Caltech for a year, in the hope that Weidel would establish a German See of the Phage Church.

Élie and I rode the Santa Fé's El Capitan to Chicago, sitting up for two days in coach class, feasting on a picnic basket Odile had packed for us with delicious rations and plenty of wine. We prepared the talks we were going to give. I would present a summary of our experiments on the tryptophan activation mechanism, while Élie would give a summary of Lwoff's recent rehabilitation of lysogenic bacteria, the belief in which had formerly been declared a heresy by Pope Max. It was more than fitting for Élie to make this presentation, since his bacteriologist parents, Eugène and Elizabeth Wollman, found some indications already in the 1930s that lysogenic bacteria perpetuate the phages they carry in a noninfectious state. The elder Wollmans, prominent staff members of the Institut Pasteur, had been impervious to the mortal danger they were in as Jews (just as my own father had long been) and kept on working in their lab at the Institut Pasteur even after the Germans had occupied Paris. They were arrested at work by the Gestapo's French *Hilfspolizei*, deported, and killed in one of the Nazi extermination camps. After the war, Lwoff continued the elder Wollman's pioneering studies and provided a more solid experimental foundation for his murdered colleagues' (in the event correct) view of lysogeny.

Some of the people at the meeting, such as Luria, Renato Dulbecco, and Jim Watson, I knew from Cold Spring Harbor. Others, such as Alfred Hershey and his student Raquel Rotman (the sole woman present), as well as Leo Szilard and his young colleague, Aaron Novick, had been no more than disembodied names to me. Max's German import, Wolf Weidel, with his long hairdo, horn-rimmed spectacles, and floor-length Wehrmacht-style trenchcoat, reminded me of the Krauts in the FIAT document translators' pool.

The meeting lasted two days. Each participant (except the neophytes Weigle and Weidel) made an informal presentation of recent research results, followed by hard, or even brutal discussions dominated by Max, without, however, the slightest trace of rancor. In accord with Phage Group etiquette, the projection of lantern slides was forbidden. Aside from Élie's brief summary of Lwoff's recent demonstration of the reality of lysogenic bacteria, which created a sensation, Hershey's report was probably the most important of the meeting. His and Rachel Rotman's findings showed that there occurs an exchange of genes between two or more phage particles infecting the same bacterial host cell. Hence viruses can engage in genetic recombination, which, since the rediscovery at the turn of the twentieth century of Gregor Mendel's "premature" 1865 paper, had been regarded as the exclusive prerogative of sexually reproducing creatures. Hershey and Rotman's finding opened the way for the reform of "classical" (Mendelian) genetics and its central concept of the gene as a "bead-on-a-string" This molecular biological reform would redefine the gene as a stretch of DNA in which the structure of a protein molecule is encoded.

Leo Szilard was the eldest and most famous person present. He had studied physics in Berlin in the early 1930s and taken out a patent on the design of an atomic pile (like the one eventually built during the war at the "Metallurgical Laboratory" in Chicago) in which a controlled chain reaction of nuclear fission might be harnessed as an energy source. In 1939, by which time he had moved to the States, he persuaded Albert Einstein to send Franklin D. Roosevelt a letter Szilard had drafted. In this letter, the president was apprised of the possibility that Otto Hahn's recent discovery in Berlin of uranium fission—which, according to Max, his inept assistance had greatly delayed—had opened the possibility of devising an atomic bomb of immense destructive power. Einstein/Szilard warned FDR that the Nazis were probably already trying to develop such a bomb. The Einstein/Szilard letter instigated the Manhattan District Project, which, six years later, culminated in the detonation over Hiroshima of the first nuclear weapon.

After the war, Szilard and Aaron Novick, his Manhattan District sidekick, joined the throng of physical scientists whose research interests were directed toward new horizons by Schrödinger's *What Is Life?* They took the Cold Spring Harbor phage course and set up a biophysics laboratory at the University of Chicago in an abandoned synagogue just south of the Midway, a couple of blocks from my alma mater, Hyde Park High. For their joint presentation of the results of their recent phage experiments, Szilard and Novick seemed to have taken Groucho and Chico Marx as their expositor role models, constantly interrupting each other with contradictions, corrections, and emendations. This made their talk hard to take seriously, which was a great pity. They had found, quite unexpectedly (as well as "prematurely"), that the progeny of a joint infection of one bacterium with two genetically different phages include some that carry the genes of one parent and yet act as if their behavior were determined by the genes of the other parent. Max and the rest of us considered this phenomenon a mere curiosity, overinterpreted by its overenthusiastic discoverers. But it would have shortened by three years our quest to fathom the mechanism of phage self-reproduction, had we properly appreciated Szilard and Novick's discovery at the Bloomington meeting.

My presentation of our findings on the tryptophan activation mechanism was one of the last talks of the meeting. I sensed that our colleagues thought that we were doing a really good job on a problem that needed to be straightened out, but one so dull that they wouldn't dream of touching it. Max praised me implicitly, by sparing me his standard postlecture comment that my talk was the worst he'd ever heard.

At the farewell party at Luria's house, everyone was in high spirits. Phage research seemed to be moving forward. I felt proud to be a member of this coterie of first-class scientists who were also personally good guys, and to be on Max's team. I was happy to have seen my buddies from Cold Spring Harbor and to have made new friends, such as Al Hershey, Leo Szilard, and Aaron Novick. How could I have been so lucky as to fall in with this crowd?

After the Bloomington meeting, I headed for New York, to see Taeko. After having had no love life since the Donna catastrophe, I could hardly wait to be with Taeko again, although I was apprehensive about seeing her: The return of her letters to their former tenor of emotional equanimity notwithstanding, I expected that she was going to try to make me declare myself regarding her. It turned out that Taeko had previously arranged to visit her elder brother, Hiroshi, in Tallahassee, where he was teaching mathematics at Florida State, just during the couple of days I was going to be in New York. But for my sake, she postponed her trip.

During my visit, Taeko and I stayed in the Queens apartment of my PRIWAKI classmate Gary Raven and his wife Inge. Taeko opened the door when I rang the bell at the Ravens,' and we fell into each others' arms, giving to and taking from each other all afternoon that for which we had been starved for so long. We talked a lot, but at an impersonal, emotionally impoverished level, devoid even of banal sweet-talk. I suppose that Taeko was reticent to start verbal romancing, and I lacked the aptitude or will for it. We spent two blissful nights at the Ravens,' although Taeko's bliss couldn't have been total, since there was no mention of her agenda's Topic A., namely our marriage. On our last day together, we had a farewell lunch at Penn Station, before Taeko took the Long Island commuter train out to Cold Spring Harbor—she was flying to Florida that evening—and I got on the Pennsylvania Limited for Chicago. Although her behavior toward me had remained as affectionate as ever, I could sense that she was deeply disappointed that our meeting had brought her no closer to her goal. I sent her a birthday telegram from Chicago, c/o Hiroshi in Tallahassee.

After her return from Florida, Taeko provided me with the kind of impersonal travelogue that sounded exactly like me, hardly making any reference to our meeting in New York. To my relief, our correspondence resumed at its erstwhile emotionally flaccid pitch. But then a strange happening at Cold Spring Harbor made Taeko lose control again:

Dearest Gun,

I wish you were here—I feel the need for someone to hold my hand and tell me that everything's going to be all right—that it's worthwhile to go on living from day to day—planning for the future and expecting the world and the future to treat me right. I find it extremely unnerving to realize suddenly that we never know what might happen in the next minute to something that one's always taken for granted—like health.

Do you remember Mildred O.? Caspari's technician? She's a wonderful kid—healthy as a horse. She was going to get married on June 20th, to a guy her parents disapproved of. Yesterday noon, Patsy found her on the floor of the toilet in Jones Lab. She had some kind of stroke and is completely paralyzed on her right side and probably been in the toilet for a couple of hours. She's quite conscious & rational, and today, they took her to a hospital in Manhattan, since they have to do extensive tests to determine the exact cause, if possible. They suspect a brain tumor or an aneurism, both of which are extremely serious.

Her parents arrived yesterday evening, and the mother is impossible. She's hysterical & says she's never met Mildred's fiancé and doesn't know his name, etc. She hinted to Caspari that it serves Mildred right to be sick, for trying to go against her wishes. We were going to give a shower for Mildred next week—and Dick, her fiancé, is expecting her in New Brunswick this weekend. We gave her father Dick's address, but we doubt that he ever got in touch with Dick.

It was bad enough the other day when Linda nearly had a head-on collision—but that I could accept—it's somebody's fault & it's man-made—but a sudden stroke like this—you've got nobody to blame and no way to be careful to avoid it. It sort of knocks the wind out my sails & leaves me cold & shivering when I think about it.

I guess there must be a lot of pleasant things I could think about, but right now I can't think of one—just that I wish you were here. When will you be here again? The months between seeing you seem forever. Good night, darling.

Taeko

A week later, I heard from Taeko again:

Mildred is still quite badly paralyzed—but she's getting a little better. They didn't find anything with a brain operation—no diagnosis at all as yet. She's in New York Hospital—hope I get a chance to see her soon—still makes me feel morbid.

Yet another week later, Taeko went to see Mildred in the hospital.

She looks quite well. She can walk now—but she still can't move her right arm. There's still no diagnosis. They operated and didn't find a thing. It's real queer—they think it's probably a psychosomatic, hysterical paralysis. Makes sense to me. Mildred was in an awful bind. Should she go ahead and marry Dick, or call off the wedding to please her mother?

You see, Mildred's story strikes close to home. I got a letter from my sister, and it seems Dad wrote her that he doesn't want her and her Caucasian fiancé to get married at home—in Seattle, that is—for reasons of social prestige, etc. I may sound silly, but I cried when I read that. I had built up the illusion that Dad was liberal and broad-minded—and it left me feeling completely lost. The only reason I was going to Seattle this fall was because I felt that one of us should stay home with the folks—that they are getting old—& it seemed disgraceful to me that with three children one of them couldn't be at home—but now, I don't know. If they're ashamed of what my sister is doing, then I'm not going to go home and give them a chance to be even more ashamed. I know that most of my friends will be Caucasian—and why the hell should I feel so much filial piety if they're going to react the way they have? And my brother wants to marry this Caucasian gal he met down in Florida, who's awfully sweet. It sort of knocks all my pins out from under me, all my basic security—home & family—shot to hell.

My response offered her no encouragement that I was going to be the Caucasian friend who would cause her father to lose face also on behalf of his other daughter. And so in early July, I received the first indication that Taeko was beginning to consider other possibilities.

Dearest Gun,

I wish you had decided to come here again this summer. Life would be so much wonderfuller—and so much less complicated. You see, I seem to have acquired a Shadow last week, who goes wherever I go and waits for me while I wait on tables.

Life was much simpler when my aim in life was to try to convince you that marriage and independence were not mutually exclusive. I also said once that if I found something better, I'd jump off your band wagon. But I'm afraid I'm spoiled—nothing quite measures up to you—and I find myself hanging on the slim hope that maybe some day you'll miss me enough to want to be with me as much as I want you— and yet I find that I'm not discouraging my Shadow at all. It's rather a nice feeling to have such a fuss made over me—no one's ever done it before—and it flatters my ego, but what if he gets serious? Am I crazy enough to think that a bird in the bush is worth two in the hand?

Taeko's letter reminded me of the letter that Dorothy had sent me in November 1944, telling me that Captain Tony had unexpectedly come back to the States from overseas duty and that she was going to see him, but not in preference to me. Except that there was a big difference in my emotional response between now and then: Taeko's message gave me a feeling of relief, instead of the ominous premonition incited by Dorothy's. Just as I had ignored the news of Tony's return, so did I ignore the emergence of Taeko's Shadow. Dorothy had not understood—how could she have?—that my silence was not due to my not giving a damn but to my other-worldly philosophical view of her as an autonomous agent, whose free choices were none of my business. But Taeko had understood my apparent insouciance correctly: I wasn't going to marry her.

> Dearest Gun,
>
> I accepted Shadow's fraternity pin, and I think I'll go through with marrying him. It's so wonderful to be loved and needed, for a change, that I think he's convinced me. I've never had anyone that I even liked fall in love with me and it's a wonderful sensation.
> I'm afraid it won't happen very often; so I'm taking my opportunities where I get them.

I couldn't very well ignore this news. If I congratulated Taeko on being engaged, she could interpret my response either as nasty sarcasm or as sincere relief, which would be ungallant, besides being only partly true. But if I complained, she would return the Shadow's pin, and I would be honor bound to give her mine, as soon as the TEP fraternity jeweler had filled my replacement order for the original pin that I first presented to Dorothy and then recycled to Hildegard.

So I neither congratulated nor complained. I merely asked Taeko to provide me with some information about Mr. Shadow, as a concerned parent or relative might have. She disclosed that he was a twenty-four-year-old graduate student in genetics at NYU, Wesley R., a Caucasian doctoral student who was doing his dissertation research at Cold Spring Harbor on the mutation rate of bacteria.

> Dearest Gun,
>
> Can you get here on September 10th? We're planning to be married that day in Blackford—in Blackford!—and Wes and I would very much like to have you be there.

I don't know what my parents are going to think. I wrote them last week, but I haven't heard from them yet; I'm practically a nervous wreck. You're the only one besides my parents who's had any warning about this—you should feel honored.

I sure hope that when you come to visit us in a couple of years you won't say to your friends "I wonder what I ever saw in her." I'll try to keep myself in trim—in fact, I hope the years will help me grow a little more sensible and mature.

Love Taeko

At the end of August, Taeko let me know that she had postponed her wedding because she couldn't bear the thought of causing further grief to her parents, after her brother as well as her sister had married non-Japanese. She and Wes had decided to wait with getting married until such time as her parents had become reconciled to the shame brought on them by their miscegenating children. Taeko and Wes waited four years until they were finally married, by which time he had joined the genetics faculty of one of Berkeley's sister campuses in the University of California system.

Speaking the Same Language

NO ONE IN MAX'S GROUP took much of a liking to Wolf Weidel—except me. He exuded tremendous self-confidence, which, just as Max's, could be easily mistaken for arrogance. I don't mean to say that Wolf *wasn't* arrogant, only that he was less arrogant than he seemed. Once, while we argued about our work and spun clever, though mostly false, theories during one of our daily brown-bag lunches in the lab, Wolf said, "Well, gentlemen, once you fabulists are done dreaming up your fantasies, let a biochemist tell you how things really are in the real world." Upon this remark, Jean Weigle added "biochemist" to his repertoire of obscene invectives.

Wolf let me in on a dark personal secret: His American visa had been delayed because his name had turned up in the Nazi Party roster seized by the U.S. Army in Munich at the end of the war. Eventually, Wolf managed to convince the American consul in Stuttgart that he never knew he was in the party; that he was listed in the roster only because the sailplane glider club to which he had once belonged as a student was incorporated—lock, stock, and barrel—into the SA.

As far as I was concerned, Wolf could tell his story about having been an unwitting party member to the Marines (and to the consul), but not to me, the razor-sharp FIAT investigator. Didn't he ever wonder about what *had* happened to his glider club? Didn't he ever get any mail from party headquarters? Wasn't the party after him to pay his dues? But I kept my doubts to myself. In particular, I didn't tell Élie or Jean that Wolf was not merely a generic *Boche* but a certified, card-carrying Nazi.

In view of my disbelief in Wolf's story, it may seem strange that we became such good friends. Admittedly, I never developed the affections for him that bound me to Élie or Jean. But—Nazi-Schmazi—there was something that drew me to Wolf, a young German scientist of my own generation who had spent the war years in Berlin. I suppose I regarded him as a means for my establishing the personal and professional ties with Germany that I was coveting—secretly, of course, since I would never have openly admitted such a perverse desire, especially not to Élie or Jean. And, what's more, Wolf and I shared some personality deficits rooted in our common German background, such as romanticism, anti-Semitism, and unreasonable expectations of women.

I felt free to talk with Wolf about my troubled dealings with Hildegard, Lore, and Taeko, as I had not with my sister and brother, or with any of my other friends. He said he had the same sort of problems, although he couldn't offer any concrete suggestions for their solution. But I never discussed with him my other emotional problems arising from my being a German Jew, a topic for which I felt he wouldn't be a suitable conversation partner.

Politically, Wolf was aristo-conservative. He considered the members of Hitler's National Socialist German Workers' Party low-class, leftist riffraff and took the name of the party at face value. Once, during another lunchtime conversation in Max's lab, Wolf said, "You guys are always raving about socialism, but you don't know what you're talking about. *I* know, because I had to live under socialism for twelve awful years!" Everyone present, including Max, was so stunned by this inanity that the conversation stopped. I didn't let on, of course, that when I was thirteen, I too thought that the Nazis were socialists, with the main difference between them and the Reds being that the Reds didn't hate the Jews.

Wolf did his doctoral dissertation research in Berlin under the biochemist Adolf Butenandt, who was awarded the 1939 Nobel prize for chemistry in recognition of his work on the chemistry of sex hormones. Butenandt published the results of Wolf's dissertation research in the *Naturwissenschaften* journal, listing himself as the first and Wolf as the second of three authors on the paper. Although by the time Wolf arrived in the States, his paper was considered a classic by biochemists, Phage Group people didn't know about it, and even if they *had* known, couldn't have cared less. But Beadle and his Caltech disciples were amazed when they learned that Max's latest recruit turned out to be the Weidel of the famous paper that reported the chemical identification of a specific molecule whose synthesis was directly blocked by mutation of an identified gene. Wolf's finding had provided the first tangible proof for Beadle's "one-gene-one-enzyme" theory, which asserted that each gene has only one primary function, namely directing the formation of one and only one enzyme, and thus controlling the biochemical reaction catalyzed by that enzyme. The one-gene-one-enzyme theory was one of the main conceptual way stations on the road to molecular biology.

During the war, Wolf worked at the KWI for Biochemistry in Berlin-Dahlem. He was saved from being reported missing on the eastern front because Butenandt, the institute's director, managed to cook up some chemical research projects of potential military importance for him. One of them was the development of a magic pill, which, when dropped covertly by a German secret agent into a 1,000-gallon Allied gasoline tank, would ruin the fuel. After a year or so of research, Wolf came up with a substance of which the stealthy German saboteur would have to dump several shovelfuls into the tank, presumably from a truck he had sneaked into the Allied fuel storage facility. On the intensification of Allied bombing raids on Berlin, the KWI for Biochemistry was evacuated to Tübingen.

The Kaiser Willhelm Gesellschaft was dissolved by the victorious Allies shortly after the end of the war and was reconstituted in 1947 as the Max Planck Gesellschaft. The successor Gesellschaft took over the still surviving

KWI's, which were renamed "Max-Planck-Institutes," or MPI's. Thus Butenandt's KWI in Tübingen, at which Wolf had been working, became the MPI for Biochemistry.

Max had gone to some trouble to avoid getting stuck with a Nazi. He had asked Georg Melchers, Director of the MPI for Biology in Tübingen, to recommend a young German scientist whom Melchers could give a clean bill of political health. Melchers was the only senior German biologist of whom Max was dead certain that he had never been a Nazi—a judgment for which, having known and admired Melchers for forty-five years, I would thrust my own hand into the fire. So I feel sure that during the war, Wolf must have kept quiet rather than brag about being a Party member and that Melchers hadn't suspected, let alone known, about it. Thanks to his talent for identifying promising young scientists such as Wolf, Melchers would play a major role in the first postwar decades in the (very, very slow) process of bringing German science back to its pre-Nazi, world-class rank.

Butenandt took it as a personal affront that Max had asked Melchers rather than him to designate that underling of his whom *he* thought most appropriate to send to Pasadena. (Butenandt and Melchers were *Intimfeinde*, a term implying that for the emotionally hale life a German needs an intimate enemy even more than an intimate friend.) So Butenandt informed Wolf that there would be no job waiting for him in the Tübingen MPI for Biochemistry when he returned from the States if he accepted Max's offer to spend a year at Caltech. Melchers told Wolf not to worry. He promised Wolf to hire him as an *Assistent* in the Tübingen MPI for Biology once he came back from Pasadena. This chicanery hardened Wolf's hatred of Butenandt, who, Wolf felt, had never allowed the full credit due him for his identification of the specific molecule aborted by gene mutation. Wolf confided to me that his main ambition in life was to get even with Butenandt.

By 1960, having become one of Melchers's co-directors of the Tübingen MPI for Biology and escaped from Butenandt's clutches, Wolf seemed ready for implementation of his revenge. But that same year, he was back in those old clutches when Butenandt became president of the Max-Planck-Gesellschaft, and the politically most powerful scientist in the German Federal Republic. Wolf never did achieve his goal of getting even with him.

Delbrück's Creed

CONTRARY TO MY UPBEAT IMPRESSION that phage research was on the move, Max had left the Bloomington meeting with the downbeat feeling that the group's research had lost momentum. So when, for the first time in many years, Max was not going to spend the summer in Cold Spring Harbor, because his wife, Manny, was expecting their second child, he arranged to have a few Phage Group stalwarts come out to Pasadena from back east for a month or so of brainstorming and dreaming up crucial experiments. They included Salva Luria and his minions, Renato Dulbecco and Jim Watson.

A few days after the arrival of the summer visitors, I joined three postdoc friends working at Caltech in a climb of Mt. Whitney in the Sierra Nevada, at 14,500 feet the highest peak in what were then the forty-eight United States. We went up Whitney the long way, from the Western slope of the Sierra Nevada, which took a week's hike, lugging a 50-pound backpack over seventy miles of mountain trails.

When, at the end of our hike, Wolf picked us up in my Victory Ford at the Whitney Portal roadhead on the eastern slope, he warned me that Max was furious. No sooner had I left for Mt. Whitney than Max announced the first brainstorming session of the Phage Group. But when he found out that I had absconded to the Sierra, he called it off. I was terrified by the news that I had made Max angry. Fearing punishment, I was in a dither all the way back to Pasadena. It didn't occur to me that I ought to feel honored by Max's anger if he really thought that, without having me around, there wasn't any point brainstorming with Luria, Jim Watson, and all the other Phage Group hot-shots. On reporting ultra-contrite to Max, I was immensely relieved that he didn't give me a verbal caning; he just asked me not to vanish again while the Phage Group was in town.

My universe had begun to revolve around Max. Psychobabblically speaking, he became the father figure I had never had. I liked my father—maybe even loved him for a while—but, as long as I could remember, I never respected, let alone admired him. I never took him seriously or considered him a role model, except insofar as he may have modeled a role that I tried to avoid. I sought to emulate Max's absolute moral integrity and his fabulous intelligence. For me, he was the measure of all things.

Yet I never developed the affectionate feelings for Max that tied me to my close friends, such as Élie and Jean, in whose company I always felt a warm glow and whom I sorely missed in their absence. In Max's presence, I could never relax enough to do any warm glowing, since the power of his personality simply overawed me. After Jim Watson discovered the DNA double helix, Max once complained that he had always treated Jim like a son, and now that Jim had become famous, the ingrate treated him like a father. To the very end—even when I saw Max for the last time, as he lay dying in his bed—I was always afraid that I would say something stupid. Throughout my career, one of my highest ambitions was to achieve things that would make Max think well of me—which didn't include getting academic appointments or winning prizes, for which Max had (or, at least pretended to have) no regard.

His award of the Nobel prize in 1969 eventually presented Max with a moral dilemma. Although he was certainly pleased to receive that highest of scientific honors, he realized that acceptance of the prize would negate his most important contribution, namely providing a high standard of incorruptibility. For he was a kind of Gandhi of molecular biology, who, without possessing any temporal power, was an ever-present and sometimes irksome spiritual force. He accepted the prize (donating the money to Amnesty International) partly because he felt it vindicated Bohr's faith in him as a young postdoc in Copenhagen, when many other senior physicists didn't think Max would ever amount to much, and partly because he thought that Sartre's making a big noise about *his* earlier refusal of the Nobel prize in literature was vulgar one-up-manship. But Max regretted his acceptance once he learned that the laureate in literature in the year of his own award, Samuel Beckett, neither accepted nor refused the prize; he simply ignored all mail from Stockholm.

Max's agenda for the brainstorming bull sessions that summer of 1949 was to draw up a synoptic outline of everything important that was known about phage, because he hoped that an orderly array of all the known phage facts might suggest novel working hypotheses or make evident clues hitherto overlooked. He thought we could create such an array without having to consult the literature. After all, between us, we carried in our heads all the significant, credible facts enshrined in the corpus of orally transmitted lore shared by the Phage Group. There was no need to take account of the findings of outsiders, which were neither significant nor credible.

We caucused in a small classroom with blackboards on three walls, whose space we parceled out over the diverse categories of the Phage Universe of Discourse, such as "free phage," "adsorption," "invasion," "multiplication," "mixed infection," and "lysis." People spoke up whenever the spirit moved them, as they would in a Quaker meeting, and the gist of their statements was entered in chalk under the appropriate rubric. After a couple of days, the boards were completely filled with closely scribbled notations.

Unfortunately, these bull sessions failed to bring the hoped-for inspiration or enlightenment. No one had any brainstorms for new directions, or even novel approaches to old questions. But Élie had made a hard copy of the whole ephemeral chalk outline, and his notes were to form the basis of the *Syllabus on Procedures, Facts, and Interpretations in Phage* that we published in the following year.

After we completed the synoptic outline, Max gave a tutorial on "complementarity" for the Phage Group summer conclave, a theme that, as we all knew, had once played a pivotal role in his expectation to run into other laws of physics. Although we were aware that complementarity had something to do with quantum mechanics, few of us (and certainly not I) understood just what, exactly, Max had in mind when he used that term.

Max explained that when Niels Bohr introduced the complementarity notion in the late 1920s, he did not refer to its ordinary, everyday meaning, namely two aspects of a thing that make it a whole. Rather, Bohr gave it a special, esoteric meaning, namely two theoretically irreconcilable aspects of the world whose factual irreconcilability no experiment can ever demonstrate. Bohr's prime example of complementarity was the description of electrons in terms of *both* waves *and* particles. According to quantum mechanics, when an electron travels through space, it does so as if it were a wave. Yet when it interacts with matter, it does so as of it were a particle. Although our rational intuition tells us that the paired wave/particle attributes are mutually irreconcilable, our experiments tell us that both attributes *have* to be included in any complete description of the electron's behavior. These attributes are complementary because the experimental setups under which either attribute—wave or particle—can be observed are mutually exclusive. In any given setup, you can see the wave *or* the particle attribute, but never both at the same time

Max referred to complementarity as a "Conspiracy of Nature," which ties observer and object together and prevents the drawing of a sharp line in any experimental setup that would show where the observer ends and the object begins. "'Remember,' Max quoted Bohr, 'that in the drama of existence we play the dual role of actor and observer.'"

His tutorial didn't make much of an impression on most of my Phage Group confrères, who thought that all that "other laws" stuff was a lot of baloney anyhow. (I once heard Luria say so—when he was safely out of Max's earshot). Maybe complementarity is important for atomic physics, but what relevance does it have for our line of work? None, because in phage experiments the object obviously begins at the plaque on the nutrient jelly plate and the observer ends at the eyeball that looks at the plaque.

On me, however, Max's complementarity tutorial made a tremendous impression. It was my first encounter with an idea that would be central for many of my later philosophical writings: While the human mind happens to be eminently suited for constructing a rationally coherent picture of the world at a superficial level, it leads to rational incoherences and contradictions when it is asked to address problems that lie too far beneath the surface.

Later in that summer of 1949, Max went back east to speak at the celebration of the 1000th meeting of the Connecticut Academy of Arts and Sciences. The Academy had invited one poet (Wallace Stevens), one composer (Paul Hindemith), and one scientist (Max) to come to New Haven and address its membership. Stevens read a set of his poems entitled "An Ordinary Evening in New Haven," Hindemith conducted a composition of his for trumpet and percussion, and Max presented a lecture entitled "A Physicist Looks at Biology." On his return to Pasadena, he repeated his lecture as a Caltech biology division seminar. The Kerckhoff Laboratory's large auditorium was completely packed with students and faculty, including such Caltech's notables as George Beadle, Linus Pauling, and Richard Feynman. Max was clearly stimulated by his audience and gave a brilliant performance.

To start with, he drew attention to a fundamental difference between physics and biology. He said that whereas the goal of physics is the discovery of universal laws, biologists cannot reasonably aspire to this aim, because any cell, embodying as it does the record of a billion years of evolution, represents more of a historical than a physical phenomenon. As Max put it, "you cannot expect to explain so wise an old bird in a few simple words." Just as we find features of the atom—its stability, for instance—that cannot be accounted for by descriptions in terms of the ordinary mechanics of apples falling from trees, so may we find features of the living cell that are not accounted for by descriptions in terms of the quantum mechanics of atomic nuclei and their clouds of orbiting electrons. The description of these cellular features would stand in a complementary relation to the descriptions of atomic physics, and thus may lead us to the "other laws of physics" that Max (and I) was hoping to encounter.

Max concluded his lecture with some remarks that accounted for the tendency of the physicists who had gone into molecular biology to deprecate biochemists. Physicists, Max said, are usually told that the only real access of molecular physics to biology is through biochemistry. But, according to Max, biochemistry was, in fact, unlikely to be very useful for gaining an understanding of the really important matters in biology. To him, the biochemists' agenda of explaining the simple through the complex smacked suspiciously of the pre-quantum-theory agenda of explaining atoms in terms of mechanical models. The biochemists' agenda seems sensible until paradoxes crop up and come into sharper focus, and this will not happen until the analysis of living cells has been carried into far greater detail. Max believed that physicists embarking on biological research will show the greatest zeal in uncovering these paradoxes and provide deep theoretical insights into the really important problems.

Twenty years later, in his Nobel prize acceptance speech, Max reflected on the New Haven meeting. He said that it hadn't escaped his notice at the time that Stevens's and Hindemith's contributions were enjoyed by everybody in the mixed audience of artists and scientists crowding the hall, while only scientists had come to hear Max's lecture. This "irreciprocity" Max thought was fitting, because the medium in which the scientist works does not lend itself to the listener's ear.

> The scientist may say to himself that his experiment is his musical composition, his pipette his clarinet. To others, however, his music is as silent as the music of the spheres. . . . The books of great scientists are gathering dust on the shelves of learned libraries. And rightly so. The scientist addresses an infinitesimal audience of fellow composers. His message is not devoid of universality, but its universality is disembodied and anonymous. While the artist's communication is linked forever with its original form, that of the scientist is modified, amplified, fused with the ideas and results of others, and melts into the stream of knowledge and ideas which forms our culture. The scientist has in common with the artist only this: that he can find no better retreat from the world than his work and also no stronger link with the world than his work.

Viruses Under One Hat

BY THE FALL OF 1949, there were six research fellows working in Max's lab—Élie Wollman, Jean Weigle, Wolf Weidel, Renato Dulbecco, Seymour Benzer, and me—a whopping growth since I showed up as Max's first Caltech postdoc the year before. We formed a close-knit sib, with Max as our *spiritus rector*.

Max had recruited Renato Dulbecco as a new member of his laboratory, and Luria was miffed by Max's suborning of his old friend whom he had brought over to Indiana from Italy barely two years ago. In Luria's lab, Renato had just found that ultraviolet-light-killed viruses can be brought back to life by shining visible light on host cells infected with them. This exciting discovery of reviving the dead was given maxi-coverage in the media, and two Italian cities—Catanzaro and Imperia—vied for the honor of being Renato's home town.

Seymour Benzer, who had received his doctorate in solid state physics from Purdue University, was my classmate in the 1948 Cold Spring Harbor phage course. Just like me, Seymour had been seduced by Schrödinger's *What Is Life?* and hoped to get started on finding the aperiodic crystal of heredity in the Caltech lab of the protagonist of Schrödinger's book. Within a few years, Seymour would actually refashion the fuzzy concept of the Mendelian gene of classical genetics into its precisely defined, latter-day molecular-genetic version. I have always felt it a shame that Seymour was not among the Nobel laureates honored for laying the foundations of molecular biology.

For us members of Max's research group there was no clear separation between our professional and our private lives, because Max's benevolent (or in the California-speak of the 1990s, "caring") interest in his disciples was all-inclusive. He not only guided our scientific work in the lab, but also supervised, not to say intruded in, what would normally be considered one's private, after-hours activities, such as partying, going out for concerts, plays, movies, or dinner, and, of course, camping. As our *pater familias*, Max considered it his business—if not actually his duty—to inform himself about all facets of our lives, with privacy not acknowledged as one of our constitutionally guaranteed rights.

Exchanging my anxiety-ridden sovereignty for an insouciant thralldom under which I could leave all the main decisions about my professional and private activities in Max's hands, appealed to me as an ideal shift in my way of living. It provided relief from all that freedom and personal responsibility for making choices with which I had been saddled ever since my mother's death. Being a communicant of the Phage Church was like being in the army, where every soldier, downward from the chairman of the joint chiefs of staff, takes orders from a superior authority figure who is held accountable for the commands one obeys.

Not everyone was as enthusiastic as I about taking orders from Max, but there were only occasional cases of open rebellion when some intrepid libertarian refused to go camping or insisted on making his or her own choices from the menu. Élie often grumbled about Max's authoritarian and intrusive ways of running our lives, but it was only Odile, and not he, who had the gumption to tell Max where to get off when she thought he was meddling in affairs that were none of his business. Max's response to rebellion by one of his minions was usually surprise rather than anger. He just didn't expect that there could be any opposition to his sage dispositions. In fact, he was contemptuous of toadies and preferred loyal resisters such as Odile, for whom he had great respect and liking.

Max suggested that it was time for Élie and me to start working on the publication of our results on the tryptophan activation of T4 phage. Since the papers were going to be in English, it was clear that I would have to do the most of the writing. But I had no experience in composing scientific papers other than my doctoral dissertation. So I chose as my model two urbane, closely reasoned papers on the mechanism of adsorption of phages to their bacterial host cells published in the early thirties by the Hungaro-Jewish physical chemist Martin Schlesinger. Written in unusually lucid scientific German, Schlesinger's papers presented a closely reasoned dialectic between theory and experimental results, using only the minimum of mathematics necessary for making quantitative predictions implied by his theory.

Schlesinger, on whom Max had nothing as far as incisive, quantitative thinking was concerned—at least judging by Schlesinger's papers—had begun his phage research at the Institute for Colloid Research in Frankfurt in 1930. Three years later, he fled to London, where, despondent over the rising Nazi tide, he committed suicide in 1936.

Would the birth of molecular biology have come much sooner if this early pioneer of exact experimentation on phages had not died prematurely? Even though Schlesinger was interested in DNA (as Max was not)—just before his death, he discovered that the chemical composition of phages is half protein, half DNA—I think the answer has to be no. Schlesinger couldn't have gotten to know the architecture of the phage particle or inferred the genetic role of the phage DNA much sooner than Delbrück, Luria, and Hershey did in the late 1940s. Like them, he would have had to wait for access to an electron microscope, to radioactive phosphorus, and to the information that the chemical composition of DNA varies with its biological source.

Emulating Schlesinger's crisp expository style, it took me most of my second year at Caltech to write the three papers presenting the results of our experiments and to hone the theory we had developed to account for them. I would never again devote as much effort and care to any of the couple of hundred other papers and essays I eventually published. So I consider these first three papers with Élie as my best. In accord with his usually clouded crystal ball, Max predicted that one day they would be famous classics. Alas, only a few people read them when they came out—we did get one fan letter from an immunologist in Australia—and they have long since been forgotten.

As I had feared at the outset of our research project, there was no need to invoke "other laws of physics" to explain the seemingly bizarre dynamics of the tryptophan activation phenomenon. We managed to devise a model based on conventional physico-chemical theory that accounted for all the data. It was a forerunner of the "cooperative" models of the complex interactions of small molecules with enzymes and other protein molecules that were first put forward a few years after our papers came out and have formed the basis for understanding the regulation of protein function ever since. As far as I know, no contributor to the vast literature of cooperative interactions ever cited our tryptophan activation model.

My Merck fellowship was coming to an end. Staying on in Max's lab was what I would have liked best—for the rest of my life, if possible, as one of the many perennial Caltech postdoc hangers-on. But Max, who couldn't have been unaware of my desire to stay in Pasadena, never expressed any interest in keeping me around. Joining Élie at the Institut Pasteur was my second preference. This was not based wholly on scientific considerations, such as the opportunity to continue our felicitous collaboration. Rather, my preference for Gay Paree arose from the emotional dependence on the Wollmans' parenting I had developed after the Donna debacle and the self-maneuvered shipwrecks of my romances with Hildegard, Lore, and Taeko. I just couldn't see how I was going to survive on my own. So Élie wrote to *le grand patron*, André Lwoff, to ask whether it was OK with him that I would move to his department next year to continue our collaboration.

Lwoff informed Élie that it was *not* OK with him. He had just taken on a young *assistant*, François Jacob, with whom Élie would have to share his *labo* when Élie came home. If it meant so much to Élie to have me with him, Lwoff would be willing to accommodate me a year later, in the fall of 1951. Since everybody knew that two people could do phage experiments in the same space as one, Élie believed that Lwoff turned me down because he didn't want to hear any static from any of those smart-alecky American wise-guys from Max's lab while he was still sorting out the basic facts of lysogenic bacteria. This he expected to have done within another year.

Max suggested that I go to Copenhagen, to work in the lab of the Danish biochemist Herman Kalckar. He made the same proposal to Jim Watson, who was finishing his doctoral dissertation with Luria at Indiana. Had I not known that Max was colossally impressed by Jim, I would have taken his sending me to work with a Danish biochemist as a sure sign of his rating me as a second-rate intellect. Max told me that it might do Jim and me some good to learn something about DNA chemistry from Kalckar.

Although Max had begun to think that maybe DNA did have something to do with genetic self-replication, he had not yet cottoned on to the still premature idea that DNA was the genetic material of the phage—Schrödinger's aperiodic crystal of heredity. Instead, he favored some non-coding role for DNA, although I don't think he found merit in the (with hindsight, preposterous) idea, then about to be published by Luria in *Science* magazine, that DNA is involved in the "baking" of the infective progeny virus particle in the final step of the intracellular phage growth cycle.

Kalckar and Max had overlapped as postdocs at Caltech before the war, but Max didn't seem to realize that Kalckar knew very little about DNA. His specialty was adenosine triphosphate (ATP) and its provision of free energy for driving biochemical reactions. Maybe Max thought that DNA, being composed of ATP-like nucleotides, provided the free energy for driving self-replication of proteinaceous genes, in chromosomes as well as in phage. Kalckar agreed to accept both of us. Jim applied for the Merck fellowship and I for the American Cancer Society fellowship, administered, like the Merck fellowship, by the National Research Council.

On completing the part of my NRC fellowship application form in which I was asked to detail my proposed plan of study and my qualifications for carrying it out successfully, I had the conceit that, as a member of Max's Phage Group, my chances for an award were very good. Ever the phony, my ideas about DNA were extremely hazy, but I was sure that I knew more about phages than any member of the fellowship committee. I thought that I had come a long way from the nonsense-blubbering mountebank who had presented himself to the Merck fellowship board two years earlier.

I had the best times of my life in Pasadena, except for the dearth of romance after my dire affair with Donna. So I still thought of Europe as the place where I would run into the woman of my dreams. I had yet to understand that my not having done anything about marrying Lore, the fabulous woman I *had* found in Europe, meant that the Miss Right I was looking for was not a real person, whom I would be no more likely to discover in Europe than in the States.

Although going to work in Paris would have been my preference, I was enthusiastic nevertheless about Copenhagen, of which I had fond memories from a bicycle trip I made to Denmark with two fellow members of the Schwarze Fähnlein when I was twelve—the Tivoli Gardens, the delicious open-faced sandwiches, the jelly doughnuts the Danes called "Berliners," and the yeasty smell of the Carlsberg brewery in Valby. Moreover, I had heard that Danish girls are something extra-special and that your typical foreign bachelor postdoc doing time in Copenhagen brings home a Danish bride.

At one of our lab lunches toward the end of 1949, Max let us know in a backhanded way that he was beginning to lose interest in genetics. This was at a time before the first great breakthroughs of molecular biology—the identification of the phage DNA as the genetic material and the discovery of the DNA double helix—had yet been made.

"It seems to me that phage research is now in good hands," Max announced. "All the good work that's being done these days is bound to lead to an understanding of biological self-replication before long."

I was astounded. "You mean, Max, that we'll find the solution to the self-replication problem without meeting up with any paradoxes along the way?

"Yes, Günter. I've begun to think so."

"What about those other laws of physics?"

"After phage research has solved the puzzle of self-replication," Max answered, "there'd still remain an even harder problem posed by living creatures. I mean the brain, for which reasonable physical mechanisms can't even be imagined. I bet that some other laws of physics are needed to explain the function of this most mysterious ensemble of atoms in the universe, to explain how mind arose from matter."

"What d'you propose to do about the brain?" Jean asked Max.

"I'm glad you asked. To prepare ourselves for our postphage, brain research future, we're going to run a series of seminars on sensory neurophysiology. I've drawn up list of publications, divided into chunks of three or four papers. Each of you'll pick one chunk for presentation at a seminar. That way, we'll all be brain experts by next spring."

Élie was the only mutineer who refused to take part in Max's reeducation project. "I'm sorry, Max, but I'm not interested in brain research. As a simple bacteriologist, I probably wouldn't understand the papers you picked out anyhow. Actually, as far as I'm concerned, phage and bacteria are good enough research materials for me, and I'll to continue working on them, if you don't mind."

Four of the chunks selected by Max consisted of papers on vision, and they were quickly grabbed by Jean, Seymour, Renato, and Wolf. The fifth chunk was devoted to hearing, which my false friends left for poor, ignorant me. As I soon realized, they already knew enough about sensory neurophysiology to stay away from hearing, compared to which vision, however complex its function may seem, is a breeze.

My chunk consisted of three papers published in the late 1920s in the *Physikalische Zeitschrift* under the title *"Zur Theorie des Hörens"* ["On the Theory of Hearing"] by the future (1961) Nobel laureate, Georg von Békésy, then at the Royal Hungarian Telegraph Research Laboratory in Budapest. My reading bogged down as soon as I got beyond the first sentence of Békésy's first paper, which declared grandly that "the treatment of a series of problems in telephone technology is greatly impeded by the lack of a theory of hearing." The papers were full of complex equations and circuit diagrams relating to hydrodynamics, resonance, and mechanical and electrical oscillators, while the text was couched in esoteric anatomical and physiological parlance, completely beyond my ken. This, my first encounter with the literature of neurobiology, was counterproductive from the perspective of Max's missionary goal of arousing my interest in brain research. Maybe there were some "other laws" that could be revealed by studying hearing, but there was no way in which *I* could possibly find them. Like Élie, I was going to stick with simple research on phage and bacteria.

One morning I found a copy of Hugo's *Danish Simplified* on my lab desk, with an inscription in Max's hand on the flyleaf:

Mind the Copenhagen Spirit!

This was Max's way of letting me know that I had been awarded my postdoctoral fellowship. He was lounging in the yellow leather easy chair when I went over to his office to thank him for the Danish language primer and the advance information.

"By the way, Max, you needn't worry about my minding the Copenhagen Spirit! I can hardly wait to get in on the renowned beer, women, and song of the Venice of the North."

"What the devil are you talking about, Günter? Beer, women and song? 'Copenhagen Spirit' means the esprit of friendship and cooperation that Bohr had instilled in his crew of atomic physicists. I want you to hang out at Bohr's Institute of Theoretical Physics, just around the corner from Kalckar's Institute of Cytophysiology, to take in the Copenhagen spirit at its source!"

An unexpected windfall of financial support from a private donor came Max's way at the beginning of 1950, which would add a new dimension to the work carried out in his laboratory. This new dimension wasn't the futuristic, postphage brain research for which we were supposed to be tooling up, but studies on animal viruses. I don't remember animal viruses having been a major topic of conversation in the lab prior to the sudden specter of money on the horizon, probably because they seemed unpromising experimental materials for cracking the self-replication problem.

The donor was Colonel James G. Boswell, a resident of San Marino, located just south of Pasadena and then home to the non-show-business (non-Jewish) superrich. The colonel had been told by his physician that there wasn't any effective cure for shingles, the awfully discomforting disease from which he was suffering.

"How come," the colonel asked his doctor, "they haven't found a cure for shingles, what with all those miracle drugs that have come on the market since the war?"

"Because shingles is caused by a virus, and the new antibiotics don't cure virus diseases. There are vaccines that protect against infection by a few viruses, smallpox being one, but nobody is close to a cure once you have already come down with a virus."

"Why not?"

"Because too little is known about the basic biology of viruses that afflict humans. But we might soon get a better understanding of virus diseases through some very interesting recent experiments on viruses that kill bacteria. In fact, some of the best work on bacterial viruses is being done at Caltech, just a couple of miles up the street from your house, Colonel."

The doctor then asked the colonel whether he'd be willing to donate the money needed for setting up a laboratory at Caltech specifically dedicated to applying the results of bacterial virus research to the problem of human viruses? The colonel was willing. He endowed the James G. Boswell Foundation Fund for Virus Research, and Max was put in charge of implementing the research projects to be financed by the fund.

As Max saw it, past efforts to gain effective control over viruses that cause diseases in animals and plants had not gotten anywhere because so little understanding of the life cycle of these pathogens was available. How do animal and plant viruses invade their host cells? How do they multiply? How do they interrupt and modify the normal functions of the infected cell? He thought that these questions tie virus research to the mainstream of modern biology, which aims at the analysis of cellular function.

Making his first use of Boswell Fund money, Max organized a conference at Caltech in March, which he called Viruses 1950 and to which he invited some forty people working on animal, plant, and bacterial viruses. He hoped that their discussions would reveal whether, and to what extent, all kinds of viruses could be brought under what he called "one hat."

Weeks before the start of the Viruses 1950 conference, the entire Phage Group collaborated by mail on editing, revising, and augmenting Élie's hard copy of the ephemeral synoptic chalk outline created during our bull sessions in the previous summer of all the significant facts known (to us) about phage. Max entitled this communal effort *Syllabus on Procedures, Facts and Interpretations in Phage* and distributed copies beforehand to all invited participants, as a tutorial to facilitate the discussions at the conference.

The invitees to Viruses 1950 included Wendell Stanley, as well as the British plant biologist, F. C. Bawden, who had shown that, contrary to Stanley's initial claim that tobacco mosaic virus (TMV) is a pure protein, TMV also contains RNA. In his supercilious Cantabrigian English, Bawden treated Stanley like a colonial bumpkin: He couldn't forgive the Ohio-born Stanley for having gotten the Nobel prize despite this error and for having made the public-relations-effective claim that TMV is a "living molecule." The most prominent of the animal-virologist invitees was Richard Shope, of the Merck Institute for Therapeutic Research. He was renowned for his work on a tumor virus named after him and for his identification of the swine influenza virus, suspected to be connected with the great influenza pandemic of 1918–1919. Bacterial viruses were represented by Max's resident phage crew, as well as by Hershey and Luria from back east.

Meeting Stanley at the Viruses 1950 conference not only provided me with my first personal contact with an anointed Nobel laureate but it also turned out to be a crucial encounter for my later career. In the following year, Stanley came to Copenhagen to attend the Second International Poliomyelitis Congress, and I happened to be seated next to him at a dinner that Niels Bohr gave in his residence at the Carlsberg Brewery for the galaxy of visiting star virologists. (I had been invited to the dinner as Bohr's sole scientific grandson—via Max—in virus research.) I mentioned to Stanley, who remembered meeting me in Pasadena, that I heard from Martha Baylor that he was recruiting people for his virus laboratory in Berkeley and asked him whether there might be an opening for me on his staff after my present NRC fellowship was going to run out in the following year. "There might be, Gunther," he answered. "Why don't you write me in Berkeley about it? Enclose your CV."

His offer of a research position in Berkeley, equivalent to the rock-bottom academic rank of instructor, reached me a few weeks later. On my arrival in Berkeley in the fall of 1952, I learned from my new colleagues why Stanley, an unassuming and seemingly simple man, had hired me. Although a Nobel laureate, he never missed a chance to add yet another feather to his cap, however tiny a feather, and he fancied the idea of having on his staff someone he imagined was a young intimate of Niels Bohr. My relations with Stanley were invariably amiable, yet throughout all the years, until his death in 1971, that I would be his colleague in Berkeley, during which I came to know dozens of other Nobel laureates, I never lost my awe of him. He always called me "Gunther," but I could never bring myself to address him in any way other than "Doctor Stanley."

The Viruses 1950 conference, which lasted three days, did little to achieve Max's goal of bringing all kinds of viruses under one hat. There was too little common ground between the pragmatic research style of the animal and plant virologists (conditioned, no doubt, by the difficulties they faced in designing experiments capable of giving clear-cut experimental answers to basic questions about the reproduction of their viruses) and the theory-driven, ideological, quantitative approach of the Phage Group, to allow much fruitful comparative colloquy.

Max and Luria found fault with Bawden's interpretations of his experiments concerned with the interference of one type of virus with the growth of another when the two types co-infect the same plant. Bawden interpreted this phenomenon as the result of a competition between individual viruses infecting the same cell, but Max and Luria insisted that, since there was no way to observe virus growth in single plant cells, the interference phenomenon could just as well be a cell population phenomenon occurring at the level of the whole plant.

Luria also disparaged the analysis of the physical and chemical properties of virus particles, which was a major research agenda in Stanley's virus laboratory. He asserted that it was unlikely that such work would throw much light on the fundamental problem of virus reproduction. It proved to be a great help for my later career in Berkeley that I kept my trap shut during the brouhaha that Luria provoked by insulting my future colleagues.

Max asked Wolf, who was going to return to Germany right after the meeting, to present a summary of the results he had obtained during his year in Pasadena. Wolf's lucid and rhetorically skillful talk left no doubt that, though he was a mere biochemist, he had made a substantial advance. Even Jean had to admit it grudgingly. Wolf had managed to pioneer the structural and chemical characterization of the outer cover of the bacterium, which as he showed, resembles that of a Zeppelin dirigible. It consists of a flexible, impermeable external skin stretched over a stiff internal skeleton, which provides the rigid framework that maintains the cigar shape of the bacterium. In his annual report to the Caltech president, Max referred to Wolf's as "the most startling finding of the year," while he claimed merely that Élie's and my tryptophan activation studies, "which are so far without parallel in physiology, may well turn out to be of wide significance."

Max had arranged for the conference to be followed by a two-day camping trip to Death Valley, intended to facilitate informal, postmortem discussions. On the hikes we took in Death Valley and while sitting around the campfire at night, swilling California jug wine and feasting on steaks and corn-on-the-cob cooked by Manny Delbrück and chairman Beadle, we talked about everything *but* viruses. Yet the camping trip was the more successful part of Viruses 1950. Comradely contacts established in Death Valley between the Phage Group stalwarts and plant and animal virologists laid the foundation for a future general virological community, once methodological advances *had* finally made it possible to bring all kinds of viruses under Max's one hat.

Viruses 1950 left Max with little hope of coopting some established animal virologist for the Boswell project of applying the results of bacterial virus research to the problem of human viruses. He decided that he had no option other than charging a member of his Phage Group with the reform of animal virology. He selected Renato for that assignment.

Renato went on a study tour of American laboratories at which work on animal viruses was under way. He didn't find the strategies, methods, and objectives he encountered very promising, largely because of a lack of adequately quantitative techniques. The only bright spots on the animal virus research scene he saw were some recent advances in the culture of animal cells that were potential hosts for the in vitro growth of animal viruses. On returning to Pasadena, Renato—who happened to have had some experience in cell culture as a medical student in Turin before the war—decided to work out a procedure for assaying animal virus particles by their formation of plaques on a sheet of cultured animal cells, just as phage particles form plaques on a lawn of bacterial cells.

Within a year of setting up his new lab, Renato had succeeded in devising a plaque assay method. The stage was now set for his implementing a comprehensive research program for quantitative studies on animal viruses to fathom their intracellular reproductive cycle. Before long, Renato isolated virus mutants and developed techniques for mixed infection of single animal cells with two or more genetically different mutant viruses. Thus began the era of animal virus genetics, one of the most important medical spinoffs of molecular biology. For these contributions Renato would be awarded the 1975 Nobel prize in physiology or medicine.

In retrospect, it seems ironic that the largely sterile discussions at the Viruses 1950 meeting led to a revolution in animal virology, by motivating Max to ask Renato to make it happen. And although, as far as I know, there still isn't any cure for shingles, few philanthropic supporters of medical research can have invested their money more wisely than Colonel Boswell.

Awakening

CALTECH INSTRUCTOR ARRESTED:
CHARGED WITH DENYING COMMUNIST PARTY TIES

read a headline on the front page of the *Pasadena Star News* toward the end of my stay in Pasadena. The story reported that the FBI had arrested one of our own crowd, Sidney Weinbaum, a theoretical physicist working as a research fellow for Linus Pauling in the Caltech chemistry division. The government charged him with perjury and fraud for having falsely denied under oath his past membership in the Communist Party.

During the war, Sydney worked for the Bendix Aircraft Corporation and other defense industries in the L.A. area. After the war, cleared by Army intelligence for secret work, he was hired by the Caltech Jet Propulsion Laboratory as a senior research engineer. But he was fired in July 1949, when his clearance was withdrawn on the grounds that he had been a member of the Communist Party. Sydney appealed his dismissal, swearing before an Army review board that he had never been a party member. While his appeal was pending, Linus Pauling hired Sydney to tide him over, and, in case his appeal was unsuccessful, to give him time to find a job not requiring top clearance.

Instead of having his clearance restored, Sydney was arrested at the request of the U.S. District Attorney, who alleged that, in the late 1930s, Sydney had been a member of the Caltech-centered Communist Party's Professional Unit 122, under the cover name "Sydney Empson." Sydney denied these charges. Next day, the *Star News* reported under the headline,

DR. LINUS PAULING DEFENDS WEINBAUM

that Pauling had the greatest confidence in Dr. Weinbaum, a high-minded man and a fine American, who never seemed interested in politics at all. My own impression was that Sydney's only extra-scientific interest was chess.

We all believed that the charges made against Sydney had been trumped up by Tricky Dick Nixon and his Red-baiting ilk on the House Un-American Activities Committee to smear Pauling, whom they detested because of his left-wing politics but somehow couldn't vilify directly. I had never heard of Unit 122, which no one had ever mentioned within my earshot. Maybe the whole "Unit 122" shebang was just made up by Tricky Dick.

The principal evidence put forward by the government against Sydney was the testimony of an FBI handwriting expert, who certified that the "Sydney Empson" signature in a Communist Party membership book was in Sydney Weinbaum's hand. The prosecution also produced two former Caltech graduate students who testified that Sydney had recruited them for Unit 122 in the 1930s, using the "Sydney Empson" cover name. These two witnesses had probably fingered Sydney to the FBI in the first place, to save their own skins while they had been interrogated during the open season on Reds and ex-Reds. Another admitted former member of Unit 122, J. Robert Oppenheimer's younger brother, Frank, testified that he didn't remember Sydney's having been a member. So regardless of whether Sydney had, or had not, been one of its members, Tricky Dick evidently hadn't made up the existence of Unit 122. Despite his unwavering denial of the charges, Sydney was convicted of perjury and sentenced to four years in a federal penitentiary.

As I was packing up to leave Caltech, I was frightened by the Weinbaum case. I firmly believed in Sydney's innocence and thought that if being put in prison could happen to him, it could happen to anybody. So I thought that by heading for Europe, I was making my getaway just in time. But shouldn't I stay on in Pasadena to help in the fight against Tricky Dick's anti-Red hysteria? Wasn't I a rat leaving the sinking ship? Yes, I thought, I was a rat, but then nobody, especially not Max, had asked me to stay on; this here rat didn't seem to be welcome any longer on board the sinking ship!

But I was also vaguely aware that my two years as Max's postdoc marked the beginning of a gradual lessening of my self-hatred as a German Jew. Unlike the officially sanctioned residential and social apartheid of Jews and gentiles at the University of Illinois, there was no overt anti-Semitism at Caltech. In fact, there was no more interest there in whether I was a Jew than in whether I was a skier or fond of Chinese food. It was at Caltech, in the first place I had been where my Jewishness didn't matter, that I managed to lose some of my shame of being a Jew.

It took me another twenty years to rid myself of what I hope were the last vestiges of my self-hatred as a German Jew. The final step in the abatement of my Jewish self-hatred I owe to the Israeli Defense Force, by virtue of its astounding 1967 Six-Day-blitzkrieg victory over the Arabs and occupation of the whole of Palestine, as well as of parts of Egypt and Syria. It turned out that, contrary to *Der Stürmer*'s anti-Semitic slander I imbibed as a child, Jewish blood was not lacking manly strength after all. My coreligionists had raised the world's most formidable army, backed by the world's most awesome intelligence service. Our famously anti-Semitic Prussian warrior-king, Frederick the Great, would have been mighty pleased to have had the IDF in his service!

When I visited Israel in 1968, it was the first time that I felt proud to be a Jew, a member of the invincible race that had turned the barren desert wasteland into a blossoming modern country. I felt at home in Israel, not so much because it is a Jewish state, but because I experienced it as a kind of Brandenburgian California, a country in which Prussian *Tüchtigkeit*, high culture, *Blut und Boden* ideology, and enlightened militarism are melded with a beautiful subtropical landscape and climate. The would-be Prussian in me appreciated that at the conclusion of my lecture at Israel's Weizmann Institute of Science—half Kaiser-Wilhelm-Institut, half Caltech—all the questions were asked in strict order of precedence of academic seniority and that at the institute's Purim holiday costume party, all the children were in military getups. The little boys were dressed as soldiers and the little girls as army nurses. Both scenes would have shocked, shocked my liberal, American-bred Berkeley colleagues.

By now, at the close of the twentieth century, Jewish self-hatred has largely disappeared—although not anti-Semitism. As foreseen in the 1920s by the self-hating German Jewish philosopher Theodor Lessing in his classic treatment of the phenomenon, *Der Jüdische Selbsthass*, the creation of a sovereign Jewish nation-state cured the Jews of their negative chauvinism. After the war, Jewish self-hatred not only disappeared but became a taboo subject, because admitting that one, or one's parents, once accepted the anti-Semitic view of their own racial inferiority might be taken as tacit support for the odious theory of Jewish complicity in the Holocaust.

Yet Jewish self-hatred ought to be brought out of the historical closet as a classic example of the genre. As the German émigré sociologist and MIT professor Kurt Lewin pointed out in an essay he published in 1941, Jewish self-hatred has its parallel in many underprivileged groups. According to Lewin, "one of the better-known and most extreme cases of self-hatred can be found among the American Negroes." Lewin thought that the German–Jewish experience is of direct practical relevance for many contemporary societies, where the self-hatred of youngsters belonging to reviled ethnic minorities impedes their escape from their socially and economically disadvantaged status.

We set off from Pasadena in a mini-caravan, Max and Manny in the Delbrücks' stately Packard, and I in my Victory Ford. The Delbrücks were headed for Cold Spring Harbor, and I was on my way back east, en route to Copenhagen. I was going to abandon the old clunker in Chicago as a present for Claire, and take the train to New York. My poor sister! She deserved more of my gratitude than having this junkpile dumped on her. But she had assured me, beatifically as ever, that it would do fine as the second car of which the Hines family was in need.

After a leisurely, four-day drive on unpaved roads through the scenically glorious Hopi–Navajo Indian Reservation country of northeastern Arizona, camping out under the stars every night, we crossed the San Juan River into Utah at Mexican Hat. At a combination coffee shop/gas station, just off the far side of the bridge, a day-old Salt Lake City newspaper was on sale, which carried a banner headline:

Truman Orders U.S. Air, Navy Units into Action

We had been out of touch with the world because the radio in my Ford wasn't working, and Max wouldn't turn on his Packard's radio. He was opposed, in principle, to all electronic media. There was neither a radio nor—God forbid!—a TV in the Delbrück home. So we were totally unaware that there had been any political crisis in Korea and were boggled by the news that

North Korean troops had crossed the 38th parallel and occupied Seoul. Obviously, Uncle Sam couldn't take this Stalin-ordained attack by his half of the Korean peninsula on our half lying down, and so there was bound to be war.

We sat down in the little riverside coffee shop for a late breakfast and tried to forecast what was likely to happen next. Congenitally committed to the Berlinocentric view of the universe, Max and I both thought that the next step to World War III was going to be the seizure of West Berlin by the [East] German Democratic Republic (GDR). They could probably pull this off without direct participation of Russian troops, by using the GDR's paramilitary *Volkspolizei*, recently armed by the Soviets in violation of solemn Allied agreements about the perpetual demilitarization of Germany. So much for the post-Hitler peace!

"Shouldn't I give up my plan of going to Europe, if there's going to be another World War?" I asked Max. "Maybe I could transfer my NRC fellowship to postdoc somewhere in the States rather than in Denmark?"

"Might be wiser for you to stay on this side of the Big Pond if the Communists really do snatch West Berlin," Max answered. "But you've still got a week or so to decide before you take off from New York. By then, the situation ought to have become clearer."

When we got to Denver, we found out that there weren't any indications that the Korean war would spread to Europe. Nothing particularly threatening seemed to be happening in Berlin, beyond the usual shrill demands that the Western Allies get out of West Berlin, which, so the Communists claimed, "lies in the territory of the sovereign GDR." So as we parted ways in Denver—the Delbrücks were going to drive to New York via a southern route through Indianapolis that bypassed Chicago—Max told me that I might as well keep to my original plan and head for Denmark. He and I were going to meet again in a year's time, when he (along with Wendell Stanley) would be one of the plenary speakers at the International Poliomyelitis Congress to be held in Copenhagen.

As I pulled into Chicago I recalled that ten years ago, Claire and Bob had met me at La Salle Street Station, a diffident teenage refugee, lucky to have escaped from Hitler-threatened Europe. And now I was going back to Europe as a certified member of an internationally renowned scientific elite, albeit with only modest research achievements to my credit. My stumbling into the Big Time of Science couldn't have come to pass if Claire and Bob hadn't made so many sacrifices on my behalf. But I was unable to express to them my feelings of gratitude. I signed over to Claire the title of my car, which, to my amazement, had carried me to Chicago without breaking down. In my never-ending repair work, I must have gradually fixed all the Victory Ford's major ailments, because Claire would still drive it for another, relatively trouble-free five years.

Like Taeko, Rita had finally given up on me in the previous summer. On the "at home" card that accompanied the announcement of her marriage to a German–Jewish wine importer in New York, she had added a handwritten note, asking me, as one of her oldest friends from high school days, to get in touch with them next time I'm in town. The irony of her adding this note to her wedding announcement eluded me, and so I called her as soon as I got to New York. The newlyweds took me to lunch in a posh Manhattan restaurant—several cuts above those checkered-red-and-white-tablecloth, candle-in-Chianti-bottle eateries to which Rita had introduced me in her career girl days. She looked terrific—got up in high fashion and sporting a big diamond rock. My admiration for her notwithstanding, I was comforted by this further providential relief from moral responsibility for a woman who loved me and whom I didn't love in return, especially since she appeared to have married well.

Rita gave me to understand that her husband was Mr. Big in Rhenish white, which I hoped was more factual than her wildly exaggerated description to her husband of my being one of the up-and-coming hot-shots of American bioscience.

"Is it really true, Gunther," he asked, "that you're not the marrying kind?"

For the benefit of his wife's self-esteem I repeated the lie I had fobbed off on Hildegard on the night of our first meeting:

"Yes, Henry, it's true. Rita sure had my number! Marriage and going for broke in science just don't mix."

My first stop in Europe would be London, where I was going to visit my family. From there, I would meander leisurely to Copenhagen, beginning with a two-week holiday in the Austrian Salzkammergut with Ronnie, Gabi, and my charming twelve-year-old niece, Monica. Then I would visit Jean Weigle in Geneva and my last stop before Copenhagen would be Berlin, where my father and I were going to meet to finalize our repossession of the Essener Strasse apartment building.

In London, I made my duty calls on my father and Friedl, with whom, as on my previous postwar visits to them, I was ill at ease. Their Belmont Trading Company seemed to flourish, as indicated by the well-appointed apartment into which they had moved and shared with Friedl's enormous, immaculately groomed poodle, Princess Serda. Although they offered to put me up, I preferred to stay with Ronnie and Gabi, to whom I felt closer than ever before, no doubt because our age differences, which, when I was a child, had made me think of them as part of the weird world of grown-ups, had become ever less significant.

The German railway coach Ronnie, Gabi, Monica, and I boarded, Austria-bound, at Ostende Quai betokened the—for me, at least—most important development that had taken place in Germany during the three years I was back in the States: the East–West fission of the German national railroad system. The coach's siding bore the acronym DB, for *Deutsche Bundesbahn*, the railroad of the [West] German Federal Republic. The railroad of the GDR had kept the old, all-German style, *Deutsche Reichsbahn* (DR). It seemed very strange for the Communists to hang on to the old name, since they were outdoing themselves in denying that their state was in any way a successor to the old *Reich* and that the peace-loving, anti-Fascist people of the GDR bore any responsibility for the war and the Nazi atrocities.

Although I could tell from the coded inscription on its undercarriage that our coach had been built before the war, its exterior and interior looked spanking new—evidently recently renovated—in striking contrast to the dirty, decrepit DR rolling stock that had still been in use in 1947. There were commercial advertisements in the compartments and the gangway—there had been none in 1947—an indicator of the West German economic *Wirtschaftswunder,* which began in 1948 on the introduction of the new D-Mark currency. The DB conductor who checked our tickets after the train crossed the German border was decked out in a smart blue uniform, with the traditional conductor's red patent-leather dispatch case dangling from his shoulder, an impressive sartorial advance over the sad-sack Wehrmacht-surplus hand-me-downs worn three years ago by DR train personnel. And the German passengers who boarded our coach no longer exuded the fetid body odor.

The destination for our two-week stay in the Salzkammergut was Fuschl, a hamlet on Lake Fuschl, not far from Salzburg. Staying in Fuschl would allow us to combine hiking and swimming with attending the first postwar season of the Salzburg Music and Drama Festival. In keeping with the Stensch family's traditional predilection for top-of-the-line things of the second category, both Lake Fuschl and our Hotel Seerose were smaller, more modest versions of nearby Lake St. Wolfgang and its Hotel Weisses Rössl, the setting of the operetta, *At the White Horse Inn,* which my parents took me to see in Berlin when I was in the first grade. Our record collection in Treptow had included a song from that operetta, "How can Sigismund help it that he's so handsome?" which was my runner-up favorite after the "Triumphal March" from *Aida.* In view of my middle name being Siegmund, I wanted to believe that the song was about me.

The excursions we made into the picturesque Salzkammergut included a trip on an ancient cog railway to the top of the Schafberg, where we went for a little hike circling the summit. While we were taking in the gorgeous view over the mountains and lakes of the whole Salzkammergut, against a panoramic background of the peaks of Styria and the Tyrol looming in the far distance, a tall, gaunt, mean-faced man in ex-Wehrmacht tatters came up to me and asked,

"Are you Americans? I heard you talking English."

"I'm American but my companions are Britishers. Why are you asking?"

"I haven't got any work. Just released from an ex-Waffen-SS POW camp. D'you happen to know where I can sign up to fight the Reds in Korea?"

My feelings of enchanted contentment evoked by the splendid scenery were lugubriously transmuted, first into fear and then into anger. Fear, because there was no barbed-wire fence that protected us from this monster, as there had been when I walked past the SS prisoners in their cages at Dachau. Unarmed and emaciated, he posed no real threat to us; but I still had not fully overcome my subliminal dread of the Nazis' evil power. And I was angry because this monster felt free to flaunt his past membership in the SS, evidently expecting that an American would welcome him as a credentialed brother-fighter against Bolshevism (and, maybe, also against the Jew conspiracy to rule the world). So this is how quickly the anti-Red hysteria in America, the Soviet machinations in Europe, and the Korean War had wiped out the memory of the Nazi crimes and restored respectability to their perpetrators!

Not letting on that we were Jews, I merely told him that I had no idea where they sign up volunteers for the U.S. Army, but that I was sure that the help of SS men was neither wanted nor needed. This incident spoiled the excursion for me; I was depressed not only by what I thought the Cold War betokened for the rehabilitation of Nazism but also by my cowardly failure to tell that SS bastard where to get off.

As the Arlberg Express from Salzburg to Zurich stopped in Kitzbühel, I recalled my last embrace of Lore on the station platform three years ago and how, dazed by rapture, I had gotten on the wrong train. I had a fleeting glimpse of the Grand Hotel and the summery, snowless Hahnenkamm ridge. Overcome by these remembrances of the greatest romantic moment of my life—when Lore pulled me out of the snow on the Kaser ski slope—I now longed for her again. Why, for God's sake, hadn't I made any efforts to bring her to America? If I could only find her! Now that I *have* come back to Europe, although two years later than I promised her, maybe it's not too late to marry my dream woman?

Actually, I had tried to reestablish contact with Lore just six months ago, a year-and-a-half after I failed to answer her last letter. In January 1950, while sunning myself, melancholy and lonely, on the deck of the Alta Ski Lodge above Salt Lake City, I remembered the divine ecstasy of lying with Lore on the deserted, postseason Hahnenkamm Lodge sun deck, kissing the day away. Totally out of the blue, I sent Lore a picture postcard from Alta to her old Ingolstadt address, giving Caltech as my return address:

My dear Lore,

What a pity that you're not with me here in Utah! I would have loved to show you how much my skiing has improved in the three years since you pulled me out of the snow on the Kaser run. Every day, as I am riding up the Alta ski lift, I fantasize that I am back in Kitzbühel—our Kitzbühel—sharing the lift chair with you.

Weeks later, the postcard came back to Pasadena, stamped "RETOUR" ("Return"). It bore a little printed ticket "Departed without leaving forwarding address," as well as a handwritten notation that said "Addressee emigrated USA Feb. 4th." So my card had missed her by just a few days!

Of course, it was too late now to marry Lore! How could I even toy with the idea that Lore would want to have anything further to do with me, even if I did find her? She would have to be a lunatic to want to marry a man who had never even told her that he loved her, and broken his promise to come back. I never did see Lore again. In the early 1990s, I happened to come on the listing of a Lore T. in the Berlin telephone directory. All excited by this discovery, I rang the woman—sure that she just *had* to be my Lore (by then in her seventies). Nothing doing; she turned out to be someone else.

Jean Weigle had an apartment in an eighteenth-century building in the old, picturesque part of Geneva that rises above the center of the modern city. Staying with Jean in his splendid digs—antique furniture, oriental rugs, walls lined with old, leather-bound tomes, view on the Place Grand Mézel with its baroque fountain, in a building that might have been visited by his namesake, Jean Jacques Rousseau—made me feel that I climbed another rung on the ladder to the high life.

Jean offered to take me on a rock-climb up Salève Mountain, just southwest of Geneva.

"Isn't rock-climbing terribly dangerous, Jean? One slip, and, poof, you've had it!"

"Nonsense. The Salève is the easiest climb in the Geneva region, ideal for breaking in beginners. Even toddlers can make it up the Salève."

Despite my being not unaware that Jean—the kindest of men in most respects—had a malicious streak when it came to dealing with persons of German origins, I said I'd give it a try.

A sheer rock wall faced us when we got out of Jean's car at the foot of the Salève.

"We're not going up here, Jean, are we?"

"Of course, we're going up here. Where else? What's the matter? Pissing in your pants, little boy? Don't worry! You'll be on the rope with me."

"We're going to use a *rope*?"

"Of course, we're going to use a rope! What d'you think?"

It wasn't too bad scrambling up the first rocks, but when we came to the steep part, I soon became exhausted. Over and over again, I had to lift one leg high to put it on a very tenuous foothold indicated by Jean and then, clutching tiny protuberances in the rock, pull myself to stand upright, with my back toward the abyss. My miserable condition alternated between being too tired to be aware of my fear and being too fearful to be aware of my fatigue. Then I noticed that Jean was heading for a virtually vertical rise. I was sure I would never to be able to get up to the top of that perpendicular wall, and since turning around and going down the route we had just come up was equally out of the question, I thought I was done for. And then I saw that Jean was squeezing into a hole in the rock. He yelled that I should follow him, helping me to get into the hole by pulling on the rope. Once in the hole, I saw light at the far end, and it dawned on me that we had got into a tunnel. The other side of the tunnel turned out to open onto the most beautiful alpine meadow I had ever seen. My obvious anguish on the way up notwithstanding, Jean hadn't said a word about the tunnel. I vowed to myself never to do any more rock climbing, a vow I have had no trouble keeping.

Surface travel by civilians to West Berlin through the "Zone" (as the GDR was referred to contemptuously in the West) had become even more difficult than it was in 1947. For non-Germans, it required a transit visa from the Soviet authorities, very hard to obtain for Americans, especially since the outbreak of the Korean War. The only practical, albeit rather expensive, way for me to get to Berlin was to fly in from Frankfurt.

When I got off the train from Geneva at Frankfurt Central, I saw that the station had been spruced up in the three years since I last passed through, although the roof, with its acres of glass panes blown to smithereens during the war, had not, as yet, been replaced. But except for the platforms still being *al fresco*, the bustling depot seemed normal. For me, Frankfurt Central was not just *any* train station, of course. It was there that I had cruelly abandoned my Hildegard, unconcerned about the fate to which I was consigning her, glad to have her off my hands so that I could devote my full attention to Lore. Traipsing along the platform toward the station exit, I turned around and looked back at a train just pulling out for Hanover, trying to conjure up Hildegard leaning out of a compartment window, waving good-bye with her handkerchief to me.

Only a few days ago, while the Arlberg Express stopped at Kitzbühel Station, I was fabulating about marrying Lore. Now, at Frankfurt Central, I was suddenly seized by the insight that it was Hildegard who was the right woman for me, not Lore, whose beautiful face and elegance I adored, but about whom I knew almost nothing. I would look up Hildegard in Berlin, beg her forgiveness, and ask her to marry me.

To approach the Tempelhof airport runway from the east, the Pan Am

DC-4 turned a low-altitude loop over Berlin. The bird's-eye view of the city brought me to a state of feverish frenzy. The thrill of returning to Berlin was nearly as intense as it had been on my first of homecoming four years ago. Berlin was still war-ravaged. The rubble had been cleared from all the streets, the shops were full of goods, and the replanting of the myriad of cut-down trees and shrubs had begun. Yet in the center of town there remained still more ruins than intact buildings. Obviously the year-long Soviet blockade, which had ended only fifteen months ago, put West Berlin's recovery way behind that of the cities in West Germany.

My father had already arrived in Berlin when I got there. It was his first return since his escape in 1938, at the time of Kristallnacht. We were staying with Alwin and Lina Köhler, the only family still living in Berlin who had been part of my parents' social circle before the war. (Our family friend, Martin Mendelsohn, the Theresienstadt survivor whom I discovered in Treptow in 1946, had moved away meanwhile, to join his daughter in Brussels.) Lina was a Jewish school friend of my father's younger sister, Aunt Käthe Salomon. She had married Alwin, an Aryan arch-Berliner, and they had a daughter, Henni. My father and Aunt Käthe always used to refer to Henni as my twin, since she and I were born on the same day, albeit to different mothers. Widowed since the soldier whom she had married during the war died on the eastern front, Henni was living with her parents. Lina survived the Holocaust, as had the Jewish father of my old schoolmate, Heinz Behr, because she and Alwin had raised their non-Aryan-half-breed-of-the-first-degree child in the Christian faith, as had the Behrs. Alwin had been put under heavy pressure to divorce Lina, but he stood by her, which cost him his job and saved her life.

The Restitution Court had finally ruled that title to the Essener Strasse property, which Georg and Günter Stensch had sold under duress in 1938 to the bookmaker, Erich Schmidt, be returned to its original owners. Actually, Schmidt had bought the house in good faith on the open market and had not himself exerted any duress on us. He paid the full purchase price into our blocked bank account, and it wasn't his doing that Hermann Göring confiscated our money in November 1938 as part of the $1 billion mark fine levied on the German Jews. I expressed some tiny pangs of conscience to my father about how Schmidt was now going to lose his whole investment.

"Forget about Schmidt, Günter. Did Schmidt worry about us when we were robbed? You bet he didn't. Now it's the turn of the *ganovim* to pay back their victims. Actually, I've already got a buyer for the property. Friedl knows someone in London who is going to take the house off our hands for a thousand pounds, payable in cash on the spot."

"Are you out of your mind, Dad? A thousand lousy quid for a building with 40-odd apartments? Maybe we can't get much income from the place right now, especially since D-marks are not convertible into hard currency. But once Berlin gets back on its feet, we ought to be able to sell the house for its real worth. Then you'll be financially secure in your old age. It'd be a real *meshuggas* to throw that chance away now for peanuts."

"No, Günter, I can't gamble on the future. I've got to sell the house right now. A generous lady friend of Friedl's in London lent us a thousand pounds and now wants them back. Since we haven't got the money to repay her, we'd be rotters if we stiffed her when the sale of the house gives us the chance to settle our debt."

"No, Dad. I won't agree to this ridiculous deal, for the sake of *your* financial security in old age, mind you! Haven't you always told me that there isn't any decency in business, that one's first lookout has got to be oneself? So if Friedl's friend is making unsecured loans, isn't that just too bad for her?"

My firm refusal to consent to the sale he thought he had already sown up made my father more angry than I had ever seen him before.

"All three of you—Rudi, Mausi, and you—are *Rabenkinder* (unnatural children), who think only of themselves! What've you ever done for your father?"

I was deeply offended by his tirade. Not so much on my own behalf, because, as I thought about it, I could see that, being the most like him and the least honorable of his three children, he probably had my number. But I always regarded my brother and my sister as the kindest and most responsible people in the world, the polar opposites of *Rabenkinder*. It was the first and only time in my life that there were any bad words between my father and me.

Since it seemed to me out of character for him to maledict his children just to do right by one of his creditors, I reckoned that Friedl was the actual author of this painful contretemps. She had evidently ordered him to finalize the sale, and he didn't dare face her in London without my signature consenting to the deal. So I made him an offer he couldn't very well refuse:

I would buy his equity in the property for the thousand quid in cash. That money represented my entire capital, which I had managed to save from my FIAT salary and my Merck fellowship stipend. We had our lawyer, Dr. Sachs, draw up the transfer of title, I gave my father a check drawn on my Pasadena bank account, and he went back to London, to pay off Friedl's generous lady friend, I thought. Thus I became the sole owner of Essener Strasse 20, penniless again but in possession of a substantial piece of real estate.

I asked Alwin Köhler to manage the property for me, the absentee landlord. Month after month, under Alwin's stewardship the building wound up with a loss. In the early summer of 1952, by which time I was getting ready to return to California, I decided to sell. I cleared about 40 percent more than what I had paid my father; but I knew it was still a ridiculous sum for a forty-unit apartment building. When the Wall came down in 1989 and Berlin became once again the all-German capital city, the property was worth a fortune.

The money I cleared from the sale of the house wasn't enough to buy my dream Cadillac, but I picked up a new Chevrolet convertible at a Chicago dealer on my return to the States. That Chevy turned out to be only a little less of a lemon than my old Victory Ford. When I junked it a few years later, the last vestige of the investment my father made in my mother's name a generation earlier had gone up in smoke.

Years later, on Friedl's death in 1958, I found out that she had lied to my father. There had been no loan from any generous lady friend. To save their Belmont Trading Company from insolvency, she had put some of the money she previously embezzled from the firm's till back into the business. But once the Essener Strasse house was ours again, she was determined to recapture her loot by the quickie sale. I never forgave my father for having called us *Rabenkinder*. Onward from my early teens, I often dreamt that my father was badly in need of my help, which I hadn't given him, more often than not, inadvertently, and that, because I hadn't helped him, he died. On waking up, my guilt feelings would subside once I realized that he hadn't died. After our quarrel in Berlin, I stopped having that dream.

My father died in 1977, aged ninety-one, in Nice. He had been living comfortably on the Riviera for the last fifteen years of his life with his third wife, a widowed Polish Jewess whom he had married in New York. As it turned out, the Essener Strasse property was not needed for his financial security in old age after all, since he received one of the lifelong, generous monthly compensatory pensions which the German Federal Republic (but not the GDR) paid out to most elderly German–Jewish survivors of the Nazi terror. He lies buried in the Jewish section of a cemetery in the foothills of the Alpes Maritîmes behind Nice, overlooking the Mediterranean, an exotic site of eternal rest for an arch-Berliner, albeit more scenic than the spousal plot next to my mother's grave in Weissensee that he had once intended for himself.

On this visit, I made my hometown debut as scientific lecturer. This came about because a young bacteriologist from the Robert-Koch-Institut —Berlin's answer to Paris's Institut Pasteur, built in honor of Prussia's anti-Pasteur, Robert Koch—had visited Max's lab last spring. He gave us a talk on the old-timey, pre-Delbrück-style phage experiments he was doing. A few minutes into his presentation, Max interrupted him to ask whether he had done such and such a control experiment. The visitor stood erect, at attention, heels together, hands flat on side of trousers, and replied, "Herr Professor Delbrück! At the Robert-Koch-Institut we do nothing *but* control experiments!" I don't think that the young Koch-Kraut understood why his riposte provoked so much laughter.

Identifying myself as a fellow-Berliner, I mentioned to him that I might be passing through Athens-on-Spree next August. He insisted that I visit him at the Robert-Koch-Institut and give a seminar on the important work I was doing at Caltech. When I called him after my arrival in Berlin to arrange my presentation of our findings on the tryptophan activation of T4 phage, my host said that it would be OK for me to give my talk in English. But I told him that I would speak in German. By making my presentation in Berlin idiom, I wanted to show off as the banished local boy who had made good, who had returned to his home town as a scientific starlet—a postdoctoral fellow of the National Research Council of the United States of America, no less.

I had never before given a scientific talk in German. I wasn't aware that I didn't know the German equivalents of many of the technical terms I would need, such as "incubate," "dilute," and "enumerate," or appreciate such subtleties as that the English word "number" has two semantically distinct German equivalents, *Nummer* and *Zahl*. Worst of all, I hadn't realized how laborious it was going to be to recast, sentence by sentence, the English word order of my stock tryptophan activation spiel to fit the very different German syntax. My talk turned out to be a humiliating disaster. If it showed anything at all, it was that banishing this local boy had not been a big loss for Berlin's scientific future. Even though my ability to speak technical German improved substantially in later years, I never again dared to give an off-the-cuff lecture in my mother tongue.

I did derive one small benefit from this rhetorical fiasco, though. An American bacteriologist spending his sabbatical leave at the Robert-Koch-Institut told me after my talk that American civilians visiting Berlin to participate in German artistic or scholarly activities can be authorized by the U.S. Cultural Affairs Office in Dahlem to travel on the military trains connecting Berlin with the West. So, as my very last fling in the military travel order racket, I got myself orders for riding the "Bremen Express" from Berlin through the "Zone" to the Bremerhaven port of embarkation. I would leave Berlin as I had arrived on my first homecoming, on a U.S. Army sleeper.

After getting to Berlin, I had made finding Hildegard my first priority. The name "Heinrich U." was still over the doorbell at Hildegard's flat in Wedding. Frau U. opened the door. She recognized me immediately, even though we had met only briefly four years before. She stared at me as if I were a phantom.

"Mr. Stent! You're back in Berlin! Hildegard isn't living here any longer, and I haven't got any way to contact her. But she is dropping by to see me every few days, and if you leave her a message, I'll pass it on at her next visit."

> Dear Hildegard,
>
> I'm stopping in Berlin for a few days on my way to Copenhagen and want to see you as soon as possible. I'm staying with friends near the Heidelberger Platz. Please give me a call at their apartment, at dinner time, if possible, at 875-438. In case you get this message only after I left for Denmark, can you please leave an address with my friends at which I can reach you? Because if I miss you on this visit, I'd hop down to Berlin to see you as soon as I got settled in Copenhagen.
>
> Your Günter

One afternoon, while I was still waiting for Hildegard's call, Henni took me to the huge pool built for the 1936 Olympic Games, recently opened to the general public for recreational swimming. I hadn't been at the pool since I was twelve, when I watched the finals of the high-diving competition, immortalized in some of the most gorgeous footage of Leni Riefenstahl's *Olympiad* movie. As I lay on the grass, sunning myself poolside, I thought back again to those days of my fear-filled, motherless childhood and took stock of where my journey had taken me since I fled Berlin on the night express to Cologne on Christmas Day, 1938.

As a Holocaust candidate, I had escaped the ghastly fate to which the Nazis had intended to consign me. As a frightened, self-hater, I had made some progress. The dread of annihilation by Hitler had subsided into the deeper recesses of my subconscious, from which it now surfaced only intermittently to trouble my dreams. I was still embarrassed by my German–Jewish origins, to which I preferred not to have to own up. But I was at least proud of my acquired identity as a nonreligious, free-thinking American, a haughty citizen of the world's Number One Nation.

As an aspiring scientist, I had progressed much farther than I could have possibly imagined. Despite having missed school for one-and-a-half of the cognitive-developmentally crucial teenage years and having gallivanted as a FIAT playboy across Europe for the better part of another year, I managed to get my Ph.D. when I had just turned twenty-four. Then I had the fantastic luck of falling in with Max, whose coterie bid fair to revolutionize biology in the coming years. And now, at twenty-six, I had come back to Europe to carry on my research as a National Research Council Fellow and spread the Good News as an Apostle of Max's Phage Church.

I hadn't made any earthshaking discoveries, but I was proud of our cofactor activation papers, which would be out soon in the *Biophysica et Biochimica Acta*. Max had praised our work and even given lectures on it in the States and abroad. There was so much to be done in phage research that I ought to be able to keep on publishing interesting papers and, when I had finally played out the postdoctoral fellowship game, get an academic job in a good university, maybe even in Berkeley.

But as a lover, the role in life that was of the greatest emotional significance for me, I had hardly made any progress at all. In spite of having gathered a lot of experience in erotic transactions, I had got mostly nowhere. I had not managed to establish an emotionally satisfying relation with any woman. But now I had come to my senses at last: I would marry Hildegard.

She called me at the Köhlers a couple of days before my departure. I was electrified by her voice; my Hildegard!

"Günter! You've come back! As you can imagine, I was totally bowled over when my mother gave me your message. Of course, I want to see you before you leave for Copenhagen. And I'll have a surprise for you!"

"How about meeting at the Café Möhring on the Kurfürstendamm, Hildelein? You probably don't remember, but that's where my father had once sprung on me my future stepmother. When can you meet me? I'm leaving Berlin day-after-tomorrow. What's your suprise, by the way?"

"Not remember the Café Möhring story, Günterlein? Are you kidding? How you met your stepmother was just about the only funny story among all the sad ones you told me about your childhood! How about tomorrow afternoon at two at the Möhring? No, I won't tell you now what my surprise is. You'll find out soon enough tomorrow."

Hildegard spoke to me without any trace of the rancor to which, God knows, she was more than entitled. She's really a good woman, the best, a jewel! Everything should work out, and we'll have a blissful marriage. After a little preliminary palaver at the Café Möhring about what either of us has been up to since we lost touch, I would take her hand, softly sing *Je ne suis pas curieux* into her ear, and then come right out with it. I've been thinking about the advice she gave me last year; that getting married is the only way in which I will be really happy. So will she please marry me?

I didn't give too much thought to the surprise she said she had in store for me. Maybe she got her job back at the Schiffbauerdamm Theater and was cast as Polly Peachum in a new production of the *Dreigroschenoper*. Or maybe she landed a big movie contract.

There was no need for her to reveal her surprise when she walked up to the table at which I was waiting for her on the sidewalk terrace of the café. She was wearing a splendid blue, silken maternity dress and, judging by her girth, about to become a mother momentarily. Her face was radiant, more beautiful than I remembered. She smiled when she beheld my stupefied visage.

"As you can see, Günterlein, you've come back too late. I'm living with a French officer. We are going to be married any day now, as soon as the Army gives him permission. I'm glad to see you, though, and happy to know that you haven't totally forgotten me."

"Forgotten you, Hildelein? I'll never forget you."

Hildegard saw that I was near tears. Instead of gloating over my super-deserved comeuppance, she also began to cry. I took her hand, and we just sat there in silence. I didn't ask any questions about her acting career, or about her fiancé—not even his (and her future) family name. She didn't ask me anything about my scientific or personal affairs—not even what I was going to do in Copenhagen. After a while, she said it was time for her to go. On parting, we shook hands, European style, as we had when I dropped Hildegard at her apartment house the night we met four years ago—four years that had been wonderful for me and horrible for her.

"Good-bye, Hildelein-mine. I wish you all the best."

"All the best to you too, Günterlein."

It turned out to be a farewell forever.